Big, Bold and Blue

LESSONS FROM AUSTRALIA'S MARINE PROTECTED AREAS

Editors: James Fitzsimons and Geoff Wescott

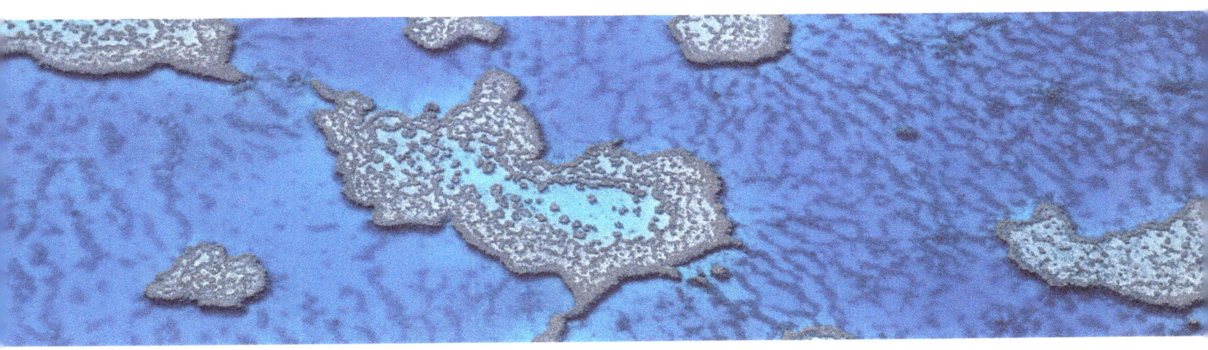

CSIRO
PUBLISHING

National Library of Australia Cataloguing-in-Publication entry

Big, bold and blue: lessons from Australia's marine
protected areas/James Fitzsimons and
Geoff Wescott, editors.

9781486301942 (paperback)
9781486301959 (epdf)
9781486301966 (epub)

Includes bibliographical references and index.

Marine resources conservation – Australia.
Marine biodiversity conservation – Australia.
Marine habitat conservation – Australia.
Endangered ecosystems – Australia.
Conservation of natural resources – Government policy – Australia

Fitzsimons, James, editor.
Wescott, G. C. (Geoff C.), editor.

333.956160994

Published by

CSIRO Publishing
36 Gardiner Road, Clayton VIC 3168
Private Bag 10, Clayton South VIC 3169
Australia

Telephone: [+613] 9545 8555
Email: csiropublishing@csiro.au
Website: www.publishing.csiro.au

Front cover: Hardy Reef, Great Barrier Reef Marine Park, Queensland (© John Carnemolla/Shutterstock).

Back cover (left to right): Macquarie Island Nature Reserve (Bob Zuur); Green sea turtle on the Great Barrier Reef (gjhamley. Available under a Creative Commons Attribution 2.0 Generic licence. For more information see https://creativecommons.org/licenses/by/2.0/. Available at https://www.flickr.com/photos/liptoniceg/8634011239); Twelve Apostles Marine National Park (Dean Ingwersen).

Set in 10.5/12 Minion Pro and ITC Stone Sans Std
Edited by Peter Storer Editorial Services
Cover design by Andrew Weatherill
Typeset by Desktop Concepts Pty Ltd, Melbourne
Index by Bruce Gillespie
Printed by Ingram Lightning Source

Feb26_RP_ILS

Foreword

Like many Australians, I grew up at the edge of our great continental island. I am a child of the littoral. The intertidal zone of influx and overlap wasn't just my home – it was the making of me. The natural world helped shape my expectations and outlook, and no part of it was more influential than the sea.

As a kid I always felt lucky to be able to do the kinds of things I loved – snorkelling, surfing, fishing and beachcombing – but it was only later in life, once I'd travelled, that I came to understand just how rare my luck had been, for I grew up on the shores of an ocean still teeming with life. I swam in clean water. And I fished and collected in what seems now to have been the last era of plenty Australia's seas were to experience since colonisation. In my family beach life was culture, and it was a rare outing indeed that didn't feature a waterway and some form of hunting and gathering. A picnic without prawn nets or crab pots or mussel buckets wasn't just pointless and boring – it was inconceivable. In the 1960s a suburban kid like me could still chip abalone off the reefs before school and bring home a bag of tailor and whiting in the evening. It was the period of domestic modernisation when the copper and mangle began to disappear from the suburban laundry to make way for the electric washing machine. The mangles were rarely seen again but the old boilers were quickly repurposed. Every fishing family I knew had a copper mounted over the barbecue for the sole purpose of cooking crustaceans because there was simply no indoor pot capacious enough to deal with the hauls of crabs, prawns and crays we brought home. A 30-litre pot was indicative of our catches and our default expectations. Some weekends it seemed we spent half the night fishing and the rest of it cooking. In the days before chest freezers were widely available, you fished and feasted and when you could eat no more you distributed the surplus around the neighbourhood. Because there was plenty to go around. That wasn't just an old-fashioned social ethic; it was our default experience – it was all we knew. Our expectation was that there would always be plenty to go around.

But by the 1970s such plenitude was waning. The estuaries we waded through for crabs were beginning to choke on algal sludge, the river prawns became scarce and catches of tailor from beaches and groynes and jetties grew patchy. At the end of that decade, what had once been commonly adjudged a modest fish soon became a whopper. As a spearfisher I found I was swimming twice as far from shore to bring back half the catch I was used to. Underwater there was less and less to see: no groper, few snapper, and the cods were tiny. The abalone reefs were already bare.

By the 1980s angling tribes like ours began to buy boats, and this was a classic example of social prosperity masking a fundamental natural impoverishment, for although many suburban fisherfolk became boaties because they felt they could afford it, the fact was we were forced to go further and further from shore in order to catch a feed. Within a decade, these first boats weren't big enough. Soon they got bigger and the average trips at sea longer.

The regular catch became lighter and the larger specimens shrank. Beach fishing became a specialty, and shore-based spearfishing an eccentric activity indeed.

By the time I was in my mid-twenties offshore fishing grounds were so hammered anglers had become secretive, suspicious and defensive. The people who'd had an aristocratic outlook in the 1960s, a mindset of generosity born of plenty, had been reduced to small-minded peasants when it came to sharing resources and information. For now there were more anglers fighting for less, and acts of bastardry like net-cutting and pot-tampering, offences once nearly unheard of, became commonplace. This culture of baffled and irritable impoverishment had overtaken commercial fishers as well. The richest fishing grounds were those least accessible. In short, the sea was at its best where it was least exposed to humans. Within a single generation Australians had begun to fish their seas bare. And yet few fishers saw themselves as part of the problem. If there was any consensus about this diminution of resources it was the certainty that it was someone else's fault.

I was in my thirties before I truly saw that I'd played a part in reducing my local marine environment to the shadow of its former self. I guess it was one long swim too many, the sense of having drawn a blank yet again, coming ashore with an empty bag and a heavy heart. In the evenings on the local jetty I watched men and women fill their buckets with the thoughtless compulsion of small children. If you asked these folks why they took so many undersized fish they told you it was their right. They were taking these fish before someone else did. These people were my age and older; they were my neighbours. This mulish, niggardly attitude and the meagre conditions it was producing were now normal; it was all they knew. And these were my people. It dawned on me that I was a member of the generation that had overspent its inheritance. By then I had three kids of my own and I was conscious that I'd contributed to the compromised estate they were born into. And I wondered what their children might inherit if the trend was not arrested and reversed. This was how I came to be an activist for marine conservation. I am still a fisher and collector, a beachcomber, boater and jetty-rat, but the simpleminded assumption of plenty is gone. The notion that the sea can take whatever we dish out has evaporated. I've sought to redeploy the suburban generosity of old and extend it to the oceans that sustain my life and hopefully restore some honour to my family culture.

For the past 20 years, as a small part of a growing movement, I have done what I could to promote areas of respite for our ocean environment. For 50 years the Australian Marine Conservation Society has campaigned for the protection of specific places like the Great Barrier Reef and against the decimation of species like tuna, sharks and turtles. We have been at the forefront of bringing marine science to public notice. But the establishment of marine protected areas has been the central focus of our conservation advocacy and the social momentum for this has steadily grown. As Australians began to travel more widely and to see the degraded state of many marine ecosystems elsewhere in the world, their appreciation for what they'd always taken for granted at home has quickened. In the 2000s many laypeople came to experience marine sanctuaries for themselves, visiting the Great Barrier Reef and Ningaloo where significant ecosystems were slowly being protected, and the notion of reserving parts of the national marine estate began to make sense, even to those who'd resisted their introduction. After all, our forebears had achieved something similar on land generations ago. And as the growing body of science on the subject began to find its way into popular discourse, Australians were increasingly aware of how vulnerable their seas were to unsustainable exploitation. Thus, an idea once viewed as outlandish is now both mainstream and bipartisan. In the lead-up to the historic announcement of a national system of marine parks in 2012, the culmination of a process more than 10 years in the making, I was heartened to see how broadly popular the initiative was. Despite the attempts of some narrow interests to undermine consensus, polling on the issue showed consistently high public support. To my mind it shows a significant move forward in our thinking, an acceptance that our relationship with the oceans needs to be redrawn.

The establishment and maintenance of MPAs will not save our oceans. On their own they cannot solve all our fisheries problems or our many pollution challenges and development pressures. But they are a necessary and fundamental step in arresting what is clearly an unsustainable trend in exploitation. Like any ecosystem, our seas need areas of respite and recovery, recruitment and increase. And they need devoted and properly-funded research. For Australians to live more consciously and intelligently in our littoral zone we need fresh ideas born of sound information and good science. A compendium like this, with contributions from leaders in the field, is not just welcome but vital.

Tim Winton

Tim Winton is a novelist and essayist. He is the patron of the Australian Marine Conservation Society.

Foreword

The Great Barrier Reef is an iconic place in the minds of many, including those who have been inspired by images on television or have been fortunate enough to visit the area. It is widely seen as synonymous with Australia's natural beauty and the magical, but delicate, diversity of our oceans.

Yet how many know that the Great Barrier Reef is just one jewel among Australia's many marine treasures, that Australia has created some of the world's largest marine protected areas, or why this protection is necessary?

Marine conservation in Australia is part of a global effort to safeguard our oceans. In recent decades, we have become increasingly aware of how these vast expanses that cover some 70% of our planet are suffering, and what needs to be done to bring them back to health.

Initially, negative impacts were the direct result of activities such as fishing and pollution. More recently, much larger and more diffuse effects have taken hold as we increase the amount of greenhouse gases in the atmosphere and change the world's climate. This is causing water temperatures to rise and the oceans to become more acidic, affecting marine species and the rich habitats that support them.

Against such trends, all nations have agreed to look after key places in the ocean by creating marine protected areas. Within these areas, efforts are made to maintain healthy ecological function and manage the worst effects of human activity. In some cases, areas are strictly protected to remove pressures altogether and to give an area time to recover. In other places, human activity is allowed but managed in a sustainable way so that the sea can continue to be a source of food and livelihood for the many people who depend on it.

Australia has been at the forefront of marine conservation, and has proved invaluable in helping promote similar efforts in many other countries.

Big, Bold and Blue: Lessons from Australia's Marine Protected Areas captures much of this experience, sharing important lessons from the Great Barrier Reef and introducing us to many other extraordinary marine protected areas across Australia. The rich content of these pages draws on leading academic expertise, perspectives on government policy, indigenous sea country management, non-governmental organisations, and commercial and recreational fishing sectors.

By documenting and sharing Australia's experience of establishing and managing marine protected areas, we anticipate this book will inspire new marine conservation successes, both in Australia and more widely. We also hope it will expand perspectives on what is possible, caution that nothing can be taken for granted and show that, through diligence and hard work, we can build a better future for the oceans, for ourselves, and for future generations.

Inger Andersen
IUCN Director General

Prof Dan Laffoley
Marine Vice Chair
IUCN World Commission on Protected Areas

Contents

Acknowledgements

Putting an edited book together takes a lot of work, and many people contribute to making it a reality – and hopefully a success. The editors would first like to thank and acknowledge the enormous efforts put in by all the authors of the chapters, which form the backbone of this book. Their individual reflections, vast experience and dedication to the task has been greatly appreciated and we are sure this will also be recognised by the book's readers. We would like to thank those authors for their responses to numerous reviews and edits and hope that our editorial efforts have contributed further to the book's quality and justifies the many hours they have put in.

In these days when high-impact-factor journal articles are the main metric (for the academic sector at least), we are particularly grateful that this eminent and experienced range of authors has contributed their important perspectives to this book. The authors often drew on colleagues and staff within state environment departments for up-to-date figures, for relevant photos and for map creation and we are appreciative of those contributions (see individual acknowledgements in specific chapters as well).

We also want to very gratefully acknowledge our respective employers – The Nature Conservancy and Deakin University – for allowing us some of their time to prepare the book and for a number of extra resources that have assisted us very significantly. In particular, we wish to acknowledge the financial support of Deakin University's Centre for Integrated Ecology and The Nature Conservancy's Science Impact Project, which has assisted with sponsoring the production of the book in full colour.

At CSIRO Publishing we are grateful for the efforts of Lauren Webb throughout the project and to the copy editor Peter Storer for catching all the things that got past us even at the end. We would like to thank CSIRO Publishing as an organisation for the overall opportunity to prepare this publication.

JF would also like to personally thank Janelle, Jarrah and Tali for their love and support, for accompanying him to marine protected areas both in Australia and abroad and for being tolerant of weekends and late nights completing this work, and to his parents for many a summer holiday exploring Wilsons Promontory Marine Park in his younger years.

GW extends his thanks to his family (Victoria and Alex) for their encouragement throughout this enterprise and hopes they enjoy the final product.

In closing, individually we thank a range of people who have supported us throughout the exercise or assisted in preparing or aiding with figures and tables, including Ben Carr, Kris Carlyon and Dean Ingwersen.

Thank you everyone and we hope you all enjoy the final outcome of your great efforts.

List of contributors

Inger Andersen
International Union for Conservation of Nature (IUCN), Rue Mauverney 28, 1196, Gland, Switzerland. Email: director.general@iucn.org

Daniel Beaver
Centre for Conservation Geography, 18 Beeson Street, Leichardt NSW 2040, Australia; Centre for Biodiversity and Conservation Science, University of Queensland, St Lucia QLD 4072, Australia; Marine and Coastal Conservation and Spatial Lab, San Francisco State University, 1600 Holloway Avenue, San Francisco, CA 94132, USA. Email: Daniel.Beaver@conservationgeography.org

Stuart Blanch
Research Institute for Environment and Livelihoods, Charles Darwin University, Darwin NT 0909, Australia; Environment Centre of the Northern Territory, GPO Box 2120 Darwin NT 0801 Australia; current address: Hunter Wetlands Centre Australia, PO Box 66, Shortland NSW 2307, Australia. Email ceo@wetlands.org.au

Simon Boag
South East Trawl Fishing Industry Association (SETFIA), PO Box 1125, Lakes Entrance VIC 3909, Australia. Email: simonboag@setfia.org.au

Pepe Clarke
BirdLife International, The David Attenborough Building, Pembroke Street, Cambridge CB2 3QZ, UK. Email: pepe.clarke@birdlife.org

Peter Cochrane
Fenner School of Environment and Society, Australian National University, PO Box 89, Waverley NSW 2024, Australia. Email: peter.cochrane@anu.edu.au

Jon Day
Australian Research Council Centre of Excellence for Coral Reef Studies, James Cook University, Townsville QLD 4811, Australia. Email: jon.day@my.jcu.edu.au

Karen Edyvane
Research Institute for Environment and Livelihoods, Charles Darwin University, Darwin NT 0909, Australia; Fenner School of Environment & Society, Australian National University, North Australia Research Unit, PO Box 41321, Casuarina NT 0811, Australia. Email: Karen.Edyvane@cdu.edu.au

James Fitzsimons
The Nature Conservancy, Suite 2–01, 60 Leicester Street, Carlton VIC 3053, Australia; School of Life and Environmental Sciences, Deakin University, 221 Burwood Highway, Burwood VIC 3125, Australia. Email: jfitzsimons@tnc.org

Lyn Goldsworthy
Fenner School of Environment & Society, Australian National University, Canberra ACT 2601, Australia. Email: lyn.goldsworthy@anu.edu.au

Lynne Zeitlin Hale
The Nature Conservancy, Graduate School of Oceanography, University of Rhode Island, South Ferry Road, Narragansett, RI 02882, USA. Email: lhale@tnc.org

Caroline Hoisington
Sydney Institute of Marine Science, 22 Chowder Bay Road, Mosman NSW 2088, Australia; Australian National Centre for Ocean Resources and Security, University of Wollongong, Wollongong NSW 2522, Australia; Current address: 37 Cutler Road, Clontarf NSW 2093, Australia. Email: hoisington@gmail.com

Vera Hughes
Department of Environment, Water and Natural Resources, GPO Box 1047, Adelaide SA 5001, Australia. Email: vera.hughes@sa.gov.au

Miya Isherwood
Land and Sea Management Unit, Torres Strait Regional Authority, PO Box 261, Thursday Island QLD 4875, Australia. Email: miya.isherwood@tsra.gov.au

Richard Kenchington
Australian National Centre for Ocean Resources and Security, University of Wollongong, Wollongong NSW 2522, Australia. Email: rkenchin@uow.edu.au

Lorne Kriwoken
Geography and Spatial Sciences Discipline, School of Land and Food, University of Tasmania, Private Bag 78, Hobart TAS 7001, Australia. Email: L.K.Kriwoken@utas.edu.au

Dan Laffoley
World Commission on Protected Areas, International Union for Conservation of Nature (IUCN), Rue Mauverney 28, 1196, Gland, Switzerland. Email: danlaffoley@btinternet.com

Gilly Llewellyn
WWF-Australia, PO Box 528, Sydney NSW 2001, Australia. Email: gllwellyn@wwf.org.au

Adrian Meder
Australian Marine Conservation Society, PO Box 5815, West End QLD 4101, Australia. Email: adrianmeder@amcs.org.au

Peter Ogilvie
Wildlife Preservation Society of Queensland, 95 William Street, Brisbane QLD 4000, Australia. Email ogilvie.house@bigpond.com

Paul Sheridan
Pew Charitable Trusts, 185 Elizabeth Street, Sydney 2000 NSW, Australia. Email: psheridan@pewtrusts.org

Chris Smyth
Chris Smyth Consulting, Williamstown VIC 3016, Australia. Email: chris.smyth36@gmail.com

Dermot Smyth
Research Institute for the Environment and Livelihoods, Charles Darwin University, Darwin NT 0815, Australia; Smyth and Bahrdt Consultants, PO Box 1202, Atherton QLD 4883, Australia. Email: dermot.smyth48@gmail.com

Mark Spalding
The Nature Conservancy, c/- Department of Zoology, University of Cambridge, Downing Street, Cambridge CB2 3EJ, UK; Department of Physical, Earth and Environmental Sciences, University of Siena, Pian dei Mantellini 44, 53100, Siena, Italy. Email: mspalding@tnc.org

Romola R Stewart
The Ecology Centre, The University of Queensland, St Lucia QLD 4072, Australia. Email: rom.stewart@gmail.com

Erika J Techera
Faculty of Law and Oceans Institute, The University of Western Australia, 35 Stirling Highway, Crawley WA 6009, Australia. Email: erika.techera@uwa.edu.au

Chris Thomas
Department of Environment, Water and Natural Resources, GPO Box 1047, Adelaide SA 5001, Australia. Email: chris.thomas@sa.gov.au

Trevor J Ward
Centre for Environmental Sustainability, School of Life Sciences, University of Technology Sydney, 15 Broadway, Ultimo NSW 2007, Australia. Email: Trevor.Ward@uts.edu.au.

Geoff Wescott
School of Life and Environmental Sciences, Deakin University, 221 Burwood Highway, Burwood VIC 3125, Australia. Email: wescott@deakin.edu.au

Barry Wilson
Western Australian Museum, Locked Bag 49, Welshpool DC WA 6986, Australia; Current address: 4 St Ives Loop, Kallaroo WA 6025, Australia. Email: murexpecten@bigpond.com

Tim Winton
Novelist and essayist, WA, Australia

Bob Zuur
WWF Antarctic and Southern Ocean Initiative, WWF-Australia, PO Box 528, Sydney NSW 2001, Australia. Email: bzuur@wwf.org.nz

PART 1

Setting the international and national scene

Previous page: Cape Byron Marine Park. (Photo: James Fitzsimons)

Chapter 1

Introduction: the expansion of Australia's marine protected area networks

James Fitzsimons and Geoff Wescott

Australia was among the first countries in the world to establish a national parks system on land, with the Royal National Park near Sydney declared in 1879. Less well known is that the Royal National Park included an area of intertidal and close inshore marine areas, constituting what could be the oldest marine protected area (MPA) in the world (Brown 2002).

Although the national park system grew in popularity on land and expanded progressively since that early declaration, it was not until the late 1960s and early 1970s when there were proposals to drill for oil on the Great Barrier Reef that serious attention was paid to establishing protected areas in marine waters in Australia. The declaration of the Great Barrier Reef Marine Park, commencing in 1975, made Australia a world leader at the time. The more recent development of MPA systems by governments in state, territory and national (Commonwealth) jurisdictional waters has again placed Australia in the global spotlight in MPA development.

Australia is an island continent with an ocean territory that is the third largest on Earth and Australians are mostly coastal dwellers – 85% of Australians live within 50 km of the coast (Trewin 2004) (Fig. 1.1). This connection to the coastal marine environment may well be a contributory factor to the development of a vast suite of marine reserves. In 2015, Australia has more than 327 790 000 ha of sea declared as MPAs, comprising almost 36% of its waters. This represents a nearly five-fold expansion from 7.1% in 2002 (Taylor *et al.* 2014). The global average is 1.6% of waters reserved. Marine national parks and other highly protected areas (IUCN Categories I–III) grew nearly seven-fold from 2% to 13.5% of Australian waters over the last decade (Taylor *et al.* 2014) (Fig. 1.2).

However, the establishment of marine protected areas has not been without controversy and, in some cases, a decline in political support. Despite the significant activity, discussion and debate around MPAs in Australia, there have been no published volumes bringing together the history, current status and future directions of each of the Australian MPA networks (national and state) nor perspectives from different sectors on these MPA networks more broadly. This book aims to fill this void.

As most countries around the world seek to build their marine protected area networks, in response to commitments they have made under the Convention on Biological Diversity, the lessons from Australia are likely to be even more timely. Past compilations of

Fig. 1.1. Visitors exploring rockpools in Bunurong Marine National Park, Victoria. (Photo: James Fitzsimons)

differing perspectives of MPA development and implementation have either been at a more general level (e.g. Claudet 2011; Day *et al.* 2015) or are not particularly recent (e.g. Beumer *et al.* 2003).

The aim of the book is two-fold. First, it seeks out common lessons, both positive and negative, acquired from the Australian experience during the recent rapid expansion of MPAs, particularly over the past 15 years. Second, it aims to aid and inform other nations how they may progress their own specific and distinctive attempts to meet internationally agreed targets set under the Convention of Biological Diversity for marine conservation, especially Aichi Target 11 (CBD 2010).

We have done this by inviting a range of eminent and experienced authors from government, non-government organisations, academia and industry to describe in their own words various aspects of MPAs in Australia. Before outlining the chapters that follow, we would like to make it clear that the focus of this book is not to review the ecological theory behind MPAs (although some chapters do touch on this topic); rather it is a book about real world examples of trying to establish MPAs, and to enhance marine conservation more generally, written by a range of experienced planners, managers and academics, representing different departments, organisations, and communities.

The definition of a marine protected area is 'a clearly defined geographical space, recognised, dedicated and managed, through legal or other effective means, to achieve the long-term conservation of nature with associated ecosystem services and cultural values' (Dudley 2008, p. 8). Although the term marine protected area is the overarching term for a

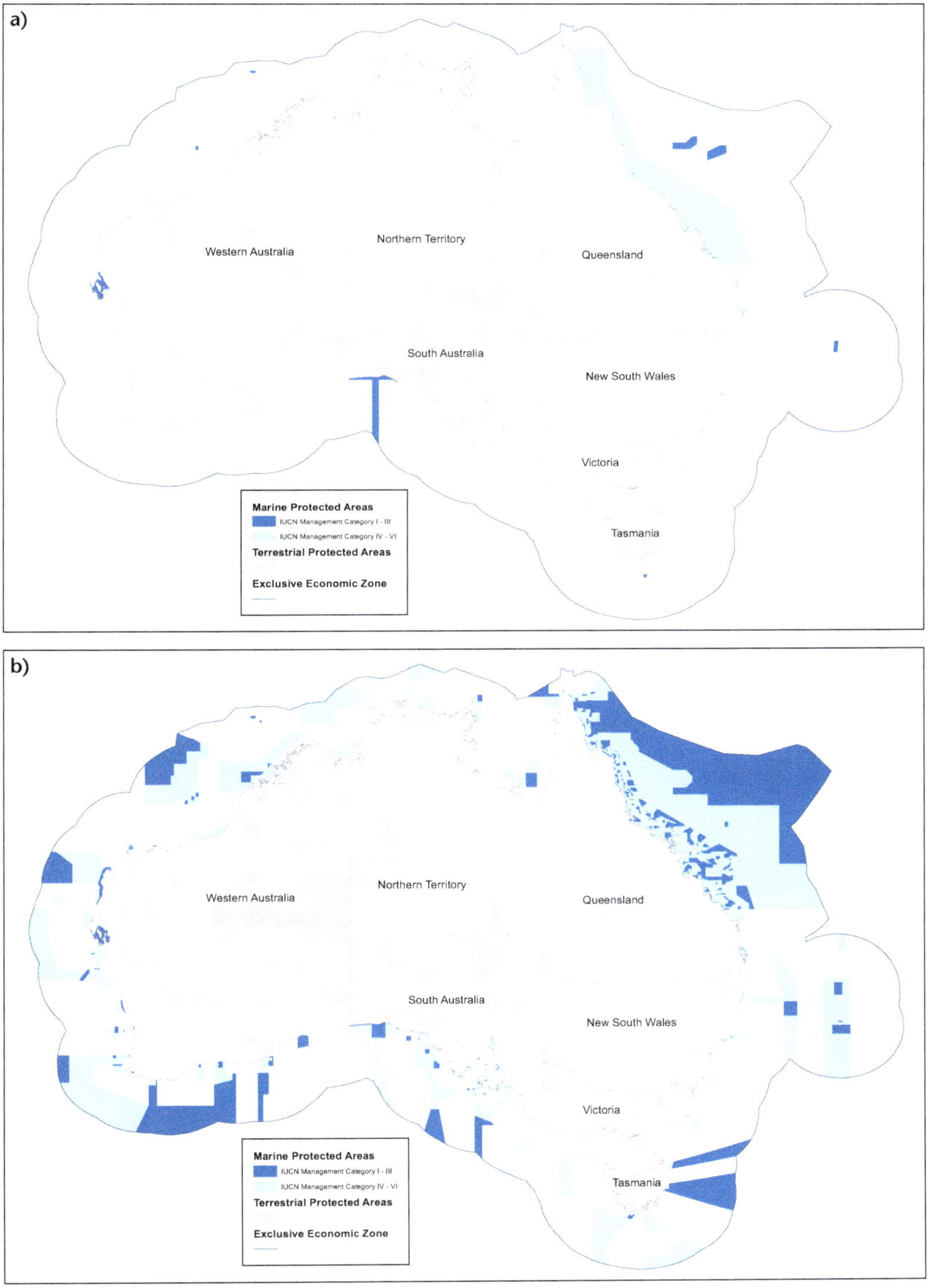

Fig. 1.2. Australia's marine protected area estate in (a) 1997 and (b) 2014. The external waters of Heard and McDonald Islands, Macquarie Island, Norfolk Island, Christmas Island and Cocos (Keeling) Islands are not shown.

range of protected areas covering marine environments, each Australian jurisdiction has their own names, legislation and categories, and these are what are presented in the chapters.

This book contains five main sections:

- Part 1 begins with an introductory outline on the purpose of the book, the broad international movement towards MPA establishment and an historical perspective on MPAs in Australia – it 'sets the scene' and places in context the sections and chapters that follow.
- Part 2 outlines the history, current extent and key lessons acquired in assembling MPA networks in each Australian jurisdiction. The first three chapters deal with waters under Commonwealth/Australian Government jurisdiction (Australia's Exclusive Economic Zone beyond 3 nautical miles of the coast, the Great Barrier Reef and Australia's Antarctic and sub-Antarctic waters). The remaining seven chapters cover each of the states and Northern Territory coastal waters within 3 nautical miles.
- Part 3 presents reviews of the science, economics, regulatory frameworks and reservation levels of MPAs in Australia.
- Part 4 provides a range of different perspectives on MPAs, from recreational and commercial fishing sectors, Indigenous use and involvement, public perceptions, campaigning experiences and the need for MPAs to sit within a context of an integrated approach to ocean planning and management, all from leaders in these different fields and sectors.
- Part 5 concludes the book by providing a synthesis of the common themes, key lessons and conclusions from all of the previous chapters.

We hope readers find this book enjoyable, as well as stimulating and challenging, and are able to place the various lessons easily into their own political and cultural context. This would further enhance marine protected areas around the world as a vital component of efforts to manage and conserve the seas that occupy over 70% of the surface area of our blue Earth.

Biographies

James Fitzsimons is the Director of Conservation for The Nature Conservancy's Australia Program where he oversees the organisation's conservation planning, science and policy functions. He is an Adjunct Professor at the School of Life and Environmental Sciences, Deakin University, with particular research interests in the fields of protected area policy, practical conservation planning and mechanisms to integrate conservation outcomes on public and private lands. He has previously worked in the fields of conservation policy and planning for government environment departments and agencies, and for non-government environment organisations.

Geoff Wescott is Associate Professor of Environment at Deakin University's Melbourne Campus. He is a previous Chair of the Marine and Coastal Community Network who he also represented on the National Oceans Advisory Group. He was a Deputy Chair of Parks Victoria and has served on the steering committee of the past three Nature Conservation Reviews of the Victorian National Parks Association. He is current President of the Australian Coastal Society, a Director of Zoos Victoria and in mid-2015 was appointed to the

Victorian Environmental Assessment Council. He has published over 100 articles on marine and coastal matters, including a popular field guide to rocky shores in south-eastern Australia.

References

Beumer JP, Grant A, Smith DC (Eds) (2003) *Aquatic Protected Areas: What Works Best and How Do We Know? World Congress on Aquatic Protected Areas.* Australian Society for Fish Biology, Perth.

Brown I (2002) Oldest MPA in the world: Royal National Park, New South Wales, Australia. Designated 1879. *MPA News* **3**(6), 5.

CBD (2010) *Strategic Plan for Biodiversity 2011–2020, Including Aichi Biodiversity Targets.* Convention on Biological Diversity Secretariat, Montreal.

Claudet J (Ed.) (2011) *Marine Protected Areas: A Multidisciplinary Approach.* Cambridge University Press, Cambridge, UK.

Day JC, Laffoley D, Zischka K (2015) Marine protected area management. In *Protected Area Governance and Management.* (Eds GL Worboys, M Lockwood, A Kothari, S Feary and I Pulsford) pp. 609–650. ANU Press, Canberra.

Dudley N (2008) *Guidelines for Applying Protected Areas Management Categories.* IUCN, Gland, Switzerland.

Taylor MFJ, Fitzsimons J, Sattler P (2014) *Building Nature's Safety Net 2014: A Decade of Protected Area Achievements in Australia.* WWF-Australia, Sydney.

Trewin D (2004) *Year Book Australia: 2004.* Australian Bureau of Statistics, Canberra.

Chapter 2

Marine protected areas: past, present and future – a global perspective

Mark Spalding and Lynne Zeitlin Hale

Introduction

Marine protected areas[1] (MPAs) are one of the most widely used tools for marine conservation. Their attraction stems from their documented benefits – when the MPA is well sited and effectively implemented – to ecosystems and species (Lester and Halpern 2008; Selig and Bruno 2010; Edgar *et al.* 2014; Mumby *et al.* 2014), as well as to people and coastal communities. They are recorded to enhance multiple ecosystem services including fisheries productivity (Goni *et al.* 2011; Harrison *et al.* 2012; Almany *et al.* 2013; Garcia *et al.* 2013; Moland *et al.* 2013; Green *et al.* 2014), recreational values (Roncin *et al.* 2008; Wielgus *et al.* 2010; Vianna *et al.* 2012) and the long-term security of ecosystems that defend coastlines from erosion, store carbon, improve water quality and provide a range of other benefits.

Long before the term 'marine protected area' was coined, traditional societies set aside sites in order to protect fish stocks or breeding grounds, similar in many ways to the global array of locally managed marine areas that still exist today (Johannes 2002). Australia's Royal National Park, near Sydney, was one of the first legally declared protected areas following the more modern, conservation-focused definitions. Established in 1879, largely as a terrestrial protected area, its boundaries deliberately included mangroves, seagrasses and a diverse benthic fauna (see also Chapter 11). Intertidal habitats remained a focus of all the early 'marine' protected areas, including the Sundarbans Forest Reserves in India and Bangladesh established in 1878, and Pelican Island National Wildlife Refuge in Florida in 1903. The earliest contenders for modern-style subtidal sites were the Sea Gardens near Nassau in the Bahamas established in 1897 (but sadly no longer protected) and Middle Morant Cay in Jamaica, established in 1907.

Despite these early beginnings, the declaration of MPAs was slow to catch on, due at least in part, to the overall lack of awareness of marine and coastal biodiversity, and its condition. The global extent of MPAs only really began to increase significantly from the late 1970s, notably with the declaration of the Great Barrier Reef Marine Park in Australia.

International spatial targets for protected areas coverage – set through the Convention on Biological Diversity (CBD) in 2006 – initially called for 10% global coverage of terrestrial surfaces by 2010, and the same for marine by 2012 (CBD 2006). While the terrestrial coverage exceeded this target by 2010 (Butchart *et al.* 2010), by 2012 MPAs still only covered 2.3% of the ocean surface (Spalding *et al.* 2013). This massive shortfall in MPA coverage compared with the CBD target led to considerable discussion as to why goals had not been met, what values of the ocean require protection, and how protection can be achieved (Spalding *et al.* 2013). This process, in turn, resulted both in a new CBD target and a modified approach, all captured under 'Aichi Target 11'. The headline target of 10% was retained for marine, the target date extended to 2020, but considerable and important nuance was added in defining what should be protected and how. Aichi Target 11 calls on nations to ensure that 'areas of particular importance for biodiversity and ecosystem services, are conserved through effectively and equitably managed, ecologically representative and well connected systems of protected areas and other effective area-based conservation measures, and integrated into the wider landscapes and seascapes' (CBD 2010).

In this chapter we consider progress towards that new target and the trends, changes, opportunities and needs we see for developing marine conservation efforts, especially through protected areas, over the next 5 years.

The increasing extent of MPAs

The headline target of encompassing 10% of marine space within protected areas has been widely cited and used to drive conservation progress. Even so, earlier authors had raised concerns that a global target of 10% coverage of the oceans by MPAs would take many decades beyond the CBD target (Wood *et al.* 2008). Since around 2008, however, a rapid acceleration of MPA declarations has led to a more recent publication (Spalding *et al.* 2013) suggesting that reaching such a figure by 2020 may now be possible. The recent headline statistics for MPAs are remarkable, with a doubling in extent of MPAs in the 6 years from 2003 to 2009 and then again in the 3 years to 2012. Global estimates of MPA coverage from August 2014 indicated over 12 000 sites were considered MPAs, covering 12 million km^2. These areas are equivalent to 3.4% of the world's ocean surface, with the great majority in areas that are under national jurisdiction (between 0 and 200 nautical miles from the shore) where coverage now stands at 8.4%. By contrast, protection of areas beyond national jurisdiction is only 0.25%, primarily in the North Atlantic and in in the Southern Ocean, where international agreements are sufficiently strong to support international cooperation in the designation and maintenance of such sites (Juffe-Bignoli *et al.* 2014).

How long such dramatic growth can be maintained is unclear, but in the short- to medium-term this trend of rapid expansion of large MPAs looks set to continue. The above statistics do not include an 833 000 km^2 expansion of the Pacific Remote Islands Marine National Monument (http://www.fpir.noaa.gov/MNM/mnm_prias.html) designated in September 2014, nor other large proposed sites such as the Cook Islands (> 1 million km^2), the Pitcairn Islands (830 000 km^2) nor other pledges from countries such as Chile, Russia, Gabon, Comoros, Madagascar, South Africa, Brazil and French Polynesia (http://worldparkscongress.org/about/promise_of_sydney_commitments.html and http://www.state.gov/e/oes/ocns/opa/ourocean/index.htm). These combined pledges amount to almost 5 million km^2 and would take the global coverage to 11.7% of areas under national jurisdiction, and 4.7% of the oceans in total.

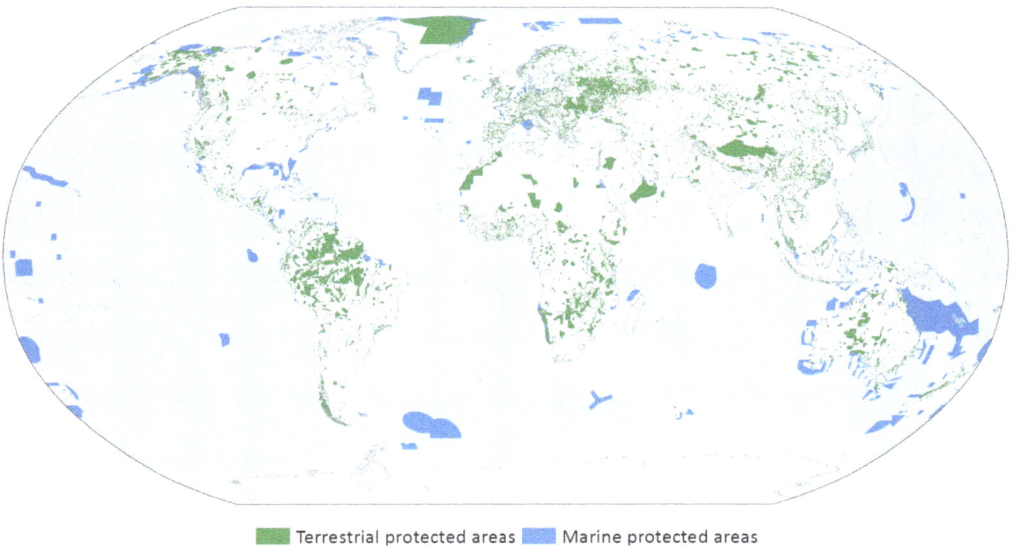

Fig. 2.1. Protected areas of the world derived from the World Database of Protected Areas (IUCN and UNEP-WCMC 2014), as published in Juffe-Bignoli *et al.* (2014), with permission.

Fig. 2.2. The British Indian Ocean Territory Marine Protected Area is the world's largest contiguous no-take MPA, including large areas of pelagic ocean as well as extensive coral reefs that house some of the highest recorded levels of fish biomass of in the entire Indian Ocean. (Photo: Mark Spalding)

The most significant contributor to the massive growth in MPA coverage, which will be re-enforced by these new declarations, is the disproportionate contribution of a small number of very large MPAs to the global statistics (Leenhardt *et al.* 2013; Spalding *et al.* 2013; Singleton and Roberts 2014). Very large MPAs make a dramatic difference to headline statistics: the 21 sites larger than 100 000 km^2 included in the 2014 statistics covered almost 8 million km^2 in total, or 75% of the global total (Table 2.1, Figs 2.1, 2.2). Such sites have been shown to be important in securing some of the last areas of remaining marine 'wilderness' (Graham and McClanahan 2013; McCauley *et al.* 2013), and larger size has been shown to be important in building effective levels of protection (Edgar *et al.* 2014). Even so, such sites can only provide a partial contribution to achieving conservation targets, which require a more systematic, representative effort to protect the ocean's biodiversity and ecosystem services (Dulvy 2013; Spalding *et al.* 2013; Butchart *et al.* 2015).

Closer scrutiny of MPA distribution shows that existing MPAs are focused on nearshore and intertidal areas, and in the tropics. Spalding *et al.* (2008) noted over 12% of a narrow coastal belt was enclosed in MPAs as far back as 2008 and more recent analyses suggest that 32% of coral reefs and 36% of mangroves are now in protected areas (Spalding *et al.* 2014a).

Looking at broad biogeographic coverage, tropical coastal realms, with the exception of the Western Indo-Pacific, have the highest levels of MPA coverage, while temperate and polar realms, with the exception of Australia and the Northern Atlantic, have much lower coverage (supplementary data in Butchart *et al.* 2015). In offshore pelagic systems, coverage is low in almost all provinces linked to the very low coverage of MPAs in areas beyond national jurisdiction. The only exceptions are those pelagic provinces that have a large extent within national jurisdictions – notably the Leeuwin Current and Non-gyral Southwest Pacific provinces off the coast of Australia (supplementary data in Butchart *et al.* 2015).

From these various observations, it is already clear that using a single global summary statistic as a metric to measure progress towards Aichi Target 11 is not enough. As described above, under Aichi Target 11, several clauses give considerable complexity and depth to the '10%' target, requiring that conservation effort not only focus on spatial extent but also on what is being protected and on how protection is achieved. It requires that: MPAs include not only areas important for biodiversity, but also areas that are important for other ecosystem services; and MPAs also be effectively and equitably managed, ecologically representative, well connected and integrated into wider landscapes and seascapes. It is therefore imperative that we assess progress not exclusively to the single metric of spatial cover, but also consider these additional elements.

Ecological representation and biodiversity

Some effort has been made to look at the ecological representation of existing MPAs. Indeed, the most widely used biogeographic classifications for coastal and ocean waters (Spalding *et al.* 2007, 2012; UNESCO 2009) were designed in large part to support such analyses. The most recent assessment in this regard has assessed both ecological representation in terms of broad biogeographic coverage as well as biodiversity coverage in terms of species ranges (Butchart *et al.* 2015). This latter study (using a global protected areas dataset from January, 2013) shows that 32% of the 232 coast and shelf ecoregions and 5% of the 37 pelagic provinces have > 10% of their area covered. In the case of coastal and shelf ecoregions, it is perhaps also noteworthy that 17 have over 75% MPA coverage, largely driven by the influence of very large MPAs, while at the opposite extreme, 65 (28%) of coast and shelf ecoregions have < 1% MPA coverage (Butchart *et al.* 2015).

Table 2.1. Marine protected areas larger than 100 000 km^2 on the World Database of Protected Areas (IUCN and UNEP-WCMC 2014), as of August, 2014

Site name and designation/s	Area (km^2)	Year of designation or most recent alteration	Includes no-take areas[c]	Country
Coral Sea Parc Naturel	1 292 967	2014	Y	France
South Georgia and South Sandwich Islands Marine Protected Area	1 070 000	2012	Y	UK
Coral Sea Commonwealth Marine Reserve	989 842	2012	Y	Australia
British Indian Ocean Territory (Chagos) Marine Protected Area	640 000	2010	Y	UK
Kermadec Benthic Protection Area	469 276	2007	Y	New Zealand
Phoenix Islands Protected Area[a]	410 500	2006		Kiribati
Papahanaumokuakea Marine National Monument[a]	362 024	2006	Y	USA
Great Barrier Reef Marine Park[a]	344 004	2004	Y	Australia
Charlie-Gibbs South High Seas Marine Protected Area (OSPAR)	324 123	2010		Areas beyond national jurisdiction
South-west Corner Commonwealth Marine Reserve	271 898	2012	Y	Australia
Marianas Trench Marine National Monument	246 608	2009		USA
Pacific Remote Islands Marine National Monument[b]	225 039	2009	Y	USA
Norfolk Commonwealth Marine Reserve	188 443	2012	Y	Australia
Reef Fish Longline and Buoy Federal Fishery Management Zone	176 525	1990		USA
Macquarie Island Commonwealth Marine Reserve	161 895	1999	Y	Australia
Motu Motiro Hiva Parque Marino	150 000	2010	Y	Chile
Argo-Rowley Terrace Commonwealth Marine Reserve	146 099	2012	Y	Australia
Galapagos Marine Reserve[a]	133 000	1996		Ecuador
Lord Howe Commonwealth Marine Reserve	110 139	2012	Y	Australia
East Florida Coast Closed Area Federal Fishery Management Zone	102 035	2000		USA

[a]These sites are also recognised as World Heritage Sites, although the declared area under this convention or other international designations for the same sites may not be identical to those areas listed here for the nationally designated MPA

[b]Pacific Remote Islands Marine National Monument was extended in September 2014 to a total of 1 057 761 km^2

[c]Although most of these sites are indicated to have some areas of no-take, the total proportion of no-take is highly variable

In terms of biodiversity, range-maps for some 4000 marine species (cartilaginous fish, selected families of bony fish, warm water reef-building corals, lobsters and crayfish, mangroves and seagrasses) were used to assess progress towards protecting representative range targets (an approximate minimum proportion of the range required to be protected to ensure the long-term security of that species, scaled according to overall range size for the species). This assessment showed that some 47% of the bony fishes and 78% of the stony corals (in both cases the samples were biased towards warm-water families) have reached their representative range targets, but that the remaining groups are all at less than 25%. (It should further be noted that the study took no account of the actual level of protection these species were getting in these protected areas, where in reality this may be limited because some sites might still allow harvest, while others may not be enforced.) Such numbers indicate that there is still a long way to go towards achieving systematic coverage of biodiversity. At the same time, they also show greater progress than for many terrestrial taxa (mammals and amphibians are at 46% and 22% covered, respectively, despite a global average of over 14% of terrestrial area coverage by protected areas) (Butchart *et al.* 2015).

Ecosystem services

The importance of conservation for human benefit has been an integral part of national and international policy and prioritisation since the first international agreements on biodiversity. However, it is only with Aichi Target 11 that ecosystem services were placed alongside biodiversity as a driving metric for prioritising conservation efforts. This focus has not been matched with specific suggestions for approaches to ensure its implementation, and indeed there are few efforts, or even methodologies, to investigate how well marine ecosystem services are being conserved.

In an earlier publication (Spalding *et al.* 2013), we observed that many ecosystem services are tightly linked to local human populations who receive direct benefits from coastal protection, food provision, tourism income and so forth (Figs 2.3, 2.4). Despite this, MPA coverage appears to show some level of inverse correlation with coastal population densities. This is perhaps not surprising: the designation and maintenance of MPAs comes at a high cost, both in dollar terms and in terms of conflicts with potential users, making designation away from areas of high populations or intensive existing use potentially 'easier'. Likewise, from a biodiversity perspective, marine ecosystems close to human populations are often more threatened, or indeed have already suffered degradation, and thus offer lower values in terms of protecting secure populations of key species or high-diversity systems. Degraded and threatened areas do not always rise to the top in biodiversity prioritisations.

Based on these observations, it has been hypothesised that the current MPA network takes little or no account of many critical ecosystem services and, given other pressures, may actually be favouring the protection of areas of low ecosystem service value (Spalding *et al.* 2013). Of course, not all ecosystem service values are closely correlated with local human beneficiaries – oxygen production and carbon storage and sequestration are clearly not – but many direct and tangible values are linked in this way. It is these services – fisheries, tourism, coastal protection and water purification – that may be being under-represented within the current portfolio of MPAs.

In order to take this argument further, there is a need to develop consistent classification, quantification and guidelines for best practice in the implementation of Aichi Target 11 with respect to conserving ecosystem services. Better models and maps at different scales would help in the development of policy and implementation. The quantification of the ecosystem services that specific marine habitats provide is beginning to advance at

Fig. 2.3. Cabo Pulmo National Park in the Gulf of California was established by a local fishing community in 1995. Spectacular recovery of fish populations followed and the MPA is now a popular dive location, providing significant economic benefits for the local population. (Photo: Mark Spalding)

Fig. 2.4. Large protected areas such as Jardines de la Reina National Park in Cuba enable the maintenance of multiple coastal habitats, and the tight ecological connections between them. This site includes vast areas of seagrass, mangrove and coral reefs, and many species such as these snapper move between habitats during their life histories. (Photo: Mark Spalding)

multiple scales (zu Ermgassen *et al.* 2013; Hutchison *et al.* 2014). New tools are helping to bring such information to much broader audiences than the 'conservation engaged' who have traditionally driven the agenda for MPA designation (Guerry *et al.* 2012; White *et al.* 2012), but much more remains to be done.

Effectiveness

The effectiveness of MPAs in conserving natural values depends on multiple factors including: levels of threat or pressure and whether those pressures and threats originate within or outside the MPA; the extent and boundary of the site relative to target ecosystems and species; the recognition and acceptance of that site by local communities; the particular regulatory regime; and the degree of compliance to and enforcement of the rules (Agardy *et al.* 2011; Fox *et al.* 2012; Edgar *et al.* 2014).

Many protected areas suffer problems of poor implementation and the problem of 'paper parks' (sites lacking any discernible management or enforcement) has been widely acknowledged (Agardy *et al.* 2011). In other cases, the design of MPAs may be inadequate in terms of size or boundary design (Claudet *et al.* 2008), or the regulatory regime may simply be inadequate to achieve conservation objectives. Many MPAs remain open to levels of use and extractive activities that reduce or prevent significant ecological benefits (Lester and Halpern 2008), while there appears to be some scaling of ecological benefits depending on the level of protection from the lowest levels up to full exclusion of extractive activities (Bartlett *et al.* 2009a; McClanahan 2010; Harrison *et al.* 2012; Sala *et al.* 2013).

There is no global overview of overall effectiveness of MPAs in meeting their stated objectives. Burke *et al.* (2011) included a rapid survey of coral reef protected areas managers and other experts and generated basic scores of ecological effectiveness for over 1000 coral reef protected areas, grouping them by number, with the outcome being: 15% effective, 38% partially effective and 47% ineffective. More recently Edgar *et al.* (2014) used field data to identify five factors that contributed most to improved conservation outcomes. They found that sites that were closed to extractive activities (no-take), effectively enforced, old, large and isolated (far removed from human population centres), or that had at least four of these five features, showed distinct improvements in terms of ecological conditions of fish populations (Edgar *et al.* 2014).

Equitability

Not all MPAs – especially some older MPAs—were designated with significant participation or buy-in from key local stakeholders (De Santo 2013; Leenhardt *et al.* 2013). In such cases, compliance has been difficult to achieve, often making such places protected areas in name only (Agardy *et al.* 2011). Fortunately, the need to engage communities in all aspects of MPA design and designation is now recognised. Globally there is a trend of increasing community-led or supported efforts, including many smaller sites designated under local or provincial rather than national regulatory regimes (Bartlett *et al.* 2009b; Jones *et al.* 2011). The term 'locally managed marine area' (LMMA) is now widely used, particularly in the Indo-Pacific, to describe areas that are under local ownership or management (Fig. 2.5). Many LMMAs include areas and management regimes that generate considerable conservation benefits, but it is not clear that all such areas meet the definition of MPAs.

Substantial public engagement in MPA selection, design and development has also become more widespread, including the extensive consultation around the expansion of Australia's marine reserve system in Commonwealth Government-controlled waters lead-

Fig. 2.5. A Locally Managed Marine Area established by the local community at Pere, southern Manus Island, Papua New Guinea, protects tropical reefs. (Photo: James Fitzsimons)

ing to designations in 2012 (Wescott and Fitzsimons 2011; see also Chapter 4) and the UK Marine Conservation Zones consultation exercise, which commenced in 2010 (Jones 2012). In the US, locally driven marine protection recently received a boost with the announcement that for the first time in two decades, NOAA is inviting communities across the nation to nominate their most treasured places in US marine and Great Lakes waters for consideration as national marine sanctuaries (NOAA 2014).

Concurrent with these trends, there remains much controversy around the designation of MPAs, especially with commercial and recreational fishers. As noted in other chapters of this book (e.g. Chapters 20 and 21), this is true in Australia and many other countries (Bennett and Dearden 2014; Weigel *et al.* 2014). This controversy remains despite multiple efforts to document MPA benefits to these groups by both conservation and fisheries agencies (Gaines *et al.* 2010; FAO 2011; Harrison *et al.* 2012; Garcia *et al.* 2013; Sala *et al.* 2013; Ballantine 2014; Green *et al.* 2014). Continuing and enhancing efforts to engage with these important stakeholders remains a major need and challenge for advancing MPAs.

Other effective area-based conservation measures

Although the IUCN definition of a protected area can include sites that are not legally gazetted (sites are to be secured by 'legal or other effective means' (Dudley 2008)) it does require that nature conservation be a primary management objective. Aichi Target 11 is still more broadly inclusive, calling for conservation targets to be achieved not only

Fig. 2.6. Asilomar State Marine Reserve, a no-take MPA in California, USA, is embedded in Monterey Bay National Marine Sanctuary, a larger MPA which allows multiple use. (Photo: James Fitzsimons)

through protected areas, but also 'other effective area-based conservation measures'. Although further guidelines are not specified, the inclusion of this language came from recognition that considerable conservation benefits are derived from other management measures, including many fisheries controls, and from local ownership and management regimes such as LMMAs. These are places where nature conservation may not be stated as a primary management objective, but may still be an important outcome. There remains a pressing need to ascertain what should, or should not, be included in this terminology.

Connected and integrated into wider landscapes and seascapes

MPAs cannot achieve their objectives if they exist in 'a sea of destruction' – the fluid nature of the oceans means that *ex situ* conditions have a strong influence on any site, through the movement of pollutants to the impact of adjacent overfishing on fish stocks and larval supply. Hence, there is increasing recognition that MPAs need be embedded in more comprehensive approaches to ocean management (Agardy *et al.* 2011). Such approaches recognise that society has multiple objectives for the use of ocean spaces and resources and that multi-sector, multi-objective management can provide opportunities for resolving competing uses and encourage sustainable use and management of ocean resources at large, socially and ecologically meaningful scales (see also Chapter 19). At the same time, MPAs and MPA networks are important components of multi-objective sea-use plans. A growing number of places – at scales from bays to full Exclusive Economic Zones – are now developing multi-objective sea use plans; however, full implementation of such plans is still limited (Agardy *et al.* 2011) (Fig. 2.6).

Challenges

While MPAs may be a rapidly growing use of marine and coastal space, they are not alone. Other activities are also fast expanding including: oil and gas exploration and development; extractive industries from aggregate extraction to deep ocean mining; shipping and ports; marine-based renewables such as wind, wave and tidal energy; traditional and novel mariculture in both coastal and offshore waters; and the continuing growth of tourism and recreation in marine settings from activities as diverse as cruise traffic to whale-watching to kite-surfing.

There are also growing *ex situ* pressures coming from land-based activities: the expansion and intensification of agriculture brings increased nutrient loading; novel chemicals from pesticides to pharmaceuticals are appearing; and the rapid development of coastal lands, including the conversion of vast areas of intertidal land to industrial, urban and agricultural landscapes continues.

Climate change and increased carbon dioxide levels are bringing additional changes to the ocean and challenges for marine conservation and MPAs. In the highest latitudes, decreasing ice cover is leading to rapid ecological change (Ducklow *et al.* 2013) and encouraging new frontiers for human expansion and ocean use. With the rapid pace of likely development, identification and designation of Arctic and Antarctic MPAs is proving challenging (Cressey 2013) (see also Chapter 6). Challenges abound in the tropics as well, where coral bleaching is having significant impacts. In all areas – including temperate waters – dramatic changes in species' spatial distributions are occurring, making the fixed boundaries of many MPAs a challenge. Ocean acidification represents a new and still poorly understood threat, which could dramatically alter patterns of ocean productivity and entire ecosystem characteristics. The role of MPAs in tackling this threat remains unclear.

A rich body of literature is building around the development of resilient protected area networks to help reduce the long-term impacts of climate change around coral reefs (Game *et al.* 2009; Maynard *et al.* 2010; Mumby *et al.* 2010, 2014; The Nature Conservancy 2010), but also around other ecosystems (Côté and Darling 2010; Montefalcone *et al.* 2011; Brock *et al.* 2012). Rising seas could impact some intertidal communities, but protected areas could be designed to allow a natural migration (Titus *et al.* 2009) where possible. More attention is being directed to actions that enable recovery and restoration of ecosystems including the active restoration of damaged systems such as coral reefs and shellfish beds (Boorman and Garbutt 2012; Schrack *et al.* 2012), as well as even more extreme measures such as the translocation of more robust genotypes between regions (Hoegh-Guldberg *et al.* 2008).

Climate change also presents an opportunity for MPAs, potentially increasing support for protecting previously ignored habitats and locations. As sea level continues to rise and storms become more intense, the important role that mangroves, coral reefs, saltmarshes and bivalve reefs play in reducing risks from erosion, inundation and wind damage is becoming better known and appreciated (Spalding *et al.* 2014b). This presents opportunities to create protected areas with risk reduction of coastal communities being a key objective alongside biodiversity conservation (Beck 2014). Similarly, as the role of the ocean and critical habitats within it for carbon sequestration become better known and understood, this value can become a motivator for, and joint objective of, MPAs.

Future directions

MPAs have played, and will continue to play, a central role in conserving marine waters and habitats, whether for their inherent wonder, biodiversity or for the many diverse

Fig. 2.7. Cinque Terre Marine Protected Area, Italy, protects marine ecosystems in the Mediterranean Sea and, with the adjoining terrestrial Cinque Terre National Park, is a World Heritage Area. (Photo: James Fitzsimons)

benefits they provides to human kind. While terrestrial protected areas now account for some 15% of the land surface of the planet, MPAs are at 3.4%. Even so, designations are accelerating and growth in this estate will continue for the foreseeable future. Given this, there is a clear need to direct and to steer the process of MPA designation and management, to maximise the efficacy of MPAs in delivering benefits, not only in helping to protect nature, but also in enhancing the benefits of nature for people and enriching the lives of the billions of people who live in, rely on or visit, coastal areas (Fig. 2.7).

The details contained within Aichi Target 11 contain many elements of such a vision.

Maximising the benefits to biodiversity

Conservation is a primary aim of protected areas and considerable efforts have been undertaken to enumerate different metrics, including measures of species richness and endemism, biogeographic uniqueness, key habitats and biomes, and functional properties such as productivity. Such measures are already considered at finer scales of reserve design, including minimum sizes of individual MPAs and MPA network design properties.

Maximising the benefits to people

In contrast to biodiversity prioritisation, we are only just beginning to develop the science around ecosystem service quantification and this has yet to be built into conservation planning in a systematic manner. This must represent a key priority, not only for the direct benefits that will accrue to human society, but because the beneficiaries will themselves become proponents of marine conservation, potentially fuelling an increase in demand for better natural resource management in the sea.

Targeting key threats

The term protected areas implies protection from something and many references have shown the tremendous value that MPAs can have in combatting threats, notably direct *in situ* threats such as over-extraction and destructive fishing. Land-based sources of pollution, and land-based changes to freshwater and sediment flows, are another major threat in coastal waters, and linking MPAs to adjacent terrestrial conservation efforts can greatly help reduce these threats. There is also a growing body of science supporting the utilisation of MPAs to build long-term resilience to future climate impacts (Brock *et al.* 2012; Billé *et al.* 2013; Mumby *et al.* 2014). One implication of such observations is that protected areas may be of particular value where threats are already high, or likely to become so. Further, there is a strong need to enhance MPA design to deal with *ex situ* threats, which requires consideration of upstream and land-based sources of threats, and of placing MPAs in wider spatial management frameworks where such threats may be controlled under integrated management approaches.

Protecting the high seas

The biggest gap in terms of MPA coverage on the global map is in areas beyond national jurisdiction: the high seas. Attention needs to target these areas, which have been widely over-fished and are now also subject to growing pressures from shipping and deep sea mineral exploitation (Ban *et al.* 2014; Rogers *et al.* 2014). Although there is a strong call for MPAs in this setting, there may also be opportunities for strengthening existing management systems such as the United Nations Convention on the Law of the Sea (UNCLOS) and regional fisheries management organisations towards improved sustainability (Global Ocean Commission 2014).

Embracing different forms of management

Aichi Target 11 effectively blurred the distinction between MPAs and other forms of marine spatial management, both in its call for 'other effective area-based conservation measures' and by calling for such areas to be integrated into wider landscapes and seascapes. In some ways, this expansion of vision beyond traditional MPAs came in recognition of the extremely different levels of protection provided by MPAs and of the fast-growing application of other forms of spatial management in the oceans. It will be valuable to recognise such different measures in their entirety. Holistic coastal and ocean planning will be a key tool moving forward, and spatial targets towards such planning should tend towards 100% of the ocean. Within this, protected areas will also need to increase, but a diversity of approaches will be important. The benefits provided by strict protection might be used to set a sub-target for no-take MPAs, to ensure the development of a network of sites delivering highest possible value in terms of biodiversity conservation, and in some settings providing important side benefits for fisheries enhancement and increased recreational value. Alongside, new approaches for achieving protection must be considered, from community ownership to licensing and quota-shares, to a broader consideration of management tools where conservation may not be a primary purpose, but may nonetheless be a positive consequence.

Perhaps the most important idea that needs to become more generally accepted as we consider MPAs moving forward is the following. For conservation to succeed, not only in the ocean, but across the globe, we must show that it is essential to making people's lives better. If this can be demonstrated, at places, and to nations and communities, the demand for MPAs and support for existing MPAs, and indeed for wider and wiser marine conservation and management, will only grow.

Acknowledgements

We thank UNEP-WCMC for the provision of data and permission to use the map in Fig. 2.1. The Lyda Hill Foundation is gratefully acknowledged for providing partial support for the authors as these ideas have been developed.

Biographies

Mark Spalding is a senior marine scientist with The Nature Conservancy. He has worked extensively on global scale assessments of marine and coastal biodiversity, biogeography, threats and conservation efforts, including a series of papers looking at the global coverage of MPAs. Under the banner of 'mapping ocean wealth', Mark is currently leading new efforts to understand, model and map the global distribution of marine ecosystem service values.

Lynne Zeitlin Hale transitioned out of her position as The Nature Conservancy's (TNC) Managing Director for Oceans in January, 2015. She currently serves as a Senior Marine Conservation Fellow focused on a range of projects in the Asia Pacific and Oceania regions. She played a central role in launching TNC's Great Southern Seascapes project in Australia along with serving as a Senior Advisor to the Coral Triangle and Pacific programs. She is based in the Graduate School of Oceanography, University of Rhode Island.

Endnote

1 The most widely used definition of MPAs derives from the broader definition of a protected area, as 'a clearly defined geographical space, recognised, dedicated and managed, through legal or other effective means, to achieve the long-term conservation of nature with associated ecosystem services and cultural values' (Dudley 2008, p. 8); while the definition of marine is taken from an earlier definition of marine protected areas, which included the phrase 'any area of intertidal or subtidal terrain together with its overlying water…' (Kelleher 1999, p. 98). Combined, such a definition includes all protected areas that include any intertidal or subtidal waters, with the proviso that the marine components are indeed a target of the conservation efforts.

References

Agardy T, di Sciara GN, Christie P (2011) Mind the gap: addressing the shortcomings of marine protected areas through large scale marine spatial planning. *Marine Policy* **35**, 226–232. doi:10.1016/j.marpol.2010.10.006.

Almany G, Hamilton RJ, Bode M, Matawai M, Potuku T, Saenz-Agudelo P, *et al.* (2013) Dispersal of grouper larvae drives local resource sharing in a coral reef fishery. *Current Biology* **23**, 626–630. doi:10.1016/j.cub.2013.03.006.

Ballantine B (2014) Fifty years on: lessons from marine reserves in New Zealand and principles for a worldwide network. *Biological Conservation* **176**, 297–307. doi:10.1016/j.biocon.2014.01.014.

Ban NC, Bax NJ, Gjerde KM, Devillers R, Dunn DC, Dunstan PK, *et al.* (2014) Systematic conservation planning: a better recipe for managing the High Seas for biodiversity conservation and sustainable use. *Conservation Letters* 7, 41–54. doi:10.1111/conl.12010.

Bartlett CY, Manua C, Cinner J, Sutton S, Jimmy R, South R, *et al.* (2009a) Comparison of outcomes of permanently closed and periodically harvested coral reef reserves. *Conservation Biology* **23**, 1475–1484. doi:10.1111/j.1523-1739.2009.01293.x.

Bartlett CY, Pakoa K, Manua C (2009b) Marine reserve phenomenon in the Pacific islands. *Marine Policy* **33**, 673–678. doi:10.1016/j.marpol.2009.01.004.

Beck MW (Ed.) (2014) *Coasts at Risk: An Assessment of Coastal Risks and the Role of Environmental Solutions.* United Nations University – Institute for Environment and Human Security, Bonn, Germany, The Nature Conservancy, Arlington VA, and the Coastal Resources Center at the University of Rhode Island Graduate School of Oceanography, Narragansett RI, <http://www.crc.uri.edu/download/SUC09_CoastsatRisk.pdf>

Bennett NJ, Dearden P (2014) Why local people do not support conservation: community perceptions of marine protected area livelihood impacts, governance and management in Thailand. *Marine Policy* **44**, 107–116. doi:10.1016/j.marpol.2013.08.017.

Billé R, Kelly R, Biastoch A, Harrould-Kolieb E, Herr D, Joos F, *et al.* (2013) Taking action against ocean acidification: a review of management and policy options. *Environmental Management* **52**, 761–779. doi:10.1007/s00267-013-0132-7.

Boorman LA, Garbutt A (2012) Restoration strategies for intertidal saltmarshes. In *Treatise on Estuarine and Coastal Science.* Vol. 10. (Eds L Chicharo and M Zalewski) pp. 189–216. Elsevier, Amsterdam.

Brock RJ, Kenchington E, Martínez-Arroyo A (Eds) (2012) *Scientific Guidelines for Designing Resilient Marine Protected Area Networks in a Changing Climate.* Commission for Environmental Cooperation: Montréal, Canada, <http://www.cec.org/Storage/141/16690_MPA_Scientific_guidelines_en_web.pdf>

Burke L, Reytar K, Spalding M, Perry AL (2011) *Reefs at Risk Revisited.* World Resources Institute, The Nature Conservancy, WorldFish Center, International Coral Reef Action Network, UNEP World Conservation Monitoring Centre and Global Coral Reef Monitoring Network, Washington DC.

Butchart SHM, Walpole M, Collen B, van Strien A, Scharlemann JPW, Almond REA, *et al.* (2010) Global biodiversity: indicators of recent declines. *Science* **328**, 1164–1168. doi:10.1126/science.1187512.

Butchart SHM, Clarke M, Smith B, Sykes R, Scharlemann JPW, Harfoot M, *et al.* (2015) Shortfalls and solutions for meeting national and global conservation area targets. *Conservation Letters* **8**, 329–337. doi: 10.1111/conl.12158.10.1111/conl.12158

CBD (2006) *COP 8 – Eighth Ordinary Meeting of the Conference of the Parties to the Convention on Biological Diversity. Curitiba, 20–31 March 2006. Decision VIII/15: Framework for Monitoring Implementation of the Achievement of the 2010 Target and Integration of Targets into the Thematic Programmes of Work. Annex 4.* United Nations Environment Programme, Nairobi.

CBD (2010) *COP 10 – Tenth Meeting of the Conference of the Parties to the Convention on Biological Diversity. Nagoya, Japan 1–29 October 2010. Decision X/2. Strategic Plan for Biodiversity 2011–2020.* United Nations Environment Programme, Nairobi.

Claudet J, Osenberg C, Benedetti-Cecchi L, Domenici P, García-Charton J-A, Pérez-Ruzafa A, *et al.* (2008) Marine reserves: size and age do matter. *Ecology Letters* **11**, 481–489. doi:10.1111/j.1461-0248.2008.01166.x.

Côté IM, Darling ES (2010) Rethinking ecosystem resilience in the face of climate change. *PLoS Biology* **8**, e1000438. doi:10.1371/journal.pbio.1000438.

Cressey D (2013) Third time unlucky for Antarctic protection bid. *Nature* <http://www.nature.com/news/third-time-unlucky-for-antarctic-protection-bid-1.14085>

De Santo EM (2013) Missing marine protected area (MPA) targets: how the push for quantity over quality undermines sustainability and social justice. *Journal of Environmental Management* **124**, 137–146. doi:10.1016/j.jenvman.2013.01.033.

Ducklow HW, Fraser WR, Meredith MP, Stammerjohn SE, Doney SC, Martinson DG, *et al.* (2013) West Antarctic Peninsula: an ice-dependent coastal marine ecosystem in transition. *Oceanography* **26**, 190–203. doi:10.5670/oceanog.2013.62.

Dudley N (2008) *Guidelines for Applying Protected Areas Management Categories.* IUCN, Gland, Switzerland.

Dulvy NK (2013) Super-sized MPAs and the marginalization of species conservation. *Aquatic Conservation: Marine and Freshwater Ecosystems* **23**, 357–362. doi:10.1002/aqc.2358.

Edgar GJ, Stuart-Smith RD, Willis TJ, Kininmonth S, Baker SC, Banks S, *et al.* (2014) Global conservation outcomes depend on marine protected areas with five key features. *Nature* **506**, 216–220. doi:10.1038/nature13022.

FAO (2011) *Fisheries Management. 4. Marine Protected Areas and Fisheries.* Food and Agriculture Organization of the United Nations, Rome.

Fox HE, Mascia MB, Basurto X, Costa A, Glew L, Heinemann D, *et al.* (2012) Reexamining the science of marine protected areas: linking knowledge to action. *Conservation Letters* **5**, 1–10. doi:10.1111/j.1755-263X.2011.00207.x.

Gaines SD, White C, Carr MH, Palumbi SR (2010) Designing marine reserve networks for both conservation and fisheries management. *Proceedings of the National Academy of Sciences of the United States of America* **107**, 18286–18293. doi:10.1073/pnas.0906473107.

Game ET, Bode M, McDonald-Madden E, Grantham HS, Possingham HP (2009) Dynamic marine protected areas can improve the resilience of coral reef systems. *Ecology Letters* **12**, 1336–1346. doi:10.1111/j.1461-0248.2009.01384.x.

Garcia SM, Hénichart L-M, Boncoeur J, Alban F, de Monbrison D (2013) *Marine Protected Areas in Fisheries Management. Synthesis on the State of the Art.* Commission Sous-Régionale des Pêches, Dakar-Fann, Sénégal.

Global Ocean Commission (2014) *From Decline to Recovery. A Rescue Package for the Global Ocean.* Global Ocean Commission, Oxford, UK.

Goni R, Badalamenti F, Tupper MH (2011) Fisheries – Effects of marine protected areas on local fisheries: evidence from empirical studies. In *Marine Protected Areas: A Multidisciplinary Approach.* (Ed. J Claudet) pp. 72–98. Cambridge University Press, Cambridge, UK.

Graham NAJ, McClanahan TR (2013) The last call for marine wilderness? *BioScience* **63**, 397–402. doi:10.1525/bio.2013.63.5.13.

Green AL, Fernandes L, Almany G, Abesamis R, McLeod E, Aliño PM, *et al.* (2014) Designing marine reserves for fisheries management, biodiversity conservation, and climate change adaptation. *Coastal Management* **42**, 143–159. doi:10.1080/08920753.2014.877763.

Guerry AD, Ruckelshaus MH, Arkema KK, Bernhardt JR, Guannel G, Kim C-K, *et al.* (2012) Modeling benefits from nature: using ecosystem services to inform coastal and marine spatial planning. *International Journal of Biodiversity Science, Ecosystem Services & Management* **8**, 107–121. doi:10.1080/21513732.2011.647835.

Harrison HB, Williamson DH, Evans RD, Almany GR, Thorrold SR, Russ GR, *et al.* (2012) Larval export from marine reserves and the recruitment benefit for fish and fisheries. *Current Biology* **22**, 1023–1028. doi:10.1016/j.cub.2012.04.008.

Hoegh-Guldberg O, Hughes L, McIntyre S, Lindenmayer DB, Parmesan C, Possingham HP, *et al.* (2008) Assisted colonization and rapid climate change. *Science* **321**, 345–346. doi:10.1126/science.1157897.

Hutchison J, Manica A, Swetnam R, Balmford A, Spalding M (2014) Predicting global patterns in the carbon storage of mangrove forests. *Conservation Letters* 7, 233–240. doi:10.1111/conl.12060.

IUCN and UNEP-WCMC (2014) *The World Database on Protected Areas (WDPA).* August 2014. International Union for Conservation of Nature (IUCN) and United Nations Environment Programme World Conservation Monitoring Centre (UNEP-WCMC), Cambridge, UK, <www.protectedplanet.net>.

Johannes RE (2002) The renaissance of community-based marine resource management in Oceania. *Annual Review of Ecology and Systematics* **33**, 317–340. doi:10.1146/annurev.ecolsys.33.010802.150524.

Jones PJS (2012) Marine protected areas in the UK: challenges in combining top-down and bottom-up approaches to governance. *Environmental Conservation* **39**, 248–258. doi:10.1017/S0376892912000136.

Jones P, Qui W, De Santo EM (2011) *Governing marine protected areas: getting the balance right.* Technical Report. United Nations Environment Programme, Nairobi.

Juffe-Bignoli D, Burgess N, Bingham H, Belle E, de Lima M, Deguignet M, *et al.* (2014) *Protected Planet Report 2014. Tracking Progress Towards Global Targets for Protected Areas.* UNEP World Conservation Monitoring Centre, Cambridge, UK.

Kelleher G (1999) *Guidelines for Marine Protected Areas.* IUCN-World Conservation Union, Gland, Switzerland and Cambridge, UK.

Leenhardt P, Cazalet B, Salvat B, Claudet J, Feral F (2013) The rise of large-scale marine protected areas: conservation or geopolitics? *Ocean and Coastal Management* **85**, 112–118. doi:10.1016/j.ocecoaman.2013.08.013.

Lester SE, Halpern BS (2008) Biological responses in marine no-take reserves versus partially protected areas. *Marine Ecology Progress Series* **367**, 49–56. doi:10.3354/meps07599.

Maynard J, Marshall P, Johnson J, Harman S (2010) Building resilience into practical conservation: identifying local management responses to global climate change in the southern Great Barrier Reef. *Coral Reefs* **29**, 381–391. doi:10.1007/s00338-010-0603-8.

McCauley DJ, Power EA, Bird DW, McInturff A, Dunbar RB, Durham WH, *et al.* (2013) Conservation at the edges of the world. *Biological Conservation* **165**, 139–145. doi:10.1016/j.biocon.2013.05.026.

McClanahan TR (2010) Effects of fisheries closures and gear restrictions on fishing income in a Kenyan coral reef. *Conservation Biology* **24**, 1519–1528. doi:10.1111/j.1523-1739.2010.01530.x.

Moland E, Olsen EM, Knutsen H, Garrigou P, Espeland SH, Kleiven AR, *et al.* (2013) Lobster and cod benefit from small-scale northern marine protected areas: inference from an empirical before–after control-impact study. *Proceedings of the Royal Society B: Biological Sciences* **280**, 20122679. doi:10.1098/rspb.2012.2679.

Montefalcone M, Parravicini V, Bianchi C (2011) Quantification of coastal ecosystem resilience. In *Treatise on Estuarine and Coastal Science.* Vol. 10. (Eds E Wolanski and D McLusky) pp. 49–70. Academic Press, Waltham, MA.

Mumby PJ, Elliott IA, Eakin CM, Skirving W, Paris CB, Edwards HJ, *et al.* (2010) Reserve design for uncertain responses of coral reefs to climate change. *Ecology Letters* **13**, 132–140. doi:10.1111/j.1461-0248.2010.01562.x.

Mumby PJ, Wolff NH, Bozec Y-M, Chollett I, Halloran P (2014) Operationalizing the resilience of coral reefs in an era of climate change. *Conservation Letters* 7, 176–187. doi:10.1111/conl.12047.

NOAA (2014) *Sanctuary Nomination Process.* NOAA, Silver Spring, MD, <http://www.nominate.noaa.gov/>.

Rogers AD, Sumaila UR, Hussain SS, Baulcomb C (2014) *The High Seas and Us: Understanding the Value of High-Seas Ecosystems.* Global Ocean Commission, Oxford, UK.

Roncin N, Alban F, Charbonnel E, Crec'hriou R, de la Cruz Modino R, Culioli J-M, *et al.* (2008) Uses of ecosystem services provided by MPAs: how much do they impact the local economy? A southern Europe perspective. *Journal for Nature Conservation* **16**, 256–270. doi:10.1016/j.jnc.2008.09.006.

Sala E, Costello C, Dougherty D, Heal G, Kelleher K, Murray JH, *et al.* (2013) A general business model for marine reserves. *PLoS ONE* **8**, e58799. doi:10.1371/journal.pone.0058799.

Schrack E, Beck M, Brumbaugh R, Crisley K, Hancock B (2012) *Restoration Works: Highlights from a Decade of Partnership Between The Nature Conservancy and the National Oceanic and Atmospheric Administration's Restoration Center.* The Nature Conservancy, Arlington, VA, <http://www.nature.org/ourinitiatives/habitats/oceanscoasts/restoration-works.pdf>.

Selig ER, Bruno JF (2010) A global analysis of the effectiveness of marine protected areas in preventing coral loss. *PLoS ONE* **5**, e9278. doi:10.1371/journal.pone.0009278.

Singleton RL, Roberts CM (2014) The contribution of very large marine protected areas to marine conservation: giant leaps or smoke and mirrors? *Marine Pollution Bulletin* **87**, 7–10. doi:10.1016/j.marpolbul.2014.07.067.

Spalding MD, Fox HE, Allen GR, Davidson N, Ferdaña ZA, Finlayson M, *et al.* (2007) Marine ecoregions of the world: a bioregionalization of coast and shelf areas. *BioScience* **57**, 573–583. doi:10.1641/B570707.

Spalding M, Fish L, Wood L (2008) Towards representative protection of the world's coasts and oceans – progress, gaps and opportunities. *Conservation Letters* **1**, 217–226. doi:10.1111/j.1755-263X.2008.00030.x.

Spalding M, Agostini V, Grant S, Rice J (2012) Pelagic provinces of the world: a biogeographic classification of the world's surface pelagic waters. *Ocean and Coastal Management* **60**, 19–30. doi:10.1016/j.ocecoaman.2011.12.016.

Spalding MD, Meliane I, Milam A, Fitzgerald C, Hale LZ (2013) Protecting marine spaces: global targets and changing approaches. *Ocean Yearbook* **27**, 213–248. doi:10.1163/22116001-90000160.

Spalding M, Burke L, Hutchison J, zu Ermgassen P, Thomas H, Ashpole J, *et al.* (2014a) 'Attaining Aichi Target 11: How well are marine ecosystem services covered by protected areas?' Policy brief prepared for the World Parks Congress, Sydney, November, 2014, <http://oceanwealth.org/wp-content/uploads/2015/02/Attaining-Aichi-Target-11-WPC-Discussion-Paper.pdf>.

Spalding MD, McIvor AL, Beck MW, Koch EW, Möller I, Reed DJ, *et al.* (2014b) Coastal ecosystems: a critical element of risk reduction. *Conservation Letters* **7**, 293–301. doi:10.1111/conl.12074.

The Nature Conservancy (2010) *R2 Reef Resilience: Building Resilience into Coral Reef Conservation.* The Nature Conservancy, Arlington, VA, <www.reefresilience.org/index.html>.

Titus JG, Hudgens DE, Trescott DL, Craghan M, Nuckols WH, Hershner CH, *et al.* (2009) State and local governments plan for development of most land vulnerable to rising sea level along the US Atlantic coast. *Environmental Research Letters* **4**, 044008. doi:10.1088/1748-9326/4/4/044008.

UNESCO (Ed.) (2009) *Global Open Oceans and Deep Seabed (GOODS) – Biogeographic Classification.* Intergovernmental Oceanographic Commission (UNESCO-IOC), Paris, <http://www.iode.org/index.php?option=com_oe&task=viewDocumentRecord&docID=3931>

Vianna GMS, Meekan MG, Pannell DJ, Marsh SP, Meeuwig JJ (2012) Socio-economic value and community benefits from shark-diving tourism in Palau: a sustainable use of reef shark populations. *Biological Conservation* **145**, 267–277. doi:10.1016/j.biocon.2011.11.022.

Weigel J-Y, Mannle KO, Bennett NJ, Carter E, Westlund L, Burgener V, *et al.* (2014) Marine protected areas and fisheries: bridging the divide. *Aquatic Conservation: Marine and Freshwater Ecosystems* **24**, 199–215. doi:10.1002/aqc.2514.

Wescott G, Fitzsimons J (2011) Stakeholder involvement and interplay in coastal zone management and marine protected area planning. In *Marine Resources Management*. (Eds W Gullett, C Schofield and J Vince) pp. 225–238. LexisNexis Butterworths, Sydney.

White C, Halpern BS, Kappel CV (2012) Ecosystem service tradeoff analysis reveals the value of marine spatial planning for multiple ocean uses. *Proceedings of the National Academy of Sciences of the United States of America* **109**, 4696–4701. doi:10.1073/pnas.111421510910.1073/pnas.1114215109

Wielgus J, Balmford A, Lewis TB, Mora C, Gerber LR (2010) Coral reef quality and recreation fees in marine protected areas. *Conservation Letters* **3**, 38–44. doi:10.1111/j.1755-263X.2009.00084.x.

Wood L, Fish L, Laughgren J, Pauly D (2008) Assessing progress towards global marine protection targets: shortfalls in information and action. *Oryx* **42**, 340–351. doi:10.1017/S003060530800046X.

zu Ermgassen PSE, Spalding MD, Grizzle R, Brumbaugh R (2013) Quantifying the loss of a marine ecosystem service: filtration by the Eastern Oyster in US estuaries. *Estuaries and Coasts* **36**, 36–43. doi:10.1007/s12237-012-9559-y.

Chapter 3

The evolution of marine conservation and marine protected areas in Australia

Richard Kenchington

Introduction

Since the late 19th century the understanding of human impacts on the marine environment has been in a process of transformation from a cultural and legal view of the seas and their resources as infinite, dangerous, available and too great to be affected by widespread human impacts. By the late 20th century there was growing understanding that human impacts were widespread, increasing and threatening the sustainability of biological diversity and ecosystem services. It is now internationally accepted that these threats should be mitigated for reasons ranging from food and physical security (prevention of illegal entry to, and use of, marine space and resources) and biosecurity (prevention of introduction of alien species or disease). This acceptance has reflected growing awareness of opportunities for new and increasing economic uses of marine space from the Review Committee on Marine Industries, Science and Technology (1989) to the 'Blue Economy' (UNCSD 2014).

In 1988, the International Union for Conservation of Nature (IUCN) adopted a primary goal for marine conservation: 'To provide for the protection, restoration, wise use, understanding and enjoyment of the marine heritage of the world in perpetuity through creation of a global, representative system of marine protected areas and through the management, in accordance with the principles of the World Conservation Strategy, of human activities that use or affect the marine environment' (IUCN 1988).

As a late developing, relatively wealthy country with a large area and low population, Australia was among the first to introduce measures for protection of marine and coastal natural areas. The intertidal and close inshore areas of the Royal National Park, Sydney, established in 1879 may constitute the oldest marine protected area (MPA) in the world (Brown 2002; Chapter 11). The *Great Barrier Reef Marine Park Act 1975* is recognised as the first initiative specifically covering the issues of environmental conservation and reasonable use of a large marine ecosystem (e.g. Gaines *et al.* 2010). The proclamation of the world's first network of marine reserves by the Australian Government in its jurisdictional (Commonwealth) waters in 2012 resulted from one of the first systematic processes within a very large national jurisdiction to establish a system of representative MPAs, using bioregions as a basis to guide priorities. All Australian states and the Northern Territory have established MPAs within their jurisdictional waters.

Although all Australian governments with marine jurisdiction have declared MPAs, there is a considerable range of specific objectives and conditions in terms of use and access to MPAs and of approaches to coordination or integration with management of activities and impacts originating outside their boundaries. The specifics are detailed in the chapters that follow, dealing with all states and the Commonwealth waters. The purpose of this chapter is to reflect on the development of MPAs in Australia, and place Australian MPAs in the context of the management of marine biodiversity at a time of expectations of increasing economic use of national and international marine space.

What are MPAs? Protected areas as marine conservation tools

The IUCN defines a protected area as (Dudley 2008; Day *et al.* 2012):

> *'A clearly defined geographical space, recognised, dedicated and managed, through legal or other effective means, to achieve the long-term conservation of nature with associated ecosystem services and cultural values.'*

Parties to the Convention on Biological Diversity (CBD), including Australia, agreed in 2004 to set aside 10% of marine waters as MPAs, which was most recently reaffirmed with Aichi Target 11 (CBD 2010), which aims to achieve by 2020:

> *'...10 per cent of coastal and marine areas, especially areas of particular importance for biodiversity and ecosystem services, are conserved through effectively and equitably managed, ecologically representative and well-connected systems of protected areas and other effective area-based conservation measures, and integrated into the wider landscape and seascape.'*

The IUCN has developed and refined seven categories of protected area that relate to management intent (Table 3.1). They are based on protecting the values of unique or relict islands of relatively intact habitat, or of culturally modified landscapes where the interaction of people and nature over time has produced an area of significant ecological, biological, cultural and scenic value. Their development primarily reflects the context of conserving examples of terrestrial ecosystems that are otherwise threatened by clearance, or destructive modification, although Day *et al.* (2012) have provided guidance on how they are to be applied and interpreted in the marine space.

In most legal systems, land management is enabled by defining and allocating title and primacy of land use for an area to an individual or group. To do this for the purpose of conservation involves applying an understood competitive sectoral process.

On land, people may live in highly modified environments but generally have a cultural understanding of natural ecosystems. They can observe natural environments for long periods without major technological investment. The habitats and the geographic ranges of terrestrial species usually reflect geological structure, soil properties and moisture regimes in a unidirectional connective flow from catchment to river mouth. Habitats may be quite small, isolated and have high levels of endemism. The atmosphere above them is physiologically inert but it provides connectivity through passive transport of the spores, pollen and seeds of plants and through flying and drifting animals. Almost all life in terrestrial ecosystems depends on primary productivity physically connected to the planetary

Table 3.1. IUCN protected area management categories

Category	Name	Description
Ia	Strict nature reserve	Protected areas that are strictly set aside to protect biodiversity and also possibly geological/geomorphological features, where human visitation, use and impacts are strictly controlled and limited to ensure protection of the conservation values. Such protected areas can serve as indispensable reference areas for scientific research and monitoring.
Ib	Wilderness area	Protected areas that are usually large unmodified or slightly modified areas, retaining their natural character and influence, without permanent or significant human habitation, which are protected and managed so as to preserve their natural condition.
II	National park	Large natural or near natural areas set aside to protect large-scale ecological processes, along with the complement of species and ecosystems characteristic of the area, which also provide a foundation for environmentally and culturally compatible spiritual, scientific, educational, recreational and visitor opportunities.
III	Natural monument or feature	Protected areas set aside to protect a specific natural monument, which can be a landform, sea mount, submarine cavern, geological feature such as a cave or even a living feature such as an ancient grove. They are generally quite small protected areas and often have high visitor value.
IV	Habitat/species management area	Protected areas aiming to protect particular species or habitats and management reflects this priority. Many category IV protected areas will need regular, active interventions to address the requirements of particular species or to maintain habitats, but this is not a requirement of the category.
V	Protected landscape/ seascape	A protected area where the interaction of people and nature over time has produced an area of distinct character with significant ecological, biological, cultural and scenic value, and where safeguarding the integrity of this interaction is vital to protecting and sustaining the area and its associated nature conservation and other values.
VI	Protected area with sustainable use of natural resources	Protected areas that conserve ecosystems and habitats, together with associated cultural values and traditional natural resource management systems. They are generally large, with most of the area in a natural condition, where a proportion is under sustainable natural resource management and where low-level non-industrial use of natural resources compatible with nature conservation is seen as one of the main aims of the area.

surface, and is morphologically, physiologically and behaviourally adapted to dynamics of diurnal, seasonal and annual variations in temperature and water availability that present a challenging environment for cellular life.

Adaptations to terrestrial life include seeds and spores resistant to desiccation, yolky eggs, viviparity, lactation, care for the young, elaborate social and territorial behaviours and human cultures of community, sense of place, property and effective separation of management within boundaries. The capacity to clearly define geographic space and to allocate title and primacy for effective management for biodiversity protection within clear boundaries is the core concept of *in situ* terrestrial conservation. This makes it possible to set and implement legislative conditions of use, to establish and enforce boundaries to exclude incompatible uses and provide a process for resolving cross-boundary impacts.

In international law, and within most marine jurisdictions, boundaries are defined by linking geodesic points on the surface of the sea or land. Use of marine space is a common right beyond private property in most jurisdictions although increasingly leases or allocations of space and resources may be made for managing defined purposes of use such as fishery quotas, mineral and hydrocarbon resource tenements, seabed cables, navigation infrastructure and offshore energy installations.

In marine ecosystems most primary production is provided by microscopic phytoplankton drifting in the illuminated upper layer of a thermally stable, chemically buffered water column that provides a benign environment for biological cells. Water column species range from phytoplankton and zooplankton to pelagic and migratory top predators and whales. Many of these species have no contact with the seabed. The water column community is also the nursery and larval distribution environment as well as an important food source for species whose adults live in, on or in territories defined by, the seabed. The complex dynamics of currents, tides and wind-generated waves of the water column, and freshwater runoff from land, link marine ecosystems by transporting nutrients, larvae, predators, prey and pollutants.

Diving enables brief visits to appreciate shallow habitats but, without technological intervention, humans cannot survive in or on the sea for extended periods of time. For most people, marine environments are experienced from the sea surface and most underwater ecosystems are invisible, inaccessible and intuitively alien. Terrestrial cultures of sense of place, property, title and effective separation of management by boundaries do not transfer readily to the sea. The history of protected areas in terrestrial environments provides useful precedents for MPAs but they are only partly transferrable to the marine context because of the vastly more biologically active third dimension ecosystem of the water column, which is largely independent of the geological properties of the seabed. A critical issue in planning and management of IUCN Categories I and II MPAs is integration with management of surrounding and linked areas and activities, and taking into account the needs and spatial dynamics of species with life cycle processes that extend beyond boundaries of areas designated for such high-level protection. The extent and nature of trophic and life cycle interactions of species between the water column and the seabed determine the opportunities and constraints for spatial management of marine biodiversity.

The application of IUCN categories in marine contexts is confusing and applied differently with specific conditions, depending on jurisdictional interpretations. Four approaches can be identified in descending order of restriction of use.

- **No-access, no-take MPAs**: equivalent to IUCN Category Ia strictest nature protection category of terrestrial systems. They protect intact areas serving, to the greatest extent possible, as refuges and scientific control or reference sites. Human activities are totally prohibited, apart from management activities, including monitoring and scientific research that cannot be undertaken elsewhere.
- **Limited-use, no-take MPAs**: equivalent to IUCN Category II with operational controls reflecting the primary function of permanent protection of biodiversity, ecosystem process and habitat, while allowing demonstrably compatible uses including research, education, nature-based recreation and tourism activities. This can also include **Special purpose MPAs** that may be equivalent to IUCN Category III or V that protect the biological diversity and amenity of sites or areas of high cultural, scenic or visitor value.

- **Multiple-use MPAs**: generally incorporating and buffering zones of no-take, limited use and possibly special purpose protection to provide for ecosystem-based management of human uses and impacts.

Part of the challenge of application and reporting of IUCN protected area categories in marine ecosystem is the need to appreciate the different, but linked, conservation priorities for seabed and water column habitats and communities (Hutchings and Kenchington 2015). Regardless of the sustainability of management of activities in the overlying water column, protection of seabed habitats and dependent benthic biota from activities that can cause direct and often long-term impacts is often a crucial consideration.

Figure 3.1 illustrates the geographic scale of marine life cycle strategies, which have implications for size and function of MPAs if they are to conserve the biodiversity and ecological processes of life in the oceans. At one extreme are strongly site-dependent species that live on or attached to the seabed or have a very small adult range and reproduce asexually or sexually without planktonic larval development. Effective conservation of these species may be provided by relatively small MPAs.

Other site-dependent species may live on, or be attached to, the seabed or with a very small adult range reproduce sexually with planktonic larval development. Populations may be scattered throughout a large geographic area but are genetically linked. Although the site-dependent phase can be accommodated by relatively small MPAs, systematic protection requires networks of MPAs to take account of recruitment linkages.

The nature of connectivity of marine environments is that, over the course of a year, an area of seabed and the overlying water column will provide habitat for species of all four life cycle ranges (e.g. Pineda *et al.* 2007).

Many demersal species have adult territories of a scale that cannot readily be contained within a single protected area or single jurisdiction, but MPAs may be appropriate for managing areas of particular life cycle significance and vulnerability, such as spawning aggregation sites, or juvenile growth areas that are critical sites for their conservation.

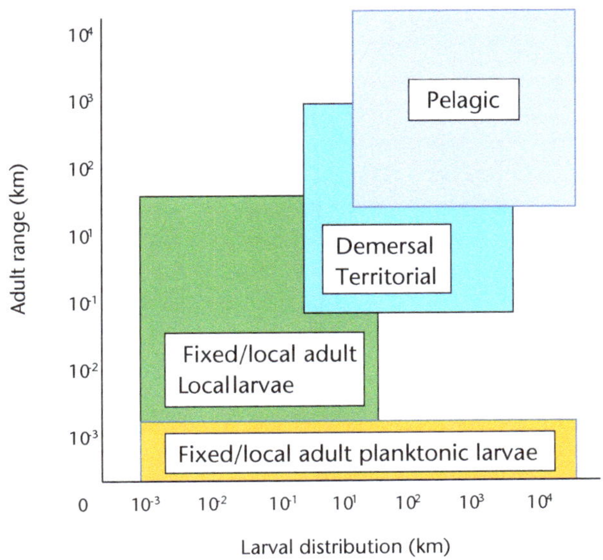

Fig. 3.1. The geographic scales of life cycle ranges of groups of marine species. (Adapted from Kenchington 1990)

At the other extreme, some pelagic and oceanic migratory species may also have spawning or birthing, sites or seasonal migratory feeding areas that can be protected by area management, but for most of their time these and wholly planktonic species are moving long distances, traversing waters of multiple jurisdictions or the high seas.

Although MPAs may have important roles for critical life cycle sites of pelagic and demersal fish, effective fishery management has a crucial and broader role in conservation because medium- to high-level predators with ranges beyond the scale of many MPAs shape the dynamics of marine biodiversity (e.g. Jackson 2008; Rice *et al.* 2012; Kenchington *et al.* 2014).

Marine protected areas in Australian jurisdictions

Despite acceptance of controls for purposes such as safety of navigation, defence areas and the management of commercial fisheries, the growing recognition of the need to conserve marine ecosystems confronts cultural assumptions of marine resources as invulnerable and common property. A three-way conflict of interest between conservation, recreation and the commercial exploitation of marine resources underlies planning and management of MPAs. The balance of those interests and their linkages and vulnerabilities to impacts from management of adjacent lands and trans-boundary marine activities are reflected differently in the legislative and administrative arrangements of Australian state and federal governments.

In the first half of the 20th century, most states had coastal and island national parks, some of which had subtidal extensions but their objectives typically addressed recreational fishing. Thus, in Queensland, Green Island was declared a recreation reserve in 1906. By 1935, nearly all the islands off the Queensland coast had been declared nature sanctuaries, some with restrictions on collection of shells and coral. In 1943, parts of Heron Island were declared a national park, while the remainder was a recreational fishing camp. In 1974, heavily used reef sites of Heron Island and Green Island were declared marine parks, which permitted recreational fishing but prohibited commercial fishing, coral and shell collecting and the removal of sand.

In 1970, major political concern about impacts and potential threats to a nationally and globally iconic marine ecosystem led to a jointly appointed Royal Commission which was established to report to the Commonwealth and Queensland governments on environmental, social and economic implications of petroleum drilling in the area of Great Barrier Reef. The report identified the need for a strategic framework to manage the Great Barrier Reef region by regulating human uses and impacts to protect and maintain the natural values of the Reef (Commonwealth of Australia 1974). It recommended establishment of a Great Barrier Reef Commission to oversee all aspects of Great Barrier Reef use and management.

The *Great Barrier Reef Marine Park Act 1975* reflected the Royal Commission's recommendations and was probably the first explicit legislation for a system of marine spatial planning and integrated management for conservation and reasonable or sustainable use at the scale of a large marine ecosystem (see Chapters 5, 12, 14). It provided an over-arching framework for strategic spatial planning and management for conservation and ecologically sustainable uses of the Great Barrier Reef region. The western boundary of the region was defined as the low water mark of mainland Queensland and the provisions of the Act over-ride any inconsistent provision of Queensland law. This created the unique situation in Australia of a single over-riding piece of legislation for management of a large marine ecosystem encompassing coastal and offshore bioregions from low water mark of a long and mostly sparsely populated coast. It provided ecosystem-based management for conservation in the no-take protected area sense of setting aside areas that would be undisturbed by humans (other than for the purposes of scientific research) and reserving other areas for appreciation and enjoyment by the public. It also provided for consistent or reasonable use through regulation of uses that exploit the resources of the reef region while minimising the effect of those activities on the Reef (see Chapter 5 for detailed discussion).

The Great Barrier Reef Marine Park (GBRMP) legislation was developed at a time of international negotiations, which led to greatly extended Commonwealth marine jurisdiction through the creation of 200 nautical mile Exclusive Economic Zones. The need for broader marine governance legislation stimulated broader policy interest in marine issues and a report on the economic potential of new technologies for development of industrial and commercial marine industries in Australia's marine jurisdictions (Review Committee on Marine Industries, Science and Technology 1989). In the early 1990s, an initiative of the Council of Australian Governments (COAG), which comprises the heads of federal, state and territory governments, resulted in the *National Strategy for Ecologically Sustainable Development* (ESDSC 1992), developed through a collaborative process by a working group that considered coastal and marine opportunities and challenges and the importance of developing integrated ESD policy frameworks to tackle them.

An outcome of the Strategy was the establishment of the National Advisory Committee on Marine Protected Areas. This identified a priority need to develop a national biogeographic classification system to form the basis for a national representative system of MPAs and a national database of MPAs. A workshop in 1995 established the basis for a marine regionalisation of Australia (Muldoon and Gillies 1995). This led to collaborative programs of state, territory and Commonwealth scientists and marine managers to develop a marine section of the Collaborative Australian Protected Area Database (CAPAD) in 1997 and an Interim Marine and Coastal Regionalisation of Australia (IMCRA) (IMCRA Technical Group 1998).

In 1996, COAG endorsed a collaborative national process to develop an Australian Oceans Policy. This process was informed by 11 commissioned research synthesis papers (e.g. Commonwealth of Australia 1997), with the intention of developing a framework for integrated oceans management with a central commitment to ecosystem-based management of Australia's marine environment (Tsamenyi and Kenchington 2012). Although the states and the private sector were involved at the start of the development of the Oceans Policy, the fuller consultation on jurisdictional integration that was intended could not be achieved by the scheduled release date so the states and territories were cut off from the Oceans Policy decision making immediately before the release of the Policy (Vince 2008).

Core governance elements of the Oceans Policy, launched in 1998 (Commonwealth of Australia 1998a,b), included a multisectoral National Oceans Ministerial Board, links to the Australian and New Zealand Environment and Conservation Council (ANZECC), a National Oceans Office in Hobart and a Commonwealth/State senior officials' coordination group to oversee the development of Regional Marine Plans.

The original objective of the Oceans Policy and Regional Marine Plans was to: implement overall ecosystem-based management; create a national representative network of MPAs; promote ecologically sustainable industries; develop integrated management of sectoral activities; work towards consistency in management across jurisdictional boundaries; and increase involvement of resource users and community at large in planning and decision making (Foster *et al.* 2005). The reduction of the role of the states from partners to consultative observers was reflected in a loss of collaborative momentum and

Table 3.2. Number and area of marine protected areas in Australian jurisdictions

(Modified from Collaborative Australian Protected Area Database 2014 (Department of the Environment 2014)[1])

Jurisdiction	Number of MPAs	Area (km²)	Proportion of jurisdiction waters in MPAs (%)
Commonwealth	62	3 102 444	36.39
New South Wales	18	3488	39.63
Northern Territory	2	2909	4.05
Queensland*	89	63 352*	52.07
South Australia	28**	28 878	48.12
Tasmania	21	1353	6.05
Victoria	30	1211	11.8
Western Australia	22	27 450	23.72
Total	**272**	**3 231 987**	

*Figures within Queensland State waters only, not Commonwealth waters of the Great Barrier Reef Marine Park
**Excludes Aquatic Reserves and Rock Lobster Reserves that are within Marine Parks

divergence of state and Commonwealth programs. In 2004, the Commonwealth process was subsumed as a sectoral environmental responsibility and the Regional Marine Planning process focused on development of the network of Commonwealth marine reserves (Tsamenyi and Kenchington 2012).

The objectives of the creation of a national network of representative MPAs and maintaining the IMCRA and CAPAD frameworks have continued. The most recently published bioregionalisation update is IMCRA 4.0 (Commonwealth of Australia 2006) and the program of accumulation and analysis of bioregional data and samples continues.

CAPAD Marine 2014 provides the most recent information on Commonwealth, state and territory MPAs, but this summary data and geospatial data contains several known errors.[1] Since the 1970s, Australia's state and federal governments have created about 300 MPAs, with the total coverage of areas and IUCN Protected Area categories summarised in Tables 3.2 and 3.3 (the different total area figures reflecting differences between gazetted areas, GIS areas and overlaps).

Table 3.3. The current extent of Australia's marine protected areas and the IUCN protected area categories (Dudley 2008) to which they have been assigned

(Source: Collaborative Australian Protected Area Database 2014)

IUCN category	Area (km²)
IA	136 740
IB	–
II	1 090 013
III	746
IV	713 669
I–IV Total	**1 941 168**
V	4373
VI	1 327 387
V–VI Total	**1 331 760**
I–VI Total	**3 272 928**

Note: This table includes overlapping areas due to the complexity involved in attributing areas with an overall IUCN category

MPAs in Australia are reported nationally and globally in terms of the IUCN categories of protection (Dudley 2008) but their design, management and allocation of categories reflects the legislative and policy framework of the jurisdictions that create them. There is confusing inconsistency between jurisdictions in the naming and allocation of categories for purposes of use and entry, which has resulted in over 30 named types of MPA (e.g. marine reserve, marine park, marine sanctuary, marine national park, etc.).

Although the marine components of Sydney's Royal National Park have been identified as the world's first MPA (Brown 2002), fishing is an accepted recreational activity, which provides an example of the challenge of applying terrestrial concepts of protected area designations to marine areas. Fitzsimons (2011) has discussed the matter of mislabelling MPAs in Australian jurisdictions, drawing particular attention to the use of IUCN Category II (National Park) for areas in which fishing is allowed despite the explicit statement by Dudley (2008): 'In marine environments, extractive use (of living or dead material) as a key activity is generally not consistent with the objectives of category II areas' (later reinforced by Day *et al.* 2012).

Differentiation between conservation priorities for seabed and water column communities and management of activities that target seabed and water column fisheries presents a particular challenge for application and reporting of IUCN protected area categories in marine ecosystems. Regardless of the sustainability or management of other activities in the overlying water column, seabed trawling and dredging cause direct, and often long-term, impacts to seabed habitats. Prevention of those activities provides substantial protection of seabed habitat and dependent benthic biota. However, although category reporting appropriately reflects that 33% of the GBRMP is zoned to provide 'no-take' protection consistent with IUCN Category I/II, a further 33% zoned to protect seabed habitat from the direct impacts of trawling or dredging is allocated to category VI: the same category as the final 33% where trawling and dredging may occur. A case could be made that zoning to protect seabed ecosystems from such activities is an important conservation measure that should appropriately be assigned as category IV: 'Habitat/species management area – Protect particular species or habitats, where management reflects this priority'.

Fitzsimons and Wescott (2008) have investigated this issue of protected area coverage of marine ecosystems from the perspective of quantifying situations with differential protection of seabed and water column. Using the example of the Tasmanian Seamounts Marine Sanctuary (now part of the Huon Commonwealth Marine Reserve) in which the area below 500 m is zoned as Category Ia, while above 500 m the water-column habitat is zoned as Category VI. They suggested that the proportions of area allocated to each category should be determined on the basis of proportion of the volume of water in each zone within the boundaries of a vertically differentiated MPA.

The complexity and diversity of MPAs and their extent in Australian waters reflects the operational constraints and opportunities in three substantially different governance contexts (Figs 3.2 and 3.3).

In state and territory waters, MPAs are effectively a new sectoral category of area-based management introduced within multi-purpose regimes that have developed since the colonies (states) were created in the 19th century. They have introduced explicit environmental objectives of biodiversity conservation, nature-based recreation and education to systems otherwise already and continuingly used and managed primarily for shipping, boating, fisheries and coastal engineering.

The GBRMP is an explicit collaborative multiple-use, strategic ecosystem-scale management scheme with over-arching conservation and sustainability objectives for an iconic large marine ecosystem. It provides for sustainable use through zoning. It includes parts of the waters of the state of Queensland but has constitutional legislative powers to over-ride

Fig. 3.2. Prince Regent River within the Lalang-garram/Camden Sound Marine Park in the Kimberley region, north-west Western Australia. This 673 000 ha MPA is bordered by Dambimangari Indigenous Protected Area to the west and Prince Regent National Park to the east, and is part of a large network of MPAs established or proposed along the Kimberley coast. (Photo: James Fitzsimons)

any inconsistent provisions of state law. Although the entire system is a designated MPA, the basis of its planning and management reflect a strategic multiple use system of marine spatial planning. It has been recognised as a role model of good practice for large marine ecosystem management (e.g. Olsson *et al.* 2008; Gleason *et al.* 2010).

The Commonwealth Marine Reserves program arose to meet the need to develop a conservation regime for very large new areas of offshore marine jurisdiction allocated to Australia through its responsibilities for the Exclusive Economic Zone and Extended Continental Shelf. It was developed on the basis of the best scientific understanding of bioregionalisation substantially on surrogacy extrapolations from seabed geology, fishery data and generally limited sampling, and analysis of seabed surface and infauna (IMCRA 4.0 – Commonwealth of Australia 2006; Kenchington and Hutchings 2012).

Conclusion

Australia has a diverse record of achievement in establishing bioregional networks of MPAs, helped by a national culture that values coastal environments and recreation. For the most part, this has been developed with and within a multi-sectoral challenge of planning ecologically sustainable development.

Despite a high level of success in progressing international targets for MPA creation, there are substantial challenges in achieving reasonable levels of protection for vulnerable

Fig. 3.3. Churchill Island Marine National Park in Western Port, Victoria: a relatively small (670 ha), highly protected no-take MPA encompassing shallow temperate waters and saltmarsh and mangroves. (Photo: James Fitzsimons)

estuarine, coastal and nearshore continental shelf bioregions. In part, this reflects a common challenge facing jurisdictions with high coastal population density and competing demands for access, use and enjoyment of coastal environments.

Resolving such issues is probably hindered by the complexity of the systems that have evolved to meet different governance arrangements and circumstances. Despite rapid and continuing development of scientific understanding and methods, there are substantial challenges in explaining MPAs to people who have not been closely involved. With 32 types of MPA in Australia, it can be challenging to answer apparently simple questions such as: What is a marine protected area? Which categories are no-take MPAs and are they the only legitimate MPAs? What other purposes can MPAs serve?

The underlying challenge for Australian MPAs is that, although many are well managed within their spatial boundaries, many of the most threatening impacts arise from activities outside their boundaries and often beyond the boundaries of the jurisdiction that established them. Thus poor water quality from land-based runoff, coastal development and climate change are continuing major threats to marine ecosystems. Effective MPAs are important for building and maintaining resilience to such impacts. Limited use no-take MPAs have the crucial role of enabling people to experience and understand and wonder at marine biodiversity, while no access no-take MPAs are crucial as sites for monitoring impacts and change in areas that are not also affected by direct uses of recreation, boating and commercial or recreational fishing.

Biography

Richard Kenchington is a marine ecologist who has studied Crown-of-thorns Starfish and undertaken large scale surveys of the Great Barrier Reef. He was a foundation member of the planning and research staff of the Great Barrier Reef Marine Park Authority in 1977, retiring as Executive Director of the Authority. He was Chair of the drafting group for the Framework for Action of the International Coral Reef Initiative. His work includes capacity building, consultancy and advice to government and international agencies. He has published on management of coral reef, coastal and marine biodiversity and his books include *Managing Marine Environments* (1990, Taylor and Francis) and *Sustainable Coastal Management and Climate Adaptation* (2012, CSIRO Publishing).

Endnote

1 The MPA data for the Collaborative Australian Protected Area Database 2014 contains numerous errors, particularly in the summary spreadsheets on the Department of the Environment's website http://www.environment.gov.au/land/nrs/science/capad/2014. These errors included counting the total area of mainly terrestrial protected areas that adjoined the coast or islands in Western Australia, Tasmania, Victoria and South Australia, inflating both the numbers and areas of MPA in those jurisdictions. The geospatial data associated with CAPAD 2014 Marine, while not having as many of these errors, has different issues such as counting Macquarie Island Marine Reserve, the marine component of Kent Group National Park and Victorian MPAs as terrestrial protected areas. In creating Table 3.2, these errors have been rectified, and doubling counting avoided where overlapping MPAs occur, but it is not to imply the numbers in the table are error free. Areas for Commonwealth, New South Wales, Northern Territory and Queensland appear to be error free in CAPAD 2014 and this is the source for the data in Table 3.2.

References

Brown I (2002) Oldest MPA in the world: Royal National Park, New South Wales, Australia. Designated 1879. *MPA News* **3**(6), 5.

CBD (2010) *COP 10 – Tenth Meeting of the Conference of the Parties to the Convention on Biological Diversity. Nagoya, Japan 18–29 October 2010. Decision X/2.Strategic Plan for Biodiversity 2011–2020.* United Nations Environment Programme, Nairobi.

Commonwealth of Australia (1974) *Report of Royal Commissions into Exploratory and Production Drilling for Petroleum in the Great Barrier Reef.* Australian Government Printing Service, Canberra.

Commonwealth of Australia (1997) *Australia's Oceans Policy Background Paper No. 1 – Multiple Use Management in the Australian Marine Environment, Principles, Definitions and Elements.* Commonwealth of Australia, Canberra.

Commonwealth of Australia (1998a) *Australia's Oceans Policy. Volume 1 – Caring, Understanding, Using Wisely.* Commonwealth of Australia, Canberra.

Commonwealth of Australia (1998b) *Australia's Oceans Policy. Volume 2 – Specific Sectoral Measures.* Commonwealth of Australia, Canberra.

Commonwealth of Australia (2006) A *Guide to the Integrated Marine and Coastal Regionalisation of Australia Version 4.0.* Department of the Environment and Heritage, Canberra.

Day J, Dudley N, Hockings M, Holmes G, Laffoley D, Stolton S, *et al.* (2012) *Guidelines for Applying the IUCN Protected Area Management Categories to Marine Protected Areas*. IUCN, Gland, Switzerland.

Department of the Environment (2014) *Collaborative Australian Protected Area Database Marine 2014*. Department of the Environment, Canberra, <http://www.environment.gov.au/land/nrs/science/capad/2014>

Dudley N (Ed.) (2008) *Guidelines for Applying Protected Area Management Categories*. IUCN, Gland, Switzerland.

ESDSC (1992) *National Strategy for Ecologically Sustainable Development*. Ecological Sustainable Development Steering Committee, endorsed by the Council of Australian Governments, Canberra.

Fitzsimons JA (2011) Mislabelling marine protected areas and why it matters – a case study of Australia. *Conservation Letters* **4**, 340–345. doi:10.1111/j.1755-263X.2011.00186.x.

Fitzsimons JA, Wescott G (2008) Getting the measure of marine protected areas: surface area or volume as measures for reserve system auditing? *Aquatic Conservation: Marine and Freshwater Ecosystems* **18**, 518–526. doi:10.1002/aqc.867.

Foster E, Haward M, Coffen-Smout T (2005) Implementing integrated oceans management: Australia's South East Regional Marine Plan (SERMP) and Canada's Eastern Scotian Shelf Integrated Management (ESSIM) initiative. *Marine Policy* **29**, 391–405. doi:10.1016/j.marpol.2004.06.007.

Gaines SD, Lester SE, Grorud-Colvert K, Costello C, Pollnac R (2010) Evolving science of marine reserves: New developments and emerging research frontiers. *Proceedings of the National Academy of Sciences of the United States of America* **107**, 18278–18285. doi:10.1073/pnas.0906473107.

Gleason M, McCreary S, Millerrt-Henson M, Ugoretz J, Fox E, Merrifield M, *et al.* (2010) Science-based and stakeholder-driven marine protected area network planning: a successful case study from north central California. *Ocean and Coastal Management* **53**, 52–68. doi:10.1016/j.ocecoaman.2009.12.001.

Hutchings P, Kenchington R (2015) Constraints of terrestrial protected area solutions in protecting marine biodiversity. *Australian Zoologist* doi:10.7882/AZ.2015.020

IMCRA Technical Group (1998) *Interim Marine and Coastal Regionalisation for Australia: An ecosystem-based classification for marine and coastal environments*. Version 3.3. Environment Australia, Canberra.

IUCN (1988) *17th Session of the General Assembly of IUCN and 17th Technical Meeting, San José, Costa Rica 1–10 February 1988, Proceedings*. IUCN, Gland, Switzerland, <https://portals.iucn.org/library/efiles/documents/GA-17th-011.pdf>.

Jackson J (2008) Ecological extinction and evolution in the brave new ocean. *Proceedings of the National Academy of Sciences of the United States of America* **105**, 11458–11465. doi:10.1073/pnas.0802812105.

Kenchington RA (1990) *Managing Marine Environments*. Taylor and Francis, New York.

Kenchington R, Hutchings P (2012) Science, biodiversity and Australian management of marine ecosystems. *Ocean and Coastal Management* **69**, 194–199. doi:10.1016/j.ocecoaman.2012.08.009.

Kenchington R, Vestergaard O, Garcia SM (2014) Spatial dimensions of fisheries and biodiversity governance. In *Governance of Marine Fisheries and Biodiversity Conservation*. (Eds SM Garcia, J Rice and A Charles) pp. 110–123. Wiley Blackwell, Chichester, UK.

Muldoon J, Gillies J (1995) ANZECC's marine biogeographic regionalisation workshop. In *Towards a Marine Regionalisation for Australia: Proceedings of a Workshop Held in Sydney,*

New South Wales 4–6 March 1994. (Ed J Muldoon) pp. 219–224. Great Barrier Reef Marine Park Authority, Townsville.

Olsson E, Folke C, Hughes TP (2008) Navigating the transition to ecosystem-based management of the Great Barrier Reef, Australia. *Proceedings of the National Academy of Sciences of the United States of America* **105**, 9489–9494. doi:10.1073/pnas.0706905105.

Pineda J, Hare JA, Sponaugle S (2007) Larval transport and dispersal in the coastal ocean and consequences for population connectivity. *Oceanography* **20**, 22–39. doi:10.5670/oceanog.2007.27.

Review Committee on Marine Industries, Science and Technology (1989) *Oceans of Wealth? A Report.* Australian Government Publishing Service, Canberra.

Rice J, Moksnsess E, Attwood C, Brown SK, Dahle G, Gjerde KH, *et al.* (2012) The role of MPAs in reconciling fisheries management with conservation of biological diversity. *Ocean and Coastal Management* **69**, 217–230. doi:10.1016/j.ocecoaman.2012.08.001.

Tsamenyi M, Kenchington R (2012) Australian oceans policymaking. *Coastal Management* **40**, 119–132. doi:10.1080/08920753.2012.652519.

UNCSD (2014) *Blue Economy Concept Paper.* United Nations Conference on Sustainable Development, New York, <https://sustainabledevelopment.un.org/content/documents/2978BEconcept.pdf>.

Vince J (2008) Ten years of implementing Australia's oceans policy: from an integrated approach to an environmental policy focus. *Maritime Studies* **159**, 1–10. doi:10.1080/07266472.2008.10878880.

PART 2

Australia's marine protected area systems

Previous page: Diver exploring the coral on the Great Barrier Reef Marine Park. (Source: Tchami. Available under a Creative Commons Attribution-SharAlike 2.0 Generic licence. For more information see https://creativecommons.org/licenses/by-sa/2.0/. Available at https://www.flickr.com/photos/tchami/ 15364861867.)

Chapter 4

The marine protected area estate in Australian (Commonwealth) waters

Peter Cochrane

Introduction

The coast and the ocean are core elements of Australia's identity as a nation and as a people – historically, environmentally, socially and economically. The oceans separate us from, and connect us to, the rest of the world. The assumption that these oceans were vast, limitless, resource rich and perhaps incidentally an acceptable place to dispose of wastes has increasingly given way to a recognition of the profound consequences of this approach for environmental and human health. Globally, the collapse of several major fisheries from overfishing and near disappearance of iconic species underscored the need for better understanding and control over the exploitation of marine resources. In more recent years, the condition and outlook reports for the Great Barrier Reef have highlighted the need to change land management practices to reduce offshore pollution and sediment impacts (see Chapter 5). These shifts towards greater awareness and action have been instigated and driven by three interconnected elements: a growing research effort and evidence base identifying these impacts and providing advice on their mitigation; activism by community and non-government organisations; and leadership within and outside government and the international community.

Over the last 40 years of these trends, Australia has had a generally bipartisan national commitment to better management of marine resources, embracing their protection and wise use. The first significant expression of this was the establishment of the Great Barrier Reef Marine Park in 1975 and its subsequent World Heritage listing in 1981. Australia banned whaling in 1979 and has subsequently played a leading role globally in whale conservation. The national conversation on ecologically sustainable development in the 1980s and early 1990s set the scene for a 10-year federal marine conservation program, followed by the development of a comprehensive Oceans Policy in 1998 (see Chapter 19) along with *Guidelines for Establishing the National Representative System of Marine Protected Areas* (ANZECC TFMPA 1998), and then a lengthy Marine Bioregional Planning process from 2002–2012. Although these broadly collaborative approaches were developed to a greater or lesser degree with the states and Northern Territory governments that manage Australia's coastal waters out to 3 nautical miles, the Australian Federal (i.e. Commonwealth)

Government's priorities have been its responsibilities that extend (with some exceptions) from 3 nautical miles offshore to the boundaries of the Exclusive Economic Zone (EEZ), generally extending out 200 nautical miles from the coastal baseline, but including significant continental shelf extensions, most notably in the Coral Sea. This chapter deals with the Australian Government's network of MPAs established within Australia's EEZ beyond the 3 nautical mile boundary with state and territory waters. The Great Barrier Reef Marine Park, while technically in Commonwealth waters, is dealt with separately in Chapter 5.

The first regional network of marine reserves in waters controlled by the Australian Government – the South-east Commonwealth Marine Reserve Network, was declared in 2007, and a further 40 reserves in four more regional networks (South-west, North-west, North and Temperate East) and the Coral Sea were proclaimed in late 2012.

The declaration of these forty reserves and the finalisation of management plans for them in early 2013 became contentious in the lead up to the federal election in September of that year, with the recreational fishing lobby in particular mounting a strident campaign against a perceived 'lock-out' of recreational fishing. Commercial fishing interests also expressed strong concerns about the loss of access to fishing grounds and the economic consequences for the industry. In response, a commitment made during the election campaign to revisit how these reserves will be managed commenced as a formal review in late 2014.

History and development of the Commonwealth MPA network

The first two Commonwealth marine reserves, Lihou Reef and Coringa-Herald National Nature Reserves, were declared in 1982 to protect significant reefs in the Coral Sea. Ashmore Reef National Nature Reserve off the north-west of Australia was added in 1983.

The first biogeographic approach by governments to assist planning a national representative system of marine reserves was developed in 1985 and endorsed by the then Council of Nature Conservation Ministers. Its lack of precision and detail led to further work on regionalisations, which, for Commonwealth waters, commenced in 1995, with the first national marine regionalisation prepared the following year (CSIRO 1996) and formally endorsed and published by Environment Ministers in 1998 (Environment Australia 1998).

International concern, debate and policy on improving protection of the oceans was articulated and developed momentum and government commitment to action through forums such as those on sustainable development (Rio Earth Summit 1992, Johannesburg 2002), through the UN Environment Programme, the International Union for Conservation of Nature (IUCN), the UN General Assembly and the Convention on Biological Diversity (CBD), which came into effect in 1993.

In the lead up to the Rio Summit in 1992, the Australian Government committed to expand Australia's existing marine reserve system through the establishment of a national representative system of marine protected areas (MPAs). The division of responsibilities for a wide range of environmental issues between the Commonwealth and state/territory governments, which became contentious in the early years of the Hawke Government, was delineated in an Intergovernmental Agreement on the Environment (COAG 1992).

The *National Strategy for the Conservation of Australia's Biological Diversity* (ANZECC 1996) was developed and agreed by Commonwealth, state and territory governments to meet commitments made under the Convention on Biological Diversity and the Intergovernmental Agreement on the Environment. It recommended expansion of marine parks and reserves to encompass representative examples of Australia's marine environments.

This led to the development and agreement by Commonwealth and state/territory governments to the *Guidelines for Establishing the National Representative System of Marine*

Protected Areas (ANZECC TFMPA 1998). In that same year, the Commonwealth Government set out its policy framework as Australia's Oceans Policy (Commonwealth of Australia 1998a,b). The policy included a 3-year, $50 million program for the commencement of Regional Marine Planning, including identifying current and emerging threats to ecosystem health and development of management strategies and frameworks to tackle them. A key component of the policy was to accelerate development of the National Representative System of Marine Protected Areas (NRSMPA), including development and release of a national and regional framework for planning new MPAs and improved management of existing ones. The Interim Marine and Coastal Regionalisation of Australia (IMCRA) version 3.3 was released in 1998, extending the initial regionalisation work of CSIRO in 1996.

By the end of 2000, there were 12 Commonwealth MPAs covering over 20 million ha. These MPAs were declared to protect iconic habitats such as coral reefs or key habitats for endangered species. The exception was the Great Australian Bight Marine Park, declared in 1998, which apart from protecting important whale calving grounds at the head of the Bight, extended to the limit of the EEZ to protect a representative sample of benthic habitat from the continental shelf to the abyssal plain.

At the World Summit on Sustainable Development (2002) the Australian Government committed to the establishment of a comprehensive, adequate and representative (CAR) system of MPAs by 2012.

In the period 2001–04, the Australian Government's effort on developing a representative system of marine reserves had a major focus on the south-east marine region. The then National Oceans Office produced a series of reports documenting environmental values, economic uses and values, impacts of use and pressures on conservation values in the region (e.g. National Oceans Office 2001; other reports at Department of the Environment 2001–04). This work supported the development of a draft south-east network proposal released for public comment in 2006, and the declaration of the final network of reserves in 2007, which included the existing Macquarie Island Commonwealth Marine Reserve and 13 new reserves. This network was managed under interim arrangements while the remainder of the reserve networks were being designed and finalised over the next 5 years. The management plan for the south-east network was finalised and published in 2013 (Director of National Parks 2013).

Work was continuing to improve the spatial framework for classifying Australia's marine environment into bioregions to help regional-scale planning. A national bioregionalisation extending to the edge of the EEZ was completed and released in 2005 and was published in 2006 together with an updated IMCRA 3.3 as the Integrated Marine and Coastal Regionalisation of Australia (IMCRA 4.0) (Fig. 4.1). The marine regions adopted for the development of the national representative system of marine reserves are shown in Fig. 4.2.

The Australian Government's approach to establishing a national representative system of marine reserves was more clearly articulated in the *Goals and Principles for the Establishment of the National Representative System of Marine Protected Areas in Commonwealth Waters* (Australian Government 2007).

By 2007, with the addition of the Heard Island and McDonald Islands Marine Reserve in the sub-Antarctic, the Cod Grounds Commonwealth Marine Reserve off the east coast and the 14 reserves in the South-east Network, the area of Commonwealth marine reserves had more than doubled to nearly 50 million ha in 26 reserves.

The next major steps in the development of the national representative system of marine reserves occurred between 2007 and 2012, and culminated in the declaration of 40 marine reserves in the South-west, North-west, North and Temperate East Commonwealth Marine Reserve Networks and the Coral Sea Commonwealth Marine Reserve in 2012 after a lengthy and systematic development of options and public consultation

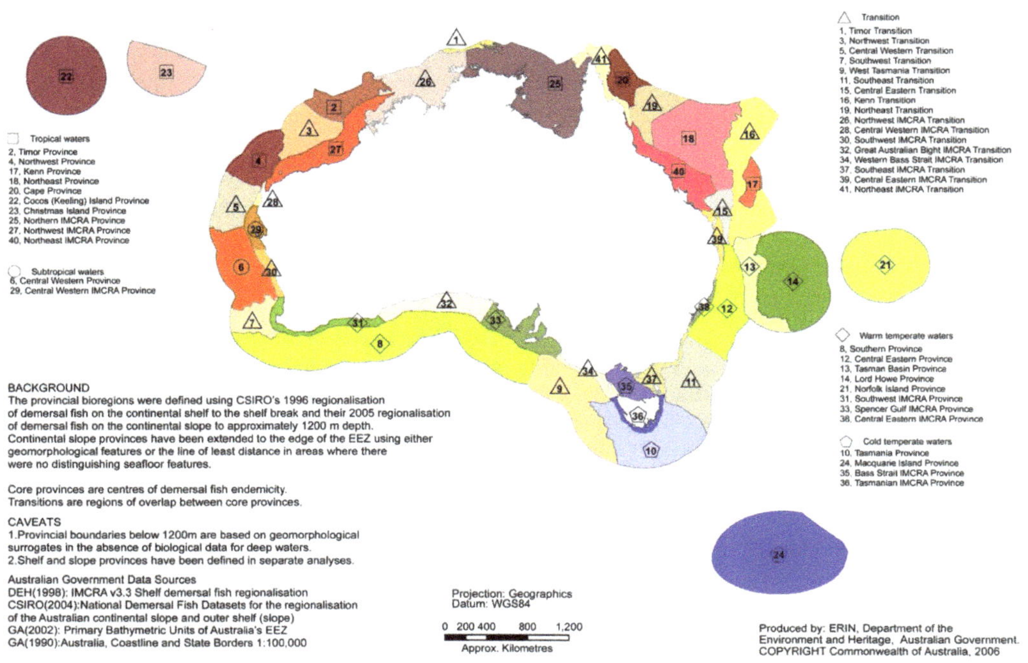

Fig. 4.1. Provincial Bioregions of the Integrated Coastal and Marine Regionalisation of Australia (IMCRA) V. 4 (2006).

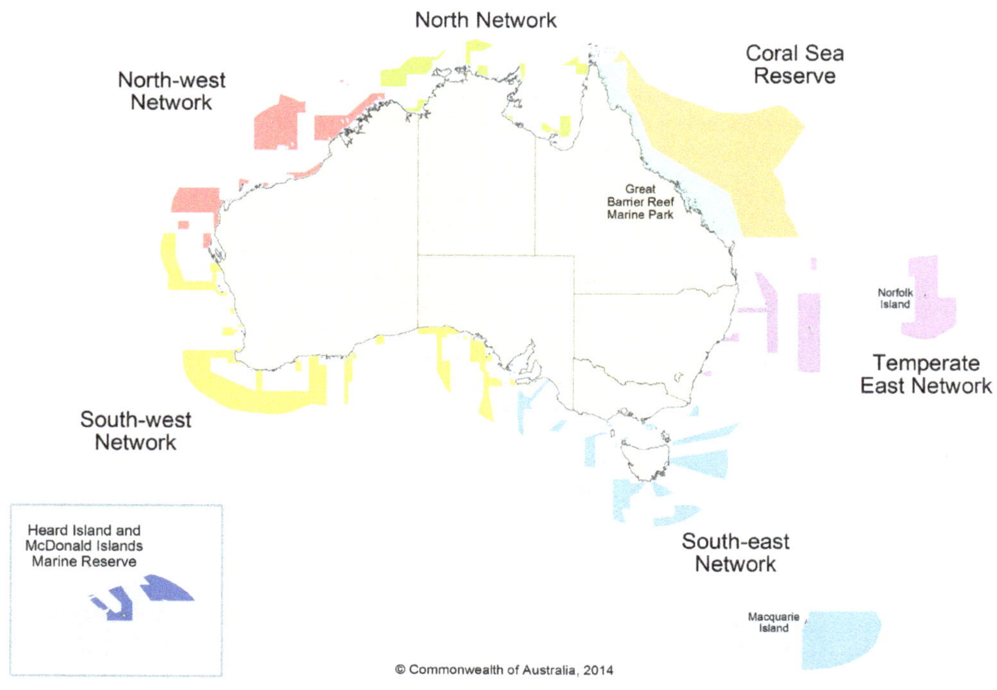

Fig. 4.2. Commonwealth Marine Reserves networks as at 2014.

(Fig. 4.2). This included the preparation and release of marine bioregional profiles describing the natural assets of each region between October 2007 and May 2009. These profiles, together with subsequent analysis and review, informed the description of 'Areas for Further Assessment' as a key step in identifying potential new reserves and these areas of interest were released progressively for public consultation from May 2009 to March 2010. Draft reserve network proposals were then developed using a computer-based decision support tool and targeted consultations with key stakeholders. A formal 90-day consultation period on each regional network proposal and the Coral Sea was held between May 2011 and February 2012, resulting in over 560 000 submissions, most being campaign submissions for and against the proposals.

Final reserve network proposals, including the Coral Sea Commonwealth Marine Reserve, were announced in June 2012, based on the public submissions, further consultations, socio-economic assessment and a regulatory impact statement by the Australian Government. These final reserve proposals were subject to a further round of public consultation, as required by the *Environment Protection and Biodiversity Conservation Act 1999* (EPBC Act), between July and September 2012. This 60-day consultation period resulted in nearly 80 000 submissions, again most being campaign submissions. The 40 new reserves in these final proposals were proclaimed by the Governor-General in November 2012. These 40 new reserves incorporated six existing reserves whose earlier proclamations were revoked in the 2012 proclamation.

While representing the outcome of a lengthy consultation process, a comprehensive compilation and consideration of scientific knowledge and expert opinion, and a delicate balancing of the various interests and sectors affected, there was not unanimous endorsement and acceptance of the final design and location of the new reserves. Comprehensiveness of coverage from a purely scientific design perspective was compromised by the accommodation of significant economic interests, notably commercial fishing, and oil and gas activity and prospectivity. There was considerable criticism by scientists of the networks (e.g. Pressey 2013). Socio-economic considerations inevitably strongly shaped final reserve design and placement in some bioregions. A particular source of aggravation to some sectors had been the appearance, relatively late in the consultation process, of several 'new' reserves, such as Dampier in the North-West bioregion (see Director of National Parks 2012). Although their inclusion was aimed at improving the comprehensiveness of the network coverage, they did not emerge from the same extensive consultation process that had shaped the remainder of the reserves. To some, this undermined the integrity of the consultations and was portrayed as 'political interference'.

The six regional networks (including the South-east network proclaimed in 2007) and the two reserves sitting outside the networks (Coral Sea, and Heard Island and McDonald Islands) now extend over an area of over 283 million ha in 59 reserves, and cover 35% of Australia's EEZ.

Targets and drivers for change

At the 10th Conference of the Parties to the CBD in 2010, the Aichi targets were negotiated and agreed, which included, in Target 11, a commitment that by 2020 at least 17% of terrestrial and inland water areas and 10% of coastal and marine areas, especially areas of particular importance for biodiversity and ecosystem services, are conserved through effectively and equitably managed, ecologically representative and well-connected systems of protected areas and other effective area-based conservation measures, and integrated into the wider landscape and seascape.

There have been many drivers for the establishment of the Commonwealth marine reserves network, domestically and internationally. Activism by concerned professionals in industry, academia and non-government organisations raised awareness and recognition by the public, industry more broadly and government that there were environmental, economic and social consequences of overuse and exploitation of marine resources, pollution and, more recently, climate change and ocean acidification. Concern over the prospect of losing iconic ocean species such as whales, turtles and sharks sparked public concern and government action to prohibit or limit harvesting, and to protect some key habitats. Regular political and media focus on whales and whaling has maintained a level of public interest in oceans for over 40 years. More recently the status and outlook for the Great Barrier Reef has been the subject of intense attention, domestically and internationally, with many of the issues of concern – climate change, land-based sources of marine pollution, ocean acidification and resilience – resonating more broadly for marine ecosystems globally.

Perhaps the key driver in recent years has become the widely shared view that if marine resources were to be available for use, appreciation and enjoyment in the future, greater attention would need to be given to conserving biodiversity and to ensuring that sustainability and resilience were key considerations in decisions on the use of marine resources. This recognition has driven major reforms in fishing policy and management, as well as the establishment of networks of MPAs with their primary focus on biodiversity conservation. The MPA paradigm, while drawing on its terrestrial counterpart, acknowledged that continued access to, and use of, marine resources as a management objective could complement better protection and conservation. This is part arose from the economic value of harvesting food from the oceans, the social cohesion of coastal communities where fishing is important and the associated political power of the fishing industry, but also from the practical regulatory perspective of how a vast ocean territory could be effectively managed with comparatively few resources – better to have responsible users that share and contribute to management objectives, than the alternative.

Generally, the approach taken to the identification and eventual declaration of marine reserves has received broad bipartisan support in successive Australian governments.

There were, and still are, however, divergent views on how to best balance conservation and wise use in a policy and practice sense. Differences in the 'what needs to change, why and how', remain at the heart of the contested space that is the design and management of MPAs, and oceans management more generally.

Environmental non-government organisations have been determined and persistent advocates for better oceans management and for biodiversity conservation as a primary management objective, through MPAs, species-specific measures and a greater focus on the sustainability of fisheries. The Australian Conservation Foundation, WWF-Australia, Pew Charitable Trusts, state conservation councils, specific organisations focused on the conservation of individual or groups of species (e.g. whales and dolphins), and together with the Australian Marine Conservation Society, have all played significant roles in raising awareness of marine conservation issues with the public, the media and politicians at the national level.

Fishing industry representatives, individual companies and businesses, oil and gas companies and their representatives, the shipping industry, local fishing communities and their representatives at a community, industry and political level have all been similarly active in advocating continuing access for economic uses of the oceans.

The current MPA estate in waters controlled by the Australian Government

The reserves that comprise the MPA estate established and managed by the Australian Government are listed in Table 4.1. Reserve locations are shown in Figs 4.2 and 4.3 (the Great Barrier Reef Marine Park is managed separately under its own legislation and agency and is dealt with in Chapter 5). These reserves have been declared and are managed under the provisions of the EPBC Act. Protected areas established under the EPBC Act are called Commonwealth reserves; all MPAs so declared have been called Commonwealth marine reserves (reserves declared previously under other designations have been renamed), and the Act requires each reserve and zones within reserves to be assigned to one of the IUCN Protected Area Categories, which are set out in the *EPBC Regulations 2000* (Department of the Environment 2015).

As a matter of Australian Government policy and practice, the Commonwealth reserves, and zones within the reserves in the current networks, are assigned to one of four IUCN Protected Area Categories (Ia, II, IV and VI), depending on the management objective for the zone (Environment Australia 2002). Although these categories and zones can be, and have been, assigned to reserves on their proclamation, the statutory management plans can reassign zones to different IUCN categories, change the management objective for the zone and therefore alter the types of activities that are allowed, and that may be allowed subject to appropriate authorisation, or prohibited (Table 4.2). The EPBC Act essentially prohibits some activities, including any commercial activity, or actions that affect native species, unless the management plan for the reserve provides otherwise, or the activity is carried out under a valid permit.

Given the key legal role played by management plans in determining the activities permitted or not in a reserve, and a political perception that management plans should be completed within the then current term of government, the preparation of management plans for the most recently declared reserves and networks commenced immediately following their proclamation in November 2012. The first 30-day consultation period on management plans closed the week before Christmas. Draft management plans were prepared and released for public comment for a further 30-day period from mid-January to mid-February 2013. Over 100 000 submissions were received in these two consultation periods. Final plans were approved by the Environment Minister in March 2013 and submitted to Parliament for a statutory 15 sitting day disallowance period.

The speed with which these plans were prepared and finalised and the tight consultation process over the major holiday season was unpopular in some sectors. Some of the content of the plans was particularly contentious with the commercial and recreational fishing sectors. Several commercial fishing businesses and their representatives foreshadowed a devastating impact on their industry from the exclusion of a range of fishing methods across the networks and the prohibition of fishing in sanctuary zones, despite government analysis showing an overall impact (loss) of 1% of annual gross value of production. Although these concerns had developed initially during the early identification phase for potential marine reserves (the broad 'areas for further assessment'), and increased as more precise boundaries were drawn on maps, this opposition to the new reserves intensified with the production of network management plans, which were the legal expression of what would, and would not, be allowed uses in the reserves and zones within the reserves. The recreational fishing lobby mounted a strident campaign against a perceived 'lock-out' of recreational

Table 4.1. Commonwealth Marine Reserves (as at 2015)

Region	Commonwealth Marine Reserve	Area (km²)
Coral Sea	Coral Sea	989 842
Coral Sea area total		**989 842**
North		
	Arafura	22 924
	Arnhem	7125
	Gulf of Carpentaria	23 775
	Joseph Bonaparte Gulf	8597
	Limmen	1399
	Oceanic Shoals	71 743
	Wessel	5908
	West Cape York	16 012
North area total		**157 483**
North-west		
	Argo-Rowley Terrace	146 099
	Ashmore Reef	583
	Carnarvon Canyon	6177
	Cartier Island	172
	Dampier	1252
	Eighty Mile Beach	10 785
	Gascoyne	81 766
	Kimberley	74 469
	Mermaid Reef	540
	Montebello	3413
	Ningaloo	2435
	Roebuck	304
	Shark Bay	7443
North-west area total		**335 437**
South-east		
	Apollo	1184
	Beagle	2928
	Boags	537
	East Gippsland	4137
	Flinders	27 043
	Franklin	671
	Freycinet	57 942
	Huon	9991
	Macquarie Island	161 895
	Murray	25 803
	Nelson	6123
	South Tasman Rise	27 704
	Tasman Fracture	42 501
	Zeehan	19 897
South-east area total		**388 355**

Region	Commonwealth Marine Reserve	Area (km^2)
South-west		
	Abrolhos	88 126
	Bremer	4472
	Eastern Recherche	20 574
	Geographe	977
	Great Australian Bight	45 926
	Jurien	1851
	Murat	938
	Perth Canyon	7409
	Southern Kangaroo Island	630
	South-west Corner	271 898
	Twilight	4641
	Two Rocks	882
	Western Eyre	57 946
	Western Kangaroo Island	2335
South-west area total		**508 605**
Temperate East		
	Central Eastern	70 054
	Cod Grounds	4
	Gifford	5829
	Hunter	6257
	Jervis	2473
	Lord Howe	110 139
	Norfolk	188 444
	Solitary Islands	152
Temperate East area total		**383 352**
Heard Island and McDonald Islands (HIMI)	Heard Island and McDonald Islands	70 953
HIMI area total		**70 953**
Grand Total		**2 834 028**

fishing from these reserves, objecting strenuously to the banning of recreational fishing from national park zones (IUCN Category II) arguing that recreational fishing had little or no impact and should therefore not be excluded (Fishing World 2012).

When Parliament was prorogued when the 2013 election was called, the statutory requirement for management plans to pass the 15-day disallowance period had not been completed, and would have to recommence with the new Parliament. During the election campaign, the opposition undertook to revisit how new marine reserves would be managed if elected (Taylor 2013). Following the change of government at the election in September 2013 the current government is implementing this commitment.

As a result of the rescission in December 2013 of the management plans for the networks in the south-west, north-west, north and temperate east and that developed for the Coral Sea, the policy position articulated before the plans were to come into effect, that is no 'on the water' changes for current users of these reserves, has been continued by the

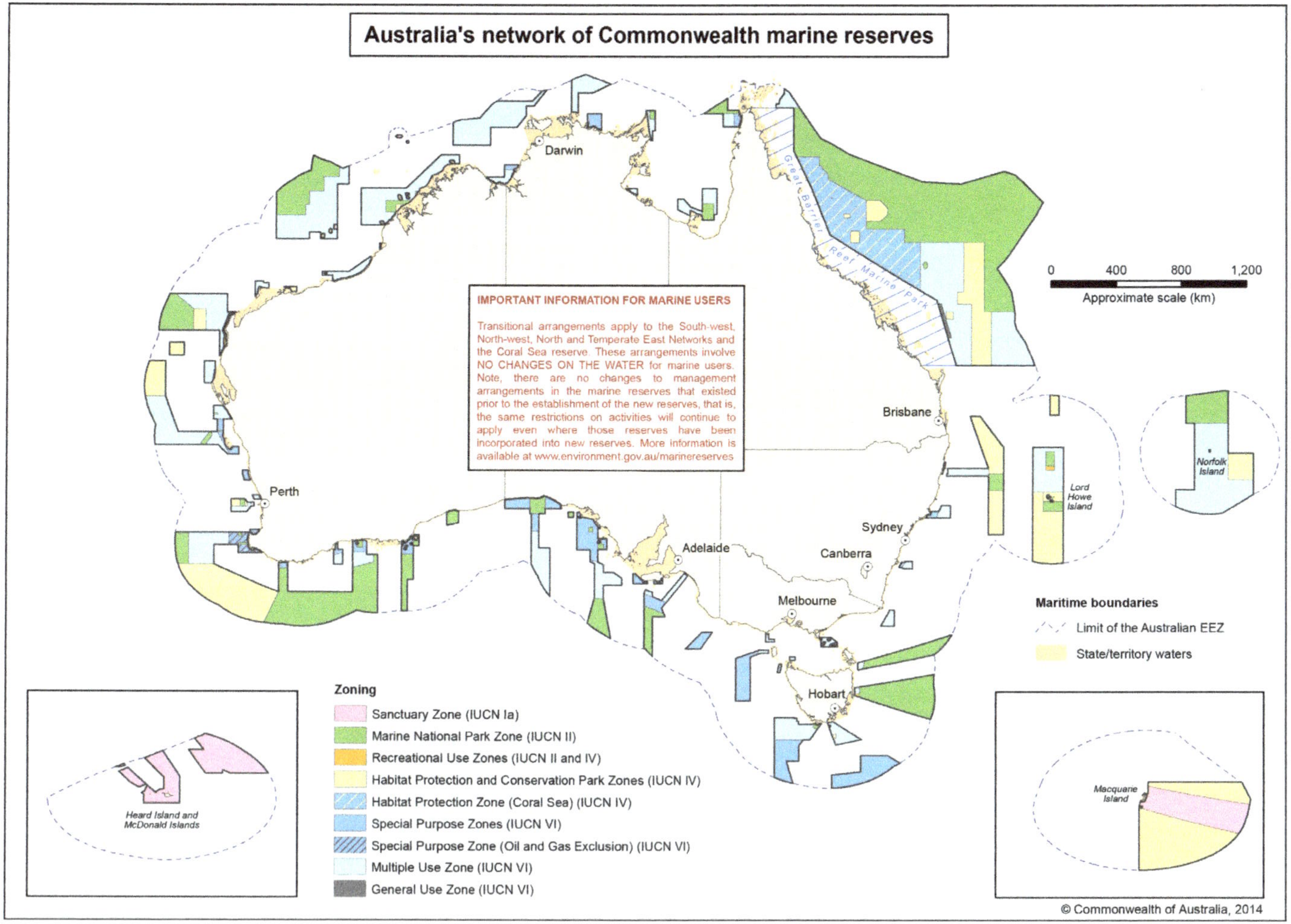

Fig. 4.3. Zoning within Commonwealth Marine Reserves as at 2014.

Table 4.2. The general approach taken to zoning activities that are allowed (A), allowed subject to a permit or other authorisation (P), and not allowed (N) in Commonwealth marine reserves as proposed in the rescinded 2013 management plans

(There were some different subcategories of IUCN VI and IUCN IV zones between networks reflecting different sets of allowed and prohibited activities, mostly related to oil and gas or commercial fishing gear types)

Activity	Multiple Use Zone (IUCN VI)	Habitat Protection Zone (IUCN IV)	National Park (IUCN II)	Sanctuary Zone (IUCN Ia)
General use				
– shipping (general transit)	A	A	A	A
– defence	A	A	A	A
Commercial fishing				
– droplines	P	P	N	N
– fish traps	P	N	N	N
– trolling and handlines	P	P	N	N
– purse seine	P	P	N	N
– lobster pots	P	N	N	N
– crab pots	P	N	N	N
– pelagic longline	P	P	N	N
– pelagic gill-net	P	P	N	N
– drift diving	P	P	N	N
– hand collection	P	P	N	N
– demersal gill-net	N	N	N	N
– demersal trawl	N	N	N	N
– demersal longline	N	N	N	N
Pearling and aquaculture	P	P	N	N
Commercial tourism				
– non-fishing related tourism	P	P	P	P
– fish-related tourism (e.g. charter fishing)	P	P	N	N
Mining	P[a]	N	N	N
Research	P	P	P	P
Indigenous activities (non-commercial)	A	A	A	A
Recreation	A	A	A	N
– scuba diving, snorkelling	A	A	A	N
– recreational fishing	P	P	N	N
– boating	A	A	A	N
– shipping	A	A	A	A

Notes: This activity table is generalised from the activity tables that had been proposed for each Commonwealth marine reserve network in the now rescinded management plans, because there are local variations and geographical differences between reserves and what is allowed in different zones. Some additional activities are allowed, or may be permitted in some zones.

[a]Mining is not allowed in any zone in the Coral Sea Commonwealth Marine Reserve

current government. This will remain the case until new management plans are in effect, which is unlikely before mid-2016.

Current policy

The policy framework articulated in the Oceans Policy (1998), the *Guidelines for Establishing the National Representative System of Marine Protected Areas* (1998) and the 2007 *Goals and Principles for the Establishment of the National Representative System of Marine Protected Areas in Commonwealth Waters* remain the core policy documents for the design and establishment of the Australian Government's MPA estate. The *Environment Protection and Biodiversity Conservation Act 1999* provides the legal framework for the development and implementation of the management plans that are required for each reserve (and/or network).

Current policy is to prepare management plans that cover each network rather than each reserve. There are five bioregional networks that include 57 marine reserves (the South-east, South-west, North-west, North and Temperate East). The Coral Sea Commonwealth Marine Reserve (Fig. 4.4) and the Heard Island and McDonald Islands (HIMI) Marine Reserve sit outside these networks.

Before developing new management plans, the Australian Government committed to consult widely with interested parties and to review the science underpinning zoning options, with a particular focus on reviewing areas of concern raised by stakeholders in previous submissions and during the 2013 election campaign.

The Australian Government's first step in reviewing the management arrangements for the new Marine Reserves established in November 2012, was to re-proclaim the South-

Fig. 4.4. Whitetip Reef Shark, Osprey Reef, Coral Sea Commonwealth Marine Reserve. (Photo: Erik Schlogl)

west, North-west, North and Temperate East Networks and the Coral Sea Commonwealth Marine Reserve in December 2013 with exactly the same description and boundaries as the November 2012 proclamation. This had the (intended) effect of legally setting aside the management plans for these reserves that had been prepared and submitted to Parliament in March 2013, and reinstituting the IUCN categories and zoning arrangements that were set out in the original proclamations of November 2012. The management plans for HIMI and the South-east network were already in effect and are unaffected by the Australian Government's review. Management arrangements for pre-existing reserves that were being included in the new networks to continue as separate reserves or incorporated into larger reserves were not changed by these decisions – that is, the same restrictions on activities in these reserves continue to apply until new management plans are finalised.

The next step of the Australian Government's review process was the establishment of a formal Commonwealth Marine Reserves Review through two parallel processes – a review of the science underpinning marine reserve management by a five member Expert Scientific Panel, and stakeholder consultation at the national and regional level through the establishment of five Bioregional Advisory Panels, led by two co-chairs, who both sit on the Expert Scientific Panel. These two parallel processes extended to the end of 2015 (Parks Australia 2015).

Previous submissions to earlier consultations on the establishment and design of the reserves networks, and their now rescinded management plans, have been reviewed and considered by the Bioregional Advisory Panels. Public submissions were invited seeking new or changed information or positions from interested parties. This produced over 13 000 written submissions. Over 260 face-to-face meetings with key stakeholders and representatives of major affected sectoral interests were held in locations around Australia over a 6-month period. In addition to the call for formal submissions, a simple online survey was developed as another avenue for interested parties to contribute their views to the Review. The Bioregional Advisory Panels have been tasked with developing zoning options that best take into account the concerns raised in submissions and through their consultations, taking into account the existing policy framework (including the *Goals and Principles for the Establishment of the National Representative System of Marine Protected Areas in Commonwealth Waters*) and the best available science and advice from the Expert Scientific Panel. This framework does not include specific area goals and targets, beyond the Aichi targets. At the time of writing, the Bioregional Advisory Panel co-chairs are evaluating and assessing all these inputs, along with a comprehensive economic assessment of the impact of potential zoning options to develop a set of recommended zoning changes in their report. The Expert Scientific Panel and the bioregional stakeholder consultation processes are working towards reporting by December 2015.

The report of the Expert Scientific Panel and the report from the co-chairs of the Bioregional Advisory Panels are expected to then inform the development of new management plans for these Reserves in 2016.

The preparation of new plans will provide interested parties with further opportunities to comment on the management arrangements during the two stages of statutory consultation that are required to be undertaken before the plans can be finalised. The new management plans will set out how the reserves will be managed, including allowed, permitted and prohibited activities in different zones, and the zoning boundaries within each reserve.

Once the consultation requirements for each management plan have been completed, the Director of National Parks will prepare reports detailing the comments received alongside the Director's views on those comments. These reports and the final management plans will be provided to the Australian Government's Minister for the Environment for approval. Once approved, the management plans will be tabled in both houses of Parliament for the required 15 sitting days during which they will be subject to disallowance.

Two significant areas of the Australian marine jurisdiction that were not included in either Bioregional Marine Planning or marine reserve consultation processes are the two Indian Ocean Territories of Christmas Island and Cocos (Keeling) Islands. It would be reasonable to conclude their omission was more a reflection of the strong domestic focus on, and long history of, developing a marine reserve system in the waters around the mainland commencing with the first marine bioregionalisation in 1996, which does not include mention of the Indian Ocean Territories. Their omission could and should be rectified in the future.

Challenges

The key challenges for the current marine reserve system are:

- dealing with the 'consultation fatigue' that many stakeholders are expressing, given the previous intensive consultations over some 7 years on the location and design of the marine reserves and on their management
- determining an appropriate balance of activities that can or may be allowed, consistent with the protection of reserve values
- making decisions with imperfect information
- prioritising and efficiently filling key information gaps
- developing an efficient and effective compliance and enforcement effort over such a large area
- ensuring there are adequate resources available for effective reserve management
- integrating reserve management into broader ocean management
- dealing with any adjustment issues that arise if commercial fishing effort is displaced from marine reserves.

The first three of these were within the scope of the Commonwealth Marine Reserves Review. The approach taken by the Bioregional Advisory Panels of reading all prior submissions and stressing that interested parties should focus on new, changed or additional information rather than restating what has been said before, appears to have alleviated some of the concern about 'consultation fatigue'.

A core task of the Review was to carefully consider the balance of interests that could be allowed, permitted or prohibited in the reserves and provide the government with recommendations for revised zoning options for the reserves. The Bioregional Advisory Panels were charged with providing advice on areas of contention and options for zoning to address these areas. They were also tasked with providing recommendations for improving the inclusion of socio-economic considerations into decision making, and suggestions for ongoing engagement of regional stakeholders. The government will consider these recommendations and respond, through the preparation of new management plans.

Imperfect information will always be a factor in managing natural resources, but it is particularly acute in the remote ocean environment where most Commonwealth reserves are located. Biodiversity information is patchy at best (see Chapters 3, 16, 17). Reserve managers will need to prioritise and support information collection and improved understanding of ocean ecology and biodiversity threats. A substantial monitoring and research effort will be required to ensure management resources are deployed efficiently and effectively. The Expert Scientific Panel assembled the best available science from published papers, and was tasked with advising on future priorities for scientific research and monitoring relating to marine biodiversity, especially any relating to the understanding of threats to biodiversity, and on options for filling the most significant information gaps hindering robust, evidence-based decisions for management.

Work is underway on the compliance and enforcement requirements of the full reserve estate. Considerable experience has been gained from the management of the South-east network, declared in 2007. To some degree, compliance and enforcement can depend on the nature and extent of allowed activities in reserves and zones. It is generally accepted that permitted users can act as 'eyes and ears' for illegal and unauthorised activities. Several past compliance and enforcement actions have been successful due to information received from authorised operators. New technologies, particularly remote sensing including from satellites, vessel monitoring systems (required for all vessels licenced to operate in federally managed fisheries), and live on-board cameras can provide some cost-effective solutions for compliance. In addition, aerial patrols by Coastwatch and on water patrols by Australian Border Force and Navy vessels will continue to play an important role in monitoring the more remote reserves. In reserves close to, and adjacent to, state and territory waters and reserves, cooperative and contractual arrangements have been negotiated between Parks Australia and relevant state agencies to include Commonwealth marine reserve management responsibilities, with examples including Ningaloo off Western Australia (Fig. 4.5), the Great Australian Bight off South Australia, the Cod Grounds and the Solitary Islands off New South Wales (Fig. 4.6), and several of the South-east network reserves off Tasmania. These arrangements are expected to continue and where relevant be extended.

Fig. 4.5. Nudibranch, Ningaloo Commonwealth Marine Reserve. (Photo: Cathy Zwick)

Fig. 4.6. Solitary Islands Commonwealth Marine Reserve. (Photo: Nigel Marsh)

The issue of adequate resources is one for governments to consider within their overall financial position and priorities. The size and national and global significance of the Commonwealth's reserve estate warrants a high priority for support.

The integration of reserve management into a broader oceans management framework should be a medium-term objective (see also Chapters 3, 19). In the first instance, collaboration and cooperation with fisheries managers and policy makers is essential, given the economic significance of the commercial fishing sector (see also Chapter 22). Inter-agency cooperation is positive and constructive at the operational and management level. There is scope for considering closer collaboration on fisheries and nature conservation policy.

The issue of dealing with displaced fishing effort can be complex. Commercial fishing effort that is restricted or prevented from operating as a result of new management arrangements in the reserves may need to be adjusted out through structural adjustment if the restrictions and impacts on operators are significant. In some cases, fishing effort can relocate, although as the management of fisheries in Commonwealth waters has progressively become more tightly focused on sustainability, the options for relocation are often limited. The previous assessment of displaced effort and the economic consequences of the now rescinded management plans had estimated an economic impact of around 1% of the Gross Value of Production (Department of Prime Minister and Cabinet 2012). This estimate, the information it was based and the methodology used has been widely criticised by fishing industry representatives and individual operators (e.g. Commonwealth Fisheries Association 2012). As the Commonwealth Marine Reserves review developed its zoning options and recommendations, social and economic impacts of potential options were carefully considered (see also Chapter 15). The real story of economic and associated social impact will not be finalised until after new management plans have been prepared, because these will be the determinants of whether there is displaced effort or not, and which fisheries and operators are affected.

Future directions

Although there has been some criticism about how comprehensive and representative the Commonwealth marine reserve networks are (Barr and Possingham 2013; Devillers *et al.* 2015) – for example, the continental shelf is poorly represented in some regions of the reserve system, perhaps most notably in the Temperate East network (see also Chapter 17) – it seems unlikely that any government for some years will consider any revisiting or extension of reserve boundaries, given the size of the current reserve system, the extensive debates over what an appropriate balance of uses and users should be to protect conservation values while allowing for current and potential economic activity, and the challenges that need to be considered within its current boundaries.

There is scope to secure better integration of marine reserve management with state and territory waters, where reserves abut across jurisdictional boundaries. This will be progressed in part through the current Commonwealth Marine Reserves Review, but will remain an ongoing issue while state and the Northern Territory governments develop their own marine reserve networks within their jurisdictions.

Several Commonwealth marine reserves overlap with native title determinations that include sea country, and several other claims that extend into reserves are yet to be determined. There is a strong desire by the relevant Indigenous communities to actively engage in the planning and management of these areas. This creates a considerable opportunity for innovative practices to emerge that respect and deliver the aspirations, expectations and needs of both these Indigenous communities and reserve managers (see also Chapter 18).

There is also scope to explore collaboration with reserve managers in other countries – perhaps most promisingly New Caledonia and New Zealand – because there are several marine reserves that reach EEZ borders that are shared with these countries, and both have an interest in collaborating across shared EEZ borders. This does not appear to be an active area of Australian Government policy interest and is more likely to arise in responding to external interest in the future.

The remaining major marine area of strong Australian and international interest in improving marine conservation outcomes and establishing a representative system of marine reserves is in Antarctic waters. Work on this continues under the Antarctic Treaty arrangements (see Chapter 6).

Conclusions

More than any other country, Australia has taken a systematic science-based approach to meeting its international commitments to establish a comprehensive, adequate and representative system of MPAs by 2012. Although it has considerably exceeded the modest headline Aichi target of 10% of nations' marine areas in MPAs by 2020, as a truly comprehensive, adequate and representative system, there are still some significant improvements to make and gaps to fill. At this early stage in the system's establishment, the Aichi target of a reserve system that is effectively and equitably managed remains to be demonstrated.

The history of establishing marine reserves in waters controlled by the Australian Government shows a progressive shift from protecting iconic habitats in a somewhat piecemeal fashion to a systematic identification and declaration of a representative system of reserves based on a steadily growing knowledge and evidence base that takes into account social and economic interests as well as the primary focus on protecting conservation values. These values include biodiversity (represented through surrogate physical and ecological features) and ecosystem processes. Grappling with the paucity of knowledge of these key values will remain a significant management priority together with improving understanding of threats to these values and developing efficient approaches to reducing these threats.

The re-proclamation by the current Australian Government of the outer boundaries of the 40 new reserves declared in 2012 reflects a bipartisan acceptance of the new estate. The differences between successive governments lie more on the management of the reserves, where the balance of interests should be struck, and the consultation, stakeholder support and science that underpins management plans and effective management. The key issue is achieving a balance of interests that allows for sound management, the effective conservation of the biodiversity within the reserves, and the realisation of the social and economic potential of these areas, consistent with conserving their conservation values.

Acknowledgements

The Secretariat of the Commonwealth Marine Reserves Review in Parks Australia provided valuable assistance with information and figures. Any inaccuracies, opinions and interpretations are the sole responsibility of the author.

Biography

Peter has over 20 years' experience in senior executive leadership and governance roles in the public and private sectors. He currently consults on environment and sustainability issues. He is a member of the Boards of Ecotourism Australia and Tangaroa Blue Foundation, and is an Adjunct Fellow at the Fenner School of Environment and Society at The Australian National University. He is a co-chair of the Bioregional Advisory Panels and a member of the Expert Scientific Panel established to conduct the Commonwealth Marine Reserves Review. He also chairs the Steering Committee for the National Environment Science Program's Marine Biodiversity Hub. Peter was Director of National Parks and head of Parks Australia from 1999 to 2013.

References

ANZECC (1996) *National Strategy for the Conservation of Australia's Biological Diversity.* Department of the Environment, Sport and Territories, Canberra.

ANZECC TFMPA (1998) *Guidelines for Establishing the National Representative System of Marine Protected Areas.* Australian and New Zealand Environment and Conservation Council, Task Force on Marine Protected Areas. Environment Australia, Canberra.

Australian Government (2007) *Goals and Principles for the Establishment of the National Representative System of Marine Protected Areas in Commonwealth Waters.* Australian Government, Canberra.

Barr L, Possingham H (2013) Are outcomes matching policy commitments in Australian marine conservation planning? *Marine Policy* **42**, 39–48. doi:10.1016/j.marpol.2013.01.012.

COAG (1992) *Intergovernmental Agreement on the Environment.* Council of Australian Governments, Canberra.

Commonwealth Fisheries Association (2012) *Commonwealth Fishing Industry Submission on the Proposed Commonwealth Marine Reserve Network.* Commonwealth Fisheries Association, Melbourne, <http://www.ntsc.com.au/wp-content/uploads/NSIA-CMRN-submission-10-September-2012-FINAL.pdf>.

Commonwealth of Australia (1998a) *Australia's Oceans Policy Volume 1: Caring, Understanding, Using Wisely.* Commonwealth of Australia, Canberra.

Commonwealth of Australia (1998b) *Australia's Oceans Policy Volume 2: Specific Sectoral Measures*. Commonwealth of Australia, Canberra.

CSIRO (1996) *Interim Marine Bioregionalisation for Australia: Towards a National System of Marine Protected Areas*. Version 1.01. CSIRO Division of Fisheries, CSIRO Division of Oceanography, Canberra, <http://www.marine.csiro.au/datacentre/ext_marlin_docs/bioregion/contents.html>.

Department of Prime Minister and Cabinet (2012) *Completing the Commonwealth Marine Reserves Network. Regulatory Impact Statement*. Department of Prime Minister and Cabinet, Canberra, <http://ris.dpmc.gov.au/2012/06/22/completing-the-commonwealth-marine-reserves-network-regulation-impact-statement-department-of-sustainability-environment-water-population-and-communities/>.

Department of the Environment (2015) *Commonwealth Marine Reserves – Legal Framework*. Department of the Environment, Canberra, <www.environment.gov.au/topics/marine/marine-reserves/overview/legal-framework>.

Department of the Environment (2001–04) *Scientific and Expert Reports on the South-East Marine Region – a Series of Reports Compiled as Part of the Establishment of the South-East Commonwealth Marine Reserves Network*. Department of the Environment, Canberra, <http://www.environment.gov.au/topics/marine/marine-reserves/south-east/resources>.

Devillers R, Pressey RL, Grech A, Kittinger JN, Edgar GJ, Ward T, *et al.* (2015) Reinventing residuals reserves in the sea: are we favouring ease of establishment over need for protection? *Aquatic Conservation: Marine and Freshwater Ecosystems* **25**, 480–504. doi:10.1002/aqc.2445.

Director of National Parks (2012) *Report of the Director of National Parks Concerning the Proposed Proclamation of 40 Commonwealth Marine Reserves*. Director of National Parks, Canberra.

Director of National Parks (2013) *South-east Commonwealth Marine Reserves Network Management Plan 2013–2023*. Director of National Parks, Canberra.

Environment Australia (1998) *Interim Marine and Coastal Regionalisation for Australia*. Department of the Environment, Canberra.

Environment Australia (2002) *Australian IUCN Reserve Management Principles for Commonwealth Marine Protected Areas*. Environment Australia, Canberra.

Fishing World (2012) 'Don't Lock Us Out' campaign launched. *Fishing World* 28 June 2012, <http://www.fishingworld.com.au/news/don-t-lock-us-out-campaign-launched>.

National Oceans Office (2001) *Pressures on Uses in the South-east Marine Region*. National Oceans Office, Hobart.

Parks Australia (2015) *Commonwealth Marine Reserves Review*. Parks Australia, Canberra, <http://www.marinereservesreview.gov.au>.

Pressey R (2013) Australia's new marine protected areas: why they won't work. *The Conversation*. 16 January 2013, <http://theconversation.com/australias-new-marine-protected-areas-why-they-wont-work-11469>.

Taylor L (2013) Coalition to set up scientific panel to review marine park boundaries. *The Guardian* 26 August 2013, <http://www.theguardian.com/world/2013/aug/26/marine-life-fishing-coalition-review>.

Chapter 5

The Great Barrier Reef Marine Park: the grandfather of modern MPAs

Jon Day

Introduction

The Great Barrier Reef (GBR), the largest coral reef ecosystem on the planet, is often referred to as one of the seven wonders of the natural world.

In addition to its global reputation, the GBR is also a significant part of the nation for most Australians. Aboriginal and Torres Strait Islander people used its coastal waters, islands and reefs for traditional resources and customary/spiritual practices for thousands of years before explorers such as Louis de Bougainville (1768), James Cook (1770), William Bligh (1789) and Matthew Flinders (1801) navigated their way through the uncharted waters.

The pioneering botanists Joseph Banks and Daniel Solander were the first to systematically document the marine and terrestrial species collected from the GBR during Cook's great voyage of discovery in 1770 (Banks 1770). Flinders, a remarkable explorer, named the Great Barrier Reef in 1801 and was one of the first to write about its underwater beauty, marvelling at the 'vivid tints of every shade betwixt green, purple, brown and white; equalling in beauty and excelling in grandeur the most favourite ... florist' (Flinders 1814).

In the late 1800s, European settlers saw the economic potential of the coastal areas and some of the islands but thought little of the reefs other than as a navigation hazard. Many coastal population centres developed over the next 150 years around industries such as agriculture, mining, commercial fishing, ports, shipping and, in more recent times, tourism.

Gradually the significance of the reefs and the adjoining marine areas was recognised and acknowledged. The GBR has long been appealing for scientists, and considerable scientific knowledge has been generated from GBR-based studies. The Yonge expedition to the Low Isles in 1928–29 was the first comprehensive scientific study of a coral reef anywhere in the world.

Tourism to several GBR islands and cays began to flourish in the 1950s, and reef-walking and shell-collecting were both popular activities. Technological advances soon followed, including glass-bottomed boats, underwater observatories and scuba diving, opening up the wonder of the reefs to rapidly increasing numbers of visitors.

Since the early 1960s, the GBR's condition and management has been subject to political controversy and conflict. This reflects the economic growth along Queensland's coast, but also the strong support that the Australian public has for the GBR.

Proposals for limestone mining by farmers and the issuing of leases for oil drilling by the Queensland Government in the 1960s and early 1970s provided a focus for a campaign by conservationists for greater protection of the GBR. This campaign subsequently led to the declaration of the Commonwealth *Great Barrier Reef Marine Park Act 1975* (the Act). A key part of the Act was the formal declaration of the Great Barrier Reef Region (GBR Region) within which sections of the Great Barrier Reef Marine Park (GBRMP) could be subsequently declared. Today the federal GBRMP covers over 99.5% of the GBR Region and extends eastward from the 'northernmost extremity of Cape York Peninsula' (Schedule 1 in the Act) to just past Lady Elliot Island in the south (Fig. 5.1). Within the GBRMP there are some 3000 separate coral reefs. Figure 5.1, however, shows an additional 550 coral reefs within the Torres Strait that are part of the same reef system that comprises the GBR. Nevertheless, the tip of mainland Australia was chosen as the northern extent of the Marine Park.

Covering 344 400 km^2, the GBRMP is equivalent in size to Italy, Japan or Malaysia. Its western boundary is generally the mean low-water mark (LWM) on the mainland coast of Queensland (see Chapter 12) and the Park extends seaward between 80 and 250 km offshore, well outside the outer reefs into deep oceanic waters.

The federal GBRMP does not include the majority of the ~1000 islands within its outer boundary, most of which are under Queensland jurisdiction (only 70 islands or parts of islands that contain lighthouses or defence training areas under federal jurisdiction are part of the GBRMP). Over half of the 1000 islands are national parks, declared and managed by the Queensland Government. The federal GBRMP also does not include 13 coastal exclusion areas around ports,[1] nor any Queensland 'internal waters'.[2] The GBRMP does, however, extend to a depth of 1000 m beneath the seabed and to a height of 915 m above sea level into the airspace above.

Queensland declared the Great Barrier Reef Coast Marine Park in November 2004, combining numerous previously small State Marine Parks into a large single State Marine Park (see Chapter 12). It covers the majority of Queensland waters between high-water mark and LWM along the mainland adjoining the GBR Region and around all Queensland islands within the outer boundaries of the federal GBRMP. This Coast Marine Park provides complementary zoning between high water and low water which 'mirrors' the adjoining federal zoning. This means today there are complementary management provisions for virtually all marine waters in the GBR Region irrespective of the jurisdiction.

Coral reefs are a major drawcard for the area (Fig. 5.2), but they comprise less than 10%[3] of the overall extent of the GBRMP today. Some 30 different reef types have been recognised and mapped but the remaining 90% of the GBRMP comprises 40 different broad-scale non-reef habitat types, ranging from shallow inshore waters to deep oceanic waters. These broad-scale habitat types were described in 1999–2000 as part of the preliminary work done to assist the rezoning; this produced a map of 70 ecologically different 'bioregions'[4] (Kerrigan *et al.* 2010).

Global significance of the Great Barrier Reef

The global significance of the GBR was recognised in 1981 when an area of 348 000 km^2 was acknowledged as being of Outstanding Universal Value[5] and was inscribed on the World Heritage List. At that time, the evaluation by the International Union for Conservation of Nature (IUCN) stated 'If only one coral reef site in the world were to be chosen for

Fig. 5.1. The Great Barrier Reef World Heritage Area (= red boundary): the World Heritage Area starts at low-water mark along the coast of mainland Queensland; the slightly smaller Great Barrier Reef Marine Park (= green boundary): the federal Marine Park does not include 13 coastal exclusions around ports or the majority of the islands within the outer boundary; and the extent of the adjoining catchment (= yellow boundary). (© Commonwealth of Australia (GBRMPA) 2015)

Fig. 5.2. Reefscape at Davies Reef, Great Barrier Reef Marine Park. (© Commonwealth of Australia (GBRMPA), Photographer: Jeffrey Maynard)

the World Heritage List, the Great Barrier Reef is the site to be chosen'. IUCN, noting the coral reefs extended 'beyond the northern boundary of the property ...', recommended the World Heritage Committee '...accept the addition of this area should it become available in the future' (IUCN 1981).

The Great Barrier Reef World Heritage Area (GBRWHA) has the same northern, eastern and southern boundaries as the GBRMP. However, it covers a slightly larger area because it includes all 1000 islands within the outer boundaries and all waters seaward of the LWM, including the ports and the internal waters of Queensland. The GBRWHA extends over 14 degrees of latitude with an expansive latitudinal and cross-shelf diversity, resulting in it containing arguably the greatest species biodiversity of any World Heritage area on the planet (Day 2015), including:

- 56% of the world's hard coral species
- one-third of the world's soft coral and sea pen species
- six of the world's seven species of marine turtles
- 54% of the world's mangrove diversity
- 23% of the world's seagrass diversity
- 13% of the world's species of starfish, sea urchins and sea cucumbers
- seabird breeding colonies on islands of world significance
- one of the world's most important populations of Dugongs.

The GBRWHA also includes spectacular seascapes and landscapes such as the Whitsunday Islands and Hinchinbrook Island, and is home to some 1625 species of bony fish, 133 species of sharks and rays, 3000 species of molluscs, 630 species of macro-algae and some 30 species of whales and dolphins (GBRMPA 2012).

This wide range of species and habitats, and their interconnectivity, make the GBR one of the richest and most complex natural ecosystems on Earth with enormous scientific, cultural and conservation significance. Twenty-six species currently listed as threatened under the *Environment Protection and Biodiversity Conservation Act 1999* (EPBC Act) occur in the GBR, including various species of turtles, sharks, whales, sawfish and seabirds (GBRMPA 2013). Dugongs are also of 'national environmental significance', being listed under two sections of the EPBC Act. Two species of inshore dolphins are of conservation concern and are likely to be declining (GBRMPA 2013).

Many of these species are also listed under the Convention on the Conservation of Migratory Species of Wild Animals (Bonn Convention), the Convention on International Trade in Endangered Species of Wild Fauna and Flora (CITES), the IUCN Red List, and various international migratory bird agreements.

The GBR was for many years the largest marine protected area (MPA) in the world; however, it lost that accolade in 2006 when several larger MPAs were declared (see Table 5.1). However, the GBR has greatly influenced many subsequent MPA efforts, particularly within Australia but also around the world. Consequently although the GBRMP is not the oldest MPA in Australia and is no longer the largest in the world, it is regarded by many as 'the grandfather' of modern MPAs.

Compared with most other MPAs or World Heritage areas, the GBR is atypical due to its large size and its complexity. However, the planning and management experience gained provides useful lessons for ecosystem-based management elsewhere, including the application of marine spatial planning as well as integrated coastal and ocean management (Day 2002a,b, 2008, 2015).

Table 5.1. Large marine protected areas (greater than 100 000 km^2) listed in order of declaration date

MPA	Country	Date declared	Area (km^2)	Size when compared with GBRMP
Great Barrier Reef Marine Park	**Australia**	**1975**	**344 400**	n.a.
Papahānaumokuākea (North-west Hawaiian Islands)	United States	2006	362 100	1.05 ×
Kermadec Benthic Protection Area	New Zealand	2007	620 500	1.80 ×
Phoenix Islands Protected Area	Kiribati	2008	408 250	1.19 ×
Marianas Trench Marine National Monument	United States	2009	250 500	0.73 ×
Chagos Marine Protected Area	British Indian Ocean Territory	2010	545 000	1.58 ×
Moto Motira Hiva Marine Park	Chile	2010	150 000	0.44 ×
Coral Sea Commonwealth Marine Reserve	Australia	2011	989 800	2.87 ×
Cook Islands Marine Park*	Cook Islands	2012	1 065 000	3.09 ×
Pitcairn Islands Marine Reserve	United Kingdom	2015	830 000	2.41 ×

*Announced in 2012 but yet to be formally declared

History and development of the Great Barrier Reef Marine Park

In the 1960s and 1970s, a group of dedicated conservationists – led by John Büsst, Judith Wright and Len Webb – battled to save Ellison Reef (near Innisfail) from limestone mining and other reefs from oil exploration. The passion and commitment of those early campaigners encouraged activists, artists, poets, ecologists and students to get involved in the first GBR campaign. That movement grew to encompass scientists, trade unionists, politicians and concerned individuals throughout Australia (Bowen and Bowen 2011).

More than a decade of heated national debate occurred over the growing threats to this globally iconic area (Lawrence *et al.* 2002). The vocal public concerns led to a Royal Commission (1970–74) that resulted in the passing of the federal Act in 1975 that in turn provided for the protection and management of the GBR.

The Royal Commission recommended '... a special statutory authority ... be established responsible to the appropriate parliament or parliaments for ecological protection and the control of research and development within the Great Barrier Reef province' (Royal Commission into Exploratory and Production Drilling for Petroleum in the Area of the Great Barrier Reef 1974). The Act therefore established the Great Barrier Reef Marine Park Authority (GBRMPA), a 'special statutory authority' responsible for functions including:

- recommending to the Minister which parts of the GBR Region should be declared as Marine Park
- preparing zoning plans for the Marine Park
- providing educational, advisory and informational services relating to the Marine Park
- management of the Marine Park.

The Act was pioneering legislation for its time, allowing for 'conservation and reasonable use' of natural resources provided that such uses did not compromise the conservation of the GBR; this effectively introduced the concept of a multiple-use marine park. The Act also provided for ocean zoning (i.e. spatial planning) in the GBR within a clear objective-based regulatory framework. This comprised a comprehensive and systematic planning process to develop spatially explicit zoning plans including mandatory requirements for public participation during specific phases of the planning.

Around the time of the Royal Commission, the Commonwealth and the states undertook negotiations resulting in the 'Offshore Constitutional Settlement', concerning jurisdiction in the waters of the territorial sea. These negotiations were followed by a challenge in the High Court and a decision in late 1975 that confirmed full jurisdiction on the part of the Commonwealth up to the low-water mark in the GBR. However, this is complicated in that the definition of 'low water' applied by the Commonwealth is 'mean low water', whereas Queensland's definition refers to 'lowest astronomical tide'.

The *Great Barrier Reef Marine Park Act 1975* is a federal Act that '... provides precedence over inconsistent provisions of almost all other federal laws, and under the Australian Constitution, federal laws have precedence over any inconsistent Queensland State laws within the GBR Region. ...' (Kenchington and Day 2011, p. 273).

The Act also states that 'a matter shall be taken to relate to the Marine Park if it relates to the use of management of an area (which may be a Queensland national park or Queensland marine park) the use or management of which would or might affect the Marine Park' (*GBRMP Act*, s. 7 (1A)). This provision still exists in 2015, along with an ability (using s. 66(2e) of the Act) to regulate or prohibit acts '... (whether in the Marine Park or elsewhere) that may pollute water in a manner harmful to animals and plants in the Marine Park'. This ability to influence activities occurring outside the GBRMP has been enhanced by Section 12 of the EPBC Act, which requires approval for any action likely to have a 'significant impact on world heritage values'.

Importantly, when the GBR Region was declared in 1975, it initially comprised only the outer boundary (albeit with a legislative framework that banned oil drilling or mining[6]) but none of the area-specific management arrangements, including zoning, were in place. Sections of the federal GBRMP were progressively declared by the Commonwealth Government within the GBR Region, starting with the relatively small Capricornia Section in October 1979.

Between 1981 and 2002, different sections of the GBRMP were consecutively declared and then zoned. Table 5.2 provides a chronological history of the GBRMP including the dates that subsequent sections of the federal GBRMP were declared. Not until 1988 was the majority of the GBRMP sequentially declared and then zoned; that is, 13 years after the initial legislation proclaimed the outer boundary of the GBR Region (Day 2015).

The GBRMP was never intended, as many believe, to be protected throughout as a no-take sanctuary or a Marine National Park. Rather, it is a large zoned multiple-use marine protected area that currently has a spectrum of seven marine zones and one terrestrial zone (i.e. Commonwealth Islands Zone) (see Table 5.3).

The spectrum of zones in the GBR ranges from the least restrictive zone at one end (i.e. the General Use Zone, which allows most reasonable uses including various types of fishing), through to the Preservation Zone at the other end (i.e. very small 'no-go' areas, set aside as undisturbed 'scientific baselines' or reference areas in which only approved scientific research is permitted).

Three of these zone types (Marine National Park Zone, Scientific Research Zone and Preservation Zone) are highly protected no-take zones. Probably the best known zone is the Marine National Park Zone (locally known as a 'green zone'), which allows public access, but prohibits any extractive activities such as fishing or collecting, similar to the concept of a national park on land (Fig. 5.3).

Fig. 5.3. Aerial of Hardy Reef, a Marine National Park zone within the Great Barrier Reef Marine Park, taken in 2014. (© Commonwealth of Australia (GBRMPA), Photographer: Jumbo Aerials)

Table 5.2. Key events in the management of the Great Barrier Reef, 1975–2015

(A more detailed list for the period up to 2005 is available as an appendix to the 2005 GBR Heritage Strategy (GBRMPA 2005))

Date	Event/comments
1975	• *Great Barrier Reef Marine Park Act 1975* passed through both Houses of Federal Parliament. The Act created the outer boundary of the GBRMP and the Great Barrier Reef Marine Park Authority comprising three members (one to be nominated by the federal government, one by the Queensland Government and one by agreement between both governments)
1976	• First meeting of the GBR Consultative Committee (the major stakeholder forum in those days)
1979	• Emerald Agreement signed outlining day-to-day management arrangements for GBRMP and establishing the joint GBR Ministerial Council comprising two Ministers from each government • First section (Capricornia) of the GBRMP was declared
1980	• Basis of Agreement signed between the Commonwealth and Queensland governments for management of GBRMP • GBR nominated for World Heritage listing
1981	• The Zoning Plan and Regulations for first small part of the GBRMP (the Capricornia Section) came into effect • GBR inscribed on the World Heritage list, meeting all four natural criteria for listing • Two other sections of the GBRMP proclaimed (Cairns and Cormorant Pass)
1983	• First permit for traditional hunting and fishing issued to Hopevale Aboriginal Community • Far Northern, Central, Townsville, Southern and Inshore Southern Sections proclaimed • GBRMP Regulations into effect • Zoning Plans for Cairns and Cormorant Pass Sections into effect.
1985	• Report of the Crown of Thorns Starfish Advisory Committee released
1986	• Zoning Plan for Far Northern Section into effect
1987	• Zoning Plan for Central Section into effect • Mackay-Capricorn Section declared • GBR Wonderland opened in Townsville; it included the GBR Aquarium, the main education focus for the GBRMPA
1988	• Zoning Plan for Mackay-Capricorn Section into effect • Zoning completed for virtually the entire GBRMP – zoning plans had been progressively developed for parts of the GBRMP since the early 1980s, but it was not until 13 years after the outer boundary of the GBRMP was declared (i.e. in 1988) that the entire GBRMP was zoned. Note, however, that at this time less than 5% of the entire GBR was in highly protected 'no-take' zones.
1989	• Major construction works began for Magnetic Quays despite objections from NGOs and the community
1990	• International Maritime Organisation declares the GBR as the world's first Particularly Sensitive Sea Area requiring compulsory pilotage in parts of the GBR
1991	• Permit assessment fees introduced to cover the cost to the GBRMPA and Queensland Parks and Wildlife Service to process permit applications (the fee is charged before assessing an application)
1992	• Revised Cairns Section Zoning Plan into effect
1993	• Community Ranger program employed eight Indigenous community-based rangers • Draft Whitsunday Plan of Management released for public comment • Environmental Management Charge introduced for tourist operations in GBR, with funds raised for research, education and management of the GBR

Date	Event/comments
1994	• Publication of the 25 year Strategic Plan for GBRWHA • Regional Marine Resource Advisory Committees established in key population centres
1995	• Significant amendments to the Act including a member of the Authority to represent Aboriginal communities, the introduction of statutory management plans, and an increase in penalties
1997	• Network of 15 Dugong Protection Areas established throughout the coastal parts of GBRWHA • Gummo Woojabuddee Section of GBRMP declared • Report *The Outstanding Universal Value of the Great Barrier Reef World Heritage Area* (Lucas *et al.* 1997) published
1998	• GBRMPA restructured to focus on four critical issue groups (tourism, fisheries, water quality and conservation) • Plans of Management for Cairns Area and Whitsundays gazetted and into effect • Representative Areas Program (RAP) commenced to improve protection of biodiversity across GBR
1999	• Local Marine Advisory Committees established in nine coastal centres • World Heritage Committee requested and endorsed two reports on the GBR (*Framework for Management* and *Focussed Recommendations*)
2000	• *GBRMP (Aquaculture) Regulations 2000* introduced • Four Expert Reef Advisory Committees established (aligned to the critical issue groups – water quality, fisheries, tourism and recreation, conservation) • Scientific experts assisted the GBRMPA to map bioregions throughout the GBR, a key underpinning for the rezoning • 18 new coastal areas, previously excluded, were added into the GBRMP
2001	• Map of 70 bioregions (30 reef and 40 non-reef bioregions) publicly released • Meetings of stakeholders and GBR experts to discuss the Representative Areas Program
2002	• Memorandum of Understanding between Australian Prime Minister and the Queensland Premier aimed at halting the decline of water quality in GBR • Zoning Plan for Gumoo Woojabuddee Section and revised Far Northern Section Zoning Plan into effect • First formal community participation phase for Representative Areas Program resulted in 10 150 public submissions
2003	• Draft zoning plan for the GBRMP released for public comment and 21 500 public submissions received • Reef Water Quality Protection Plan released by Australian and Queensland governments • Reef Guardians Schools pilot program launched
2004	• The new zoning plan covering the entire extent of the GBRMP into effect on 1 July 2004 • Eco-certification established to recognise best practice in marine tourism • First 15-year permit was issued for an accredited tourist operation • Queensland 'mirrored' the new zoning in virtually all adjoining state waters (in the GBR Coast Marine Park), ensuring complementary zoning for all federal and state waters within the GBRWHA
2005	• Regional offices established to enhance community partnerships
2006	• Major review of GBRMP Act; this resulted in greater transparency, including the requirement for a periodic Outlook Report and amendments to the zoning process but also had negative implications for the GBRMPA's independence (e.g. it removed the GBRMPA's ability to proceed unilaterally with zoning)
2007	• *Climate Change Action Plan 2007–2012* released

Date	Event/comments
2009	• First GBR Outlook Report released
2011	• Reef Guardians program expanded to include Reef Guardian Farmers and Fishers • World Heritage Committee puts the spotlight on the GBRWHA expressing 'extreme concern' about particular coastal developments; the Committee requested improved reporting, a comprehensive strategic assessment of the entire GBR and a GBR visit by a reactive monitoring mission
2012	• Joint (IUCN-UNESCO/World Heritage Centre) reactive monitoring mission visited GBR • World Heritage Committee continued to express concern, requesting an updated report 'with a view to considering, in the absence of substantial progress, the possible inscription of the GBR on the List of World Heritage in Danger' • Retrospective Statement of Outstanding Universal Value for GBRWHA approved by World Heritage Committee • First Heritage Management Plan in the GBR (Lady Elliot Island) approved
2013	• GBR *Biodiversity Conservation Strategy* released • GBR Draft Strategic Assessment released for public comment • World Heritage Committee continued to express concern, requesting an updated report '...with a view to considering... a possible 'in Danger' listing • Report *Outlook for Great Barrier Reef Coastal Ecosystems* released
2014	• *Great Barrier Reef Outlook Report 2014* released • Major restructure and downsizing of the GBRMPA occurred • World Heritage Committee welcomed progress but requested a further updated report, warning the GBR might still be listed 'in Danger' • Major public controversy over Abbot Point development proposal • Draft *Reef 2050 Long-term Sustainability Plan* released
2015	• Revised *Reef 2050 Long-term Sustainability Plan* released • World Heritage Committee decided not to list GBR as 'in Danger' but rather, to closely monitor ongoing management to ensure effective and sustained protection of the World Heritage values

The statutory objects of zoning have today evolved from the initial wording in the Act (1975) to the most recent amendments (2008) (see Day 2015 for full details). Matters such as World Heritage and the use of marine resources by Traditional Owners[7] must now be considered when zoning, and there is an explicit need to protect the full range of biodiversity of the area, rather than just keystone species or habitats.

To achieve complementarity in the adjoining state waters, Queensland subsequently declared and zoned marine parks over what they regarded as state waters ('tidal lands and tidal waters') within the outer boundaries of the major sections of the federal GBRMP, including around Queensland islands and adjacent to the major sections declared under the federal Act (i.e. along the mainland coast) (see Chapter 12).

Table 5.3 shows that the word 'conservation' is used in the objective of the General Use, Habitat Protection, Conservation Park and Commonwealth Island Zones; 'conservation' is defined in the Zoning Plans as '...allowing for its ecologically sustainable use'. In contrast, the term 'protection' is defined for the Buffer, Scientific Research and Marine National Park Zones, and refers to '...managing impacts to ensure that the natural significance [of the place] is retained'.

The fact that the words 'protection (or preservation) of the natural integrity and values' appears in the objective for five of the eight zone types is also significant, especially as the objective of the zone is one of the mandatory requirements to consider when assessing whether a permit should be granted to use or enter a zone.

As shown in Table 5.3, the different zones equate to four of the six IUCN categories of protected areas, noting there is not an exact correlation between the GBR zone objectives and the definitions for each of the IUCN categories. Using the *IUCN Guidelines for*

Table 5.3. Great Barrier Reef Marine Park zones and zone objectives

Zone name[a]	Zone colour	Statutory objective(s) of the zone	IUCN protected area category
General Use Zone	Light blue	To provide for the conservation of areas of the Marine Park, while providing opportunities for reasonable use.	VI
Habitat Protection Zone	Dark blue	(a) To provide for the conservation of areas of the Marine Park through the protection and management of sensitive habitats, generally free from potentially damaging activities; and (b) Subject to (a), to provide opportunities for reasonable use.	VI
Conservation Park Zone	Yellow	(a) To provide for the conservation of areas of the Marine Park; (b) Subject to (a), to provide opportunities for reasonable use and enjoyment, including limited extractive use	IV
Buffer Zone	Olive green	(a) To provide for the protection of the natural integrity and values of areas of the Marine Park, generally free from extractive activities; (b) Subject to (a), to provide opportunities for: (i) certain activities, including the presentation of the values of the Marine Park, to be undertaken in relatively undisturbed areas; and (ii) trolling for pelagic species	IV
Scientific Research Zone	Orange	(a) To provide for the protection of the natural integrity and values of areas of the Marine Park, generally free from extractive activities; and (b) Subject to (a), to provide opportunities for scientific research to be undertaken in relatively undisturbed areas.	Ia
Marine National Park Zone	Green	(a) To provide for the protection of the natural integrity and values of areas of the Marine Park, generally free from extractive activities; and (b) Subject to (a), to provide opportunities for certain activities, including the presentation of the values of the Marine Park, to be undertaken in relatively undisturbed areas.	II
Preservation Zone	Pink	(a) To provide for the preservation of the natural integrity and values of areas of the Marine Park, generally undisturbed by human activities.	Ia
Commonwealth Islands Zone	Cream	(a) To provide for the conservation of the natural integrity and values areas of the Marine Park above low-water mark; and (b) To provide for use of the zone by the Commonwealth; and (c) Subject to (a), to provide for facilities and uses consistent with the values of the area	II

[a]See Fig. 5.4 for the appropriate activities within each zone, also Fig. 5.6 and Table 5.4 for the spatial extent of these zones

GBRMP Zoning
(see relevant *Zoning Plans* and *Regulations* for details)

	General Use Zone	Habitat Protection Zone	Conservation Park Zone	Buffer Zone	Scientific Research Zone *	Marine National Park Zone	Preservation Zone
Aquaculture	Permit	Permit	Permit*	✕	✕	✕	✕
Bait netting	✓	✓	✓*	✕	✕	✕	✕
Boating, diving, photography	✓	✓	✓	✓	✓*	✓	✕
Crabbing (trapping)	✓	✓	✓*	✕	✕	✕	✕
Harvest fishing for aquarium fish, coral and beachworm	Permit	Permit	Permit*	✕	✕	✕	✕
Harvest fishing for sea cucumber, trochus, tropical rock lobster	Permit	Permit	✕	✕	✕	✕	✕
Limited collecting	✓*	✓*	✓*	✕	✕	✕	✕
Limited spearfishing (snorkel only)	✓	✓	✓*	✕	✕	✕	✕
Line fishing	✓*	✓*	✓*	✕	✕	✕	✕
Netting (other than bait netting)	✓	✓	✕	✕	✕	✕	✕
Research (other than limited impact research)	Permit	Permit	Permit	Permit	Permit	Permit	Permit
Shipping (other than in a designated shipping area)	✓	Permit	Permit	Permit	Permit	Permit	✕
Tourism programme	Permit	Permit	Permit	Permit	Permit	Permit	✕
Traditional use of marine resources	✓*	✓*	✓*	✓*	✓*	✓*	✕
Trawling	✓	✕	✕	✕	✕	✕	✕
Trolling	✓*	✓*	✓*	✓*	✕	✕	✕

PLEASE NOTE: This guide provides an introduction to Zoning in the Great Barrier Reef Marine Park. Relevant Great Barrier Reef Marine Park Zoning Plans should be consulted for confirmation of use or entry requirements.

* Additional restrictions / conditions apply.

ACCESS TO ALL ZONES IS PERMITTED IN AN EMERGENCY.

Fig. 5.4. Activities matrix indicating which activities can occur in which zone, which are prohibited and which activities need a permit. (© Commonwealth of Australia (GBRMPA) 2015)

Applying the IUCN Categories to Marine Protected Areas (Day *et al.* 2012), the closest equivalent IUCN category for each zone is shown. The prime reason for applying the IUCN categories is that it enables the GBR zone types to be benchmarked against an internationally recognised system for all protected areas.

The zoning network applies to all activities undertaken in the GBRMP and the activities matrix (Fig. 5.4) helps users to understand what activities can occur where, and when a permit may be required.

Contrary to what many people believed, for the first 29 years following declaration of the GBRMP, the extent of no-take zoning was less than 5% of the entire Marine Park. By the late 1990s, 15 770 km^2 (~4.6%) of the GBRMP was zoned as no-take areas (green zones), with a further 453 km^2 as 'no-go' areas (preservation or 'pink' zones).

As shown in Fig. 5.5, the locations of the green zones at this time reflected an historical focus on coral reefs, with a skewed emphasis in the more remote and 'pristine' areas (72% of the total extent of no-take in the GBRMP was in a single large cross-shelf transect of National Park Zone in the remote Far Northern Section). A further 15% of the total GBRMP was also afforded seabed protection by the Habitat Protection (dark blue) Zone which precluded bottom trawling.

The introduction of a trawl plan for the East Coast Otter Trawl Fishery in 1999 was an important management measure that subsequently had a significant impact in the GBR. This plan was subsequently reviewed in January 2001 and more stringent management arrangements introduced.[8]

The Representative Areas Program

In the mid-late 1990s, concerns were raised by scientists and NGOs that the existing zoning scheme did not adequately protect the range of biodiversity known to exist within the GBR, and that the levels of protection were inadequate to ensure that the reefs and the surrounding non-reef areas remained healthy, productive and resilient into the future (Day 2015).

Around the same time, obligations under the World Heritage Convention, and particularly the requirement to define the Outstanding Universal Value of the GBR, led the GBRMPA to commission a report by Lucas *et al.* (1997); this increase in focus on the world heritage values made it clear that many aspects of the Outstanding Universal Value needed greater protection (Richard Kenchington, *pers. comm.*).

Consequently, between 1999 and 2003, the GBRMPA undertook a complex planning and consultative program to increase the protection of biodiversity for the entire GBRMP (Day *et al.* 2002, 2004; Fernandes *et al.* 2005). The primary aim of the Representative Areas Program was to increase the protection of the range of biodiversity in the GBR, by increasing the extent of 'no-take' areas and ensuring they included 'representative' examples of all habitat types. A further aim was to minimise impacts on the existing users of the GBRMP, including fishers.

This comprehensive, but controversial, rezoning program included considerable scientific input, community involvement and innovation, and took over 5 years to complete. It included considerable political and socio-economic challenges, and the development of a wide-ranging planning and consultative program (Fernandes *et al.* 2005). During the planning process, the aims were broadened to encapsulate a complete review and rezoning of the entire GBRMP rather than solely increasing the extent of no-take areas. The final outcome was the subsequent declaration of the 2004 Zoning Plan (Fig. 5.6), which achieved a variety of aims and exceeded the expectations of many who were involved.

The 31 650 written public submissions received – 10 150 in the first formal phase of public participation and 21 500 in the second phase commenting on the draft zoning plan – were unprecedented for any previous planning programs in the GBRMP and significant for any MPA program in Australia (Wescott and Fitzsimons 2011). Many modifications were made to the draft zoning plan as a result of the detailed information provided in submissions and other information received.

Figure 5.7 shows the extent of changes between the draft plan (GBRMPA 2003) and final zoning plan (GBRMPA 2004) using, as an example, a small portion of the GBRMP offshore from Townsville. In some locations, particularly inshore coastal areas, fewer

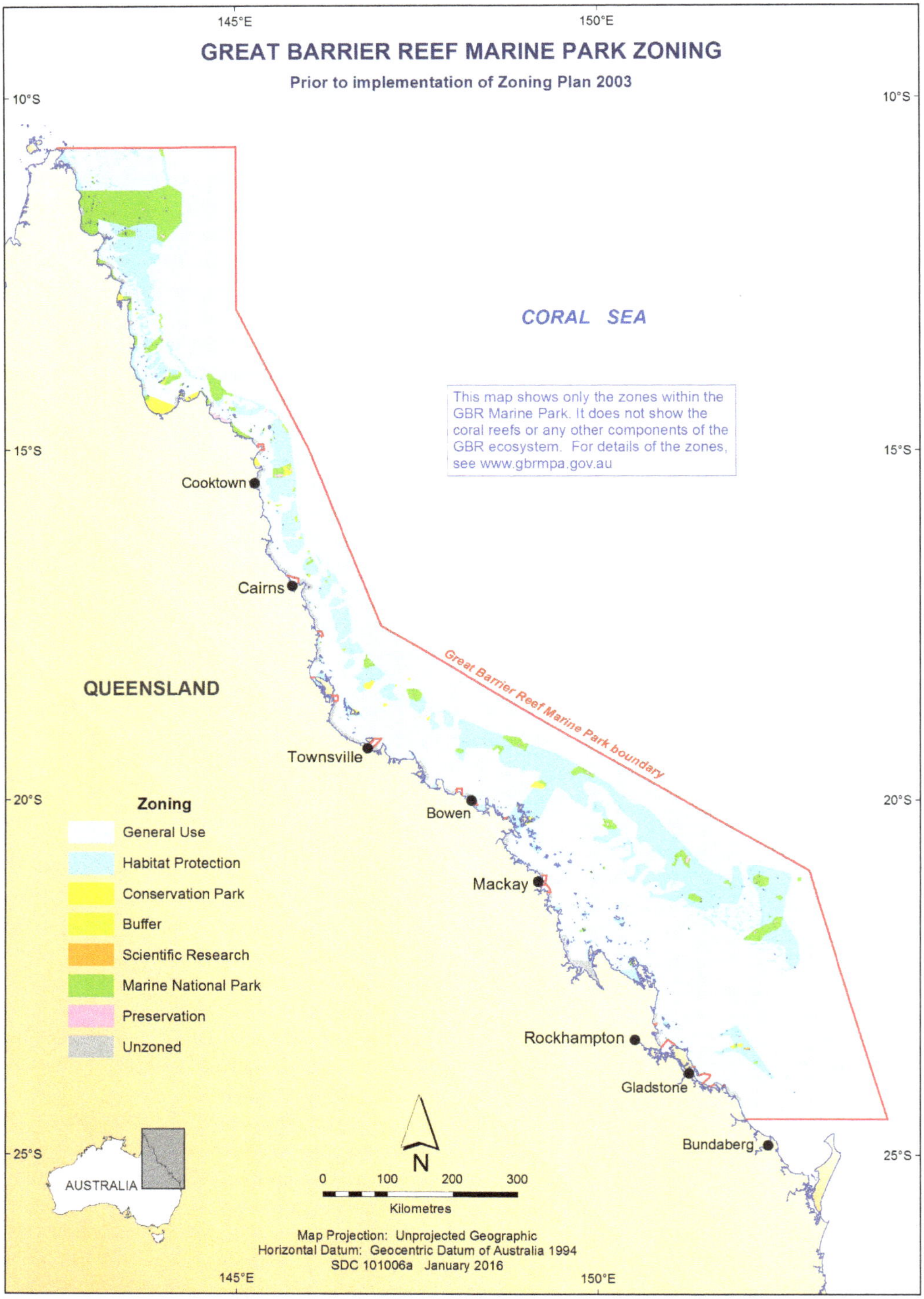

Fig. 5.5. Great Barrier Reef Marine Park zoning before implementation of the 2003 Zoning Plan. (© Commonwealth of Australia (GBRMPA) 2015)

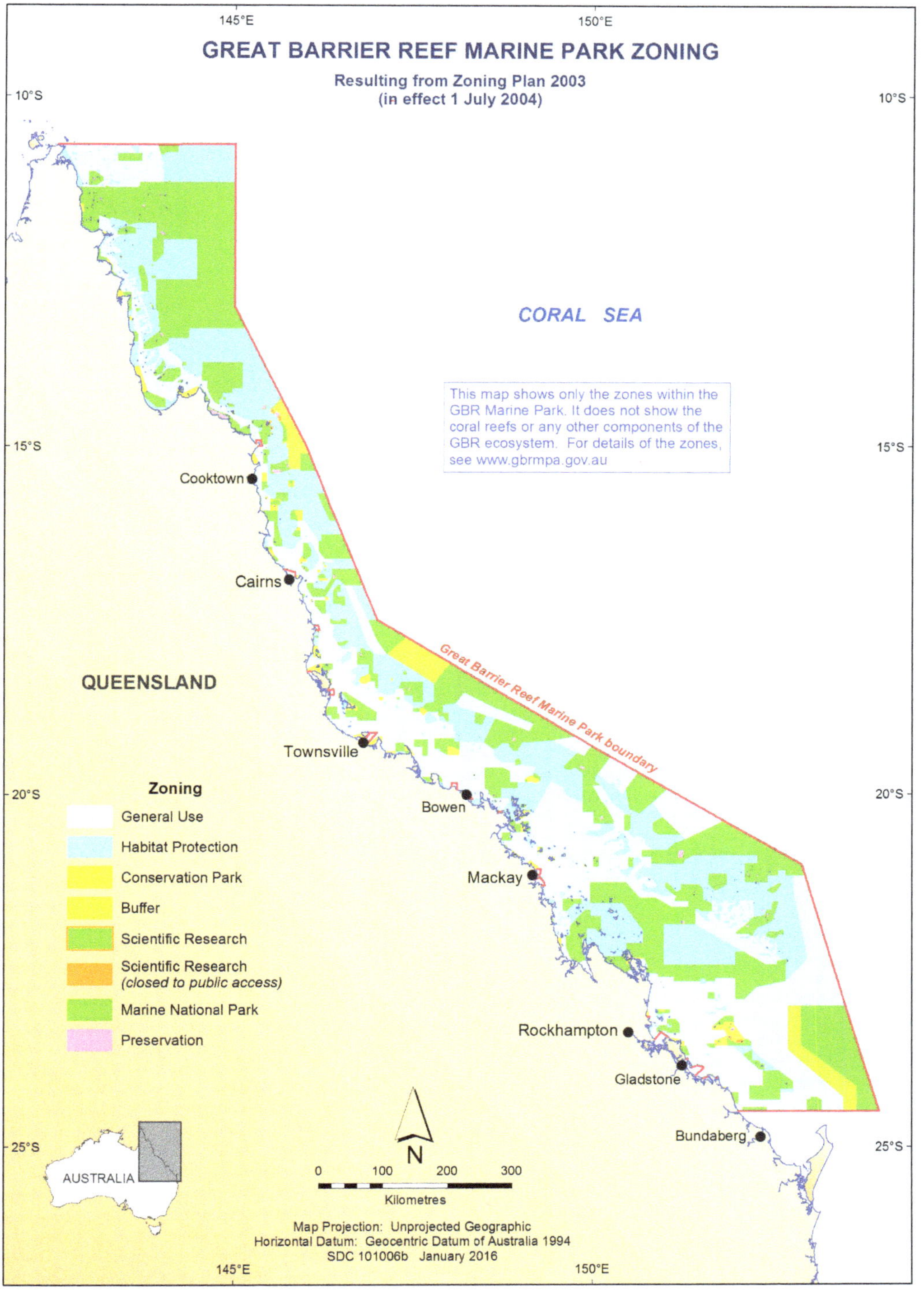

Fig. 5.6. Great Barrier Reef Marine Park zoning resulting from the 2003 Zoning Plan in effect after 1 July 2004. (© Commonwealth of Australia (GBRMPA) 2015)

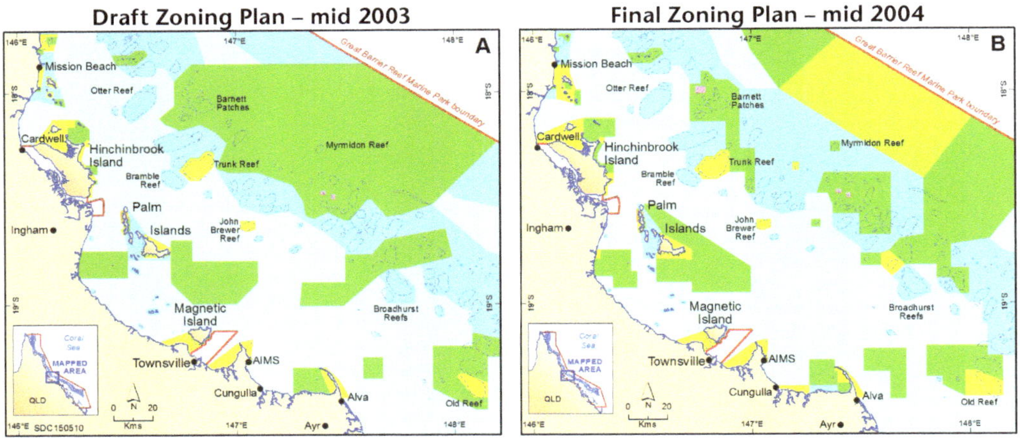

Fig. 5.7. Extent of zoning changes that occurred between (A) the draft and (B) final zoning plans to address key stakeholders concerns (example shown is a small part of the Marine Park near the city of Townsville). (© Commonwealth of Australia (GBRMPA) 2015)

options were available to modify the proposed no-take areas, while still achieving the recommended minimum levels of protection for each bioregion.

When the new zoning plan became law on 1 July 2004, the extent of no-take zones increased from 4.6% to 33.3% (or 114 530 km^2) of the GBRMP (Fig. 5.6). More important than the overall area, was the fact the new network protected representative examples of all 70 bioregions. However, changes occurred in all zones (see Table 5.4). In terms of additional biodiversity protection, there were numerous benefits beyond the new no-take zoning network; for example, the doubling of the Habitat Protection Zone was a major increase in the protection of benthic habitats.

The resulting plan maximised the protection of biodiversity while minimising the impacts on all other users, including fishers, thereby achieving the objective of the Representative Areas Program. Research and monitoring in the GBR continues to show the benefits of the new zoning network (for example, Russ *et al.* 2008; Sweatman 2008; McCook *et al.* 2010; Harrison *et al.* 2012; Emslie *et al.* 2015).

Science – both biophysical and social sciences – provided the fundamental underpinning for the rezoning (see also Chapter 16). However, the final zoning outcome was essentially a political negotiation involving considerable trade-offs rather than being science-driven. Although the best available ecological science provided the fundamentally important foundation for the new zoning, unquestionably the social, economic and management considerations were crucial when determining the final approved plan.

The planning process and the subsequent results set international benchmarks in marine conservation and marine spatial planning; key aspects of the planning process and many obstacles are discussed in more detail in Day *et al.* (2004), Fernandes *et al.* (2005, 2009) and GBRMPA (2010).

The area of each zone type and the proportion of the GBRMP are shown in Table 5.4.

Development of the 2003 Zoning Plan cost approximately A$12 million over the 5-year planning period (Osmond *et al.* 2010). This estimate includes staff salaries and operating costs but does not include the Structural Adjustment Package (SAP) that was subsequently introduced by the federal government to assist fishers, fishery related businesses, employees and communities thought to be adversely affected by the rezoning (see also Chapter 22).

Table 5.4. Extent of zones in the Great Barrier Reef Marine Park before, and after, 1 July 2004

Zone type	Area (km²)[a] Prior to 1 July 2004	Area (km²)[b] After 1 July 2004
General Use Zone	268 125 *(77.6%)*	116 530 *(33.8%)*
Habitat Protection Zone	52 380 *(15.2%)*	97 250 *(28.2%)*
Conservation Park Zone	2053 *(< 1%)*	5160 *(1.5%)*
Buffer Zone	507 *(< 1%)*	9880 *(2.9%)*
Scientific Research Zone	28 *(< 1%)*	155 *(< 1%)*
Marine National Park Zone	15 772 *(4.6%)*	114 530 *(33.3%)*
Preservation Zone	453 *(< 1%)*	710 *(< 1%)*
Islands	882 (all islands including state and Commonwealth islands) *(< 1%)*	185 (Commonwealth islands only) *(< 1%)*
Unzoned	5200 *(1.5%)*	0
Total	**345 400**	**344 400**[c]

[a]Areas from Day (2002b); the total areal determination in 2002 included all islands within the outer boundary of the Marine Park (i.e. all State and Queensland islands)
[b]Areas from Day and Dobbs (2013)
[c]Note the total area in 2013 is less than the 2002 estimate due to improved knowledge of Queensland islands (and their deletion from the list in 2013) and more accurate mapping of the Marine Park boundary. The total areal determination in 2013 included only Commonwealth islands as part of the GBRMP.

The total planning costs indicated above, while significant, can be justified in terms of an investment in the future resilience and ecological sustainability of the GBR. The size of the area being managed and its complexity warrant significant planning costs. However, more important is the achievement of far-reaching, long-lasting statutory protection for a World Heritage property that generates over A$5 billion every year for the national economy (Deloitte Access Economics 2013) and adjoins some of the fastest growing areas in Australia.

In contrast, the benefits versus the costs of the SAP have been widely debated; for example:

> '*....The roughly AUD$250 million that was spent [for the SAP] ... was an enormous and disproportionate sum to ... mitigate the impacts of an MPA program that had only relatively minor impacts on the financial interests of the commercial and recreational fishing sectors*'. (Macintosh *et al.* 2010, p. 587)

Reasons for effective management in the Great Barrier Reef

The complex legal and institutional framework that has developed since 1975 for the protection and management of the GBR is relatively effective, but not widely understood. Many think the entire GBR is a single protected area managed by a single agency. In reality, it is a complex amalgam of agencies, management tools and various approaches to management, all working together. Collectively, these assist in the implementation of both the multiple use approach and the joint management approach.

Although the *Great Barrier Reef Marine Park Act 1975* is the fundamental piece of federal legislation, it is one of many federal and state laws that are applicable to the GBR over differing jurisdictional areas as depicted in the 2014 Outlook Report (GBRMPA 2014a, Fig. 7.1, p. 196).

In addition, Australia is a signatory to various international conventions,[9] which brings international obligations. Some of these obligations have led to specific provisions in domestic law (for example, the world heritage provisions in the EPBC Act).

A key component of the success of the GBR has been the close working partnership between the federal management agency (GBRMPA), and the state of Queensland. This relationship has evolved over 40 years, and includes aspects such as complementary legislation and joint permits. For example, Queensland mirrored the federal marine park zoning plan in the adjoining state waters within 5 months of the federal zoning plan revision coming into effect (see Chapter 12).

This complementarity with zoning has occurred for various reasons. One contributing factor was that the two jurisdictions cannot agree where the boundary between state and federal waters occurs, both defining 'low water' in different ways (effectively the two jurisdictions 'agree to disagree', but this is not a problem given the complementary management approach). Furthermore, the position of the LWM constantly moves because of erosion and accretion, so trying to map it is impractical. This issue has been exacerbated because there are no clear or agreed principles for defining exactly where are the internal waters of Queensland (i.e. which parts of bays, channels, river mouths or estuaries are internal waters and hence not part of the Commonwealth Marine Park). Finally, LWM is frequently covered by water and this makes it unworkable as a boundary from an enforcement perspective, especially if the rules were different in each jurisdiction.

The Offshore Constitutional Settlement (Attorney-General's Department 2014) enables fisheries management within the GBR Region to be undertaken by the Queensland Government,[10] but all fishing activities in the GBR are subject to the federal and Queensland zoning plans.

The resulting partnership has assisted in managing the complex and interrelated mix of marine, coastal, and island issues throughout the GBR. It provides for integrated, holistic management of the entire GBR ecosystem and assists in more effective consideration of many of the issues that occur outside the area, but impact the Marine Park.

A multiple-use approach across the entire GBR has the advantage of allowing the area to be managed as an integrated whole, not just a series of isolated highly protected areas within a surrounding 'sea' of unmanaged activities. In the highly connected 'fluid' marine environment, and certainly for larger MPAs, this approach is considered preferable. Day (2002b) lists ecological social, practical and economic advantages of such a broad-scale integrated approach combined with zoning.

Consequently the GBR does not rely on a single plan of management for the entire area, but rather 'a management system' comprising a wide range of management tools and approaches of which the Zoning Plan is only one of many spatial management layers.

Over the years, various aspects of the zoning have been modified or refined (e.g. Day 2002b; Day *et al.* 2002) along with other spatial and temporal approaches to improve management of the GBR.

Today, the cornerstone of the management strategies for the GBR remains the comprehensive multiple-use statutory Zoning Plan, which is subordinate legislation covering the entire federal GBRMP.

Effective marine conservation, however, requires more than just a comprehensive spatial planning framework, especially given the range of pressures facing MPAs today. Although zoning is a well-recognised form of marine spatial planning, it is not necessarily the most effective way to manage all ocean activities or impacts (Day 2015). Many issues facing MPAs cannot be effectively managed by zoning alone; Day (2015) lists:

- **Water quality** – ocean zoning usually does not apply outside the marine area up onto the land where most water quality issues arise.
- **Fishing impacts** – zoning may assist by managing some, but not all, unsustainable fishing impacts.
- **Climate change** – zoning can help build resilience, but does not consider such factors as acidification or rising sea temperatures.
- **Coastal developments, especially ports** – zoning will not apply in ports if they are located outside the jurisdictional control of a Marine Park.
- **Shipping and pollution incidents** – a zoning plan may help by delineating shipping lanes but will not help to reduce ship groundings or marine pollution.
- **Increasing population growth and recreation** – marine zoning will not deal with key issues unless it can help to curtail use or reduce some of the consequential impacts.

Many of these pressures originate outside the marine realm and are not directly ameliorated by ocean zoning or marine spatial planning. Most pressures are better managed using management 'tools' or management approaches other than zoning (Day 2015), and various examples have been developed by other sectoral agencies using their own specific legislation (e.g. for fisheries management or shipping as shown in Fig. 5.8). However, when management tools from other agencies or sectors are applied in the GBRMP, they are still subject to the overall zoning framework and must be consistent with the underlying zone objectives.

The variety of other marine spatial planning tools that are applied as additional layers to complement the underlying zoning plan include:

- statutory plans of management for key areas
- site plans
- special management areas
- other spatial provisions (e.g. defence training areas, shipping areas, agreements with Traditional Owners).

Figure 5.7B depicts a small part of the 2003 Zoning Plan showing the range of zones in a small area off Townsville in the central part of the GBRMP. This same area is shown in Fig. 5.8 with some of the other spatial management arrangements, each with their own legislative provisions, on top of the underlying zoning.

Collectively, the zoning plan, along with other spatial and temporal tools, make up the comprehensive marine spatial planning approach (Day 2015). All these tools comprise a key part of the integrated management approach for the GBR, which has been adaptively improved over the past four decades.

Along with the wide range of management strategies, various federal and state government agencies are involved in managing the GBR. The annual operating budget of the GBRMPA in 2013/14 was around A$50 million (GBRMPA 2014b), but it is much harder to determine the total annual resources that all government agencies contribute for managing the GBR. A recent attempt was made in the *Reef 2050 Long-term Sustainability Plan* (Commonwealth of Australia 2015, Table 1, p. 54); this lists four federal agencies, six state agencies and several research programs that collectively contribute over A$200 million a year to the management of the GBR. It is unclear how much of this total figure is directly applied to the conservation and management of the GBR and how much is for associated activities (for example, the table mentioned above lists two agencies responsible for shipping, ports and navigation, expending A$49.5 million in 2004/15; this is some 24% of the total reputedly spent for managing the GBR).

Traditional Owners of the Great Barrier Reef

Today, 70 Traditional Owner clan groups (Aboriginal and Torres Strait Islanders) whose sea country includes the GBR are recognised, and their continuing social, cultural, economic and spiritual connections to the GBR Region are acknowledged (see Chapter 18). The managing agencies use a suite of sea country management arrangements, including Traditional Use of Marine Resources Agreements (TUMRAs) and Indigenous Land Use Agreements (ILUAs), to work with the Traditional Owners to protect cultural and heritage values, conserve biodiversity and enhance the resilience of the GBR. In addition, marine Indigenous Protected Areas have more recently been declared over parts of the GBRMP (see also Chapter 18).

There are several applications for Native Title over parts of the waters within the GBR (sea claims) but as of mid-2015, none had been granted. The *Great Barrier Reef Marine Park Zoning Plan 2003* acknowledges that under the *Native Title Act 1993*, Traditional Owners may undertake the traditional use of marine resources in the GBRMP. Activities include fishing, collecting, hunting or looking after cultural sites as part of Aboriginal and Torres Strait Islander people's cultures, customs or traditions, for the purpose of satisfying personal, domestic or communal needs.

In recent decades, the GBRMPA has increased its collaboration with Traditional Owners, recognising that effective and meaningful partnerships are essential to protect cultural and heritage values, and conserve biodiversity. These benefits can also enhance the resilience of the GBR to cope with other pressures. Engagement with Indigenous peoples occurs in many ways including membership on the Authority's Board and an Indigenous Reef Advisory Committee, incorporating traditional ecological knowledge into management, and compliance training for Traditional Owners.

The ecosystem-based management approach

The integrated multiple-use and multi-layered management approach has been seen as a model of ecosystem-based management (see Fig. 5.9). The GBR was for many years considered by a range of marine managers and scholars to be one of the world's better working examples of marine zoning and ecosystem-level management. Ruckelshaus *et al.* (2008, p. 59) wrote 'The Great Barrier Reef ecosystem boasts a system-wide spatial management approach that is arguably the world's most sophisticated and extensively implemented example of marine zoning'.

Arkema *et al.* (2006, p. 530) concluded 'These achievements are largely due to the GBRMPA's development of specific scientific and socioeconomic goals, use of adaptive management, and its willingness to involve the public, adopt transparency, and consult with multiple user groups'.

When considering whether the GBRMPA's approach to marine spatial planning and marine management will work elsewhere around the world, it is important to recognise that many other MPAs have differing management models and objectives. Therefore the specific political, economic, social and managerial contexts of those areas need to be considered against the situation in the GBR when translating the GBR lessons elsewhere (Day 2011).

Current policy

Although the GBRMP has a primary overriding main conservation objective of 'long-term protection and conservation of the biodiversity, environment and heritage values of the GBR Region', many people do not understand that it also has other objectives, including to

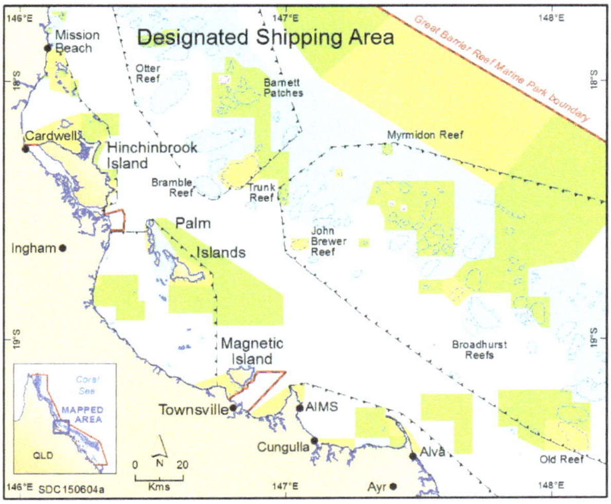

Shipping is one of many examples of marine activities that are not limited to a single zone within a multi-use zoning scheme. **Designated shipping areas** are part of the statutory Zoning Plan but, as shown, such areas where ships may transit 'as-of-right' cross over many of the underlying zones. For most users, designated shipping areas are not obvious on the publicly available zoning maps. However, all ship's crews need to be aware where the statutory shipping areas occur, with large penalties likely if they transgress.

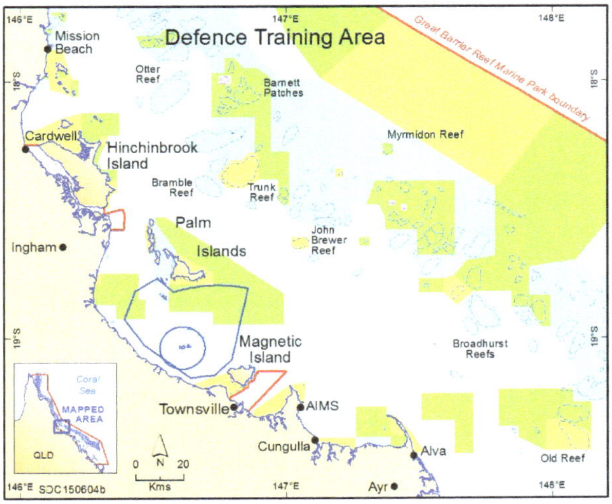

While some **Defence Training areas** in the GBR provide permanent closures to the public, others like this one are not permanent. When this area is being used for defence training purposes, it is closed to all users by Public Notice. However, for most of the year the area is open with very few restrictions other than the underlying zoning (i.e. both a spatial and a temporal arrangement).

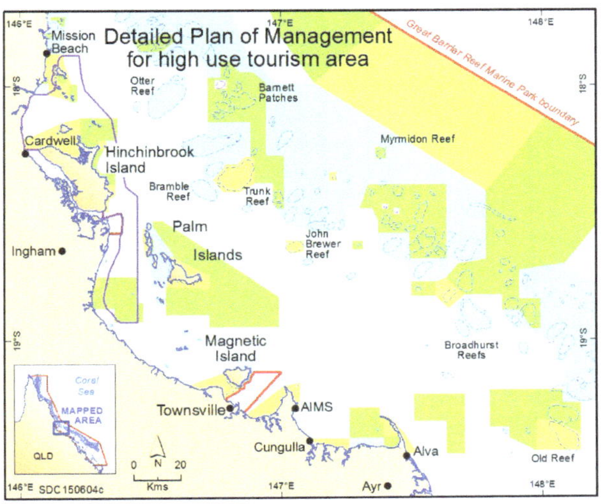

A small subset of the mapped area is highly used with potential conflicts between tourism, recreation and conservation aspects. Another statutory plan (a '**Plan of Management**') has been specifically developed to address these issues, using an overlay of spatial management (akin to a recreation-opportunity spectrum) which is irrespective of the underlying zoning.

Fig. 5.8. Small subset of the Great Barrier Reef Marine Park showing various spatial management layers on top of the underlying zoning. (Adapted from Day 2015; maps prepared by the GBRMPA)

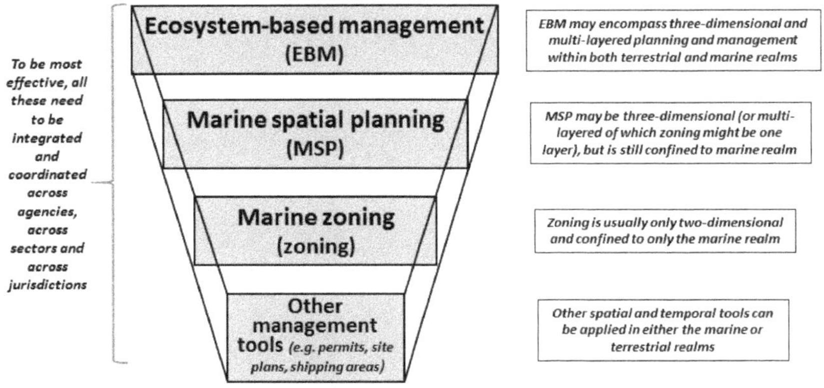

Fig. 5.9. Interrelationship of zoning, marine spatial planning and ecosystem-based management as they relate to the Great Barrier Reef Marine Park. (Adapted from Day 2015)

'allow ecologically sustainable use of the Great Barrier Reef Region'. However, section 2A of the Act provides a very clear hierarchy of objectives for the GBR stating these other objectives can only occur 'so far as is consistent with the main object'.

Today, a comprehensive suite of ecosystem-based management arrangements assist in achieving the environmental outcomes for the GBR. Guided by the principle of balancing conservation and sustainable use, the multiple-use management regime allows for a wide range of activities and uses, including various extractive industries or activities that do cause impacts, while still doing much to protect one of the world's most diverse marine ecosystems.

Since the revised zoning came into effect in 2004, public communication and education campaigns have used a range of media to raise awareness about zoning. Free zoning maps are available in a range of community information centres along the coast, including most shops supplying marine products such as bait and tackle. The zoning maps are also available on the internet along with downloadable coordinates for most of the zones. Signs with zoning information have also been located at major boat ramps along the coast since the first zoning plan came into effect, so there are many ways that users know where they can go and what they can do.

The GBRMPA continues to view education as an effective strategy to encourage compliance with the management objectives of the Marine Park, and uses a range of other communication tools to raise awareness. These include media releases, media interviews, advertising and billboards to educate visitors from outside Queensland. However, enforcement action, infringement notices and prosecution remain important deterrence tools that are available to park managers.

Policing an area the size of Italy or Japan requires a systematic and disciplined approach. The Australian and Queensland governments are strongly committed to ensuring compliance in the Marine Park, recognising that without adequate compliance the management of the park would be less successful. Boat and aircraft patrols therefore operate regularly in the Marine Park, checking on activities and monitoring ecological conditions.

There has been, however, no effective increase in the actual funding available for compliance and enforcement during the period 2005–2015. Furthermore, a recent audit report (ANAO 2015) identified shortcomings across a broad range of the GBRMPA's regulatory activities, including its assessment of permit applications, monitoring of permit compliance and response to non-compliance.

The role of education and public participation in ongoing management, however, should not be underestimated. People who regularly use the GBRMP provide the managing agencies with valuable surveillance information about suspected illegal activities and general usage patterns, and increasingly are participating in citizen science activities.

Despite major threats and complex issues, the GBR remains a good working example of the ecosystem-based management approach. The long-term success of effective management in the GBR has been attributed to various factors (Day and Dobbs 2013; Day 2015) including:

- strong political support (e.g. Box 5.1)
- a sound governance/legislative framework, with complementary legislation for all state and federal waters
- an ecosystem-level approach to management including management influence over a wider context than just the federal Marine Park
- widespread public support and consensus that the GBR is important, with many industries recognising that their future depends upon its health
- well-developed/integrated management with all relevant federal and state agencies
- planning at various scales for various purposes (and acknowledging that the Zoning Plan is only one layer of management)
- effective community engagement, good cooperation with most sectors and a strong educational focus
- effective research and monitoring programs, prioritised to provide information for management.

Box 5.1. The importance of leadership in the Great Barrier Reef

The success of the GBRMP over the decades is due to the combined efforts of many individuals, communities, NGOs, stakeholders and agencies. Individuals who have played notable leadership roles in GBR management include:

- **Graeme Kelleher:** the first permanent Chairman and CEO of the GBRMPA; Kelleher oversaw the progressive declaration of sections of the GBRMP and the development of zoning throughout most of the Park, the declaration of the GBR as the world's first Particularly Sensitive Sea Area and the inscription of the GBR on the World Heritage List.
- **Minister Robert Hill:** as the responsible Federal Minister from 1996 to 2001, Hill's determination ensured that several complex issues were considered by the Queensland and federal governments (e.g. Hill's concerns included water quality from land-based sources, greater sustainability of trawling in the GBR, increased protection for Dugongs, and a reduction in the use of high explosives for defence training).
- **The Hon. Virginia Chadwick:** during her 8 years as the GBRMPA Chair, Virginia's leadership resulted in many significant achievements, most notably the rezoning of the entire GBRMP in 2004. Virginia was also instrumental in developing better partnerships with the tourism industry, the introduction of formal agreements with Traditional Owners, the Reef Guardians program, improved community partnerships and increasing funding for field management.
- **Minister David Kemp:** as the responsible Federal Minister from 2001 to 2004, Kemp first needed to be convinced of the merits of the Representative Areas Program, but subsequently became one of the Program's biggest advocates and championed it with his colleagues to obtain agreement in federal cabinet.

Challenges for the Great Barrier Reef Marine Park

The challenges and priorities facing the GBR and its consequent management have changed considerably over the decades (Table 5.5).

In addition, a long list of direct legacy impacts[11] still affects the values of the GBR (see GBRMPA 2013, pp. 6–10). These include: the commercial harvesting of Dugongs, green turtles, crocodiles and whales; collecting pearl shell, trochus and teat fish; unsustainable activities on the reefs and adjoining waters (e.g. by-catch in shark nets, bombing during defence training) and in the adjoining catchments (such as broad-scale land clearing, changes to natural water flows) and the loss of connectivity. The differing years over which such impacts have occurred are depicted in the 2014 Outlook Report (GBRMPA 2014a, Fig. 2.1, p. 16). Today, many of these legacy impacts remain, substantially reinforcing some species' inability to recover from more recent impacts such as those in Table 5.5.

Current pressures on the GBR are caused by diverse activities including mining, agriculture, ports, shipping, tourism, fishing, urban growth and industrial development, along with natural occurrences such as cyclones, Crown-of-thorns Starfish outbreaks and disease. Although many individual decisions and management approaches to tackle these pressures may appear in themselves adequate, the cumulative impacts[12] of many decisions, on top of the legacy impacts and impending impacts of climate change, are of significant concern.

Unquestionably, the values in the GBR are declining, primarily due to human activities over which the park management agencies have limited influence (such as fishing and coastal development). Declining values are also due to other pressures beyond the mandate of the managing agencies and need to be remedied at a much broader level (such as ocean warming, increasing severity of extreme weather events and ocean acidification associated with climate change). Cumulatively, these activities have caused an estimated loss of 50% of coral cover (De'ath *et al.* 2012) and significant declines in Dugong (Marsh *et al.* 2007) and many seabird species (Condon *et al.* 2007).

There have been some positive trends in recent years, such as some limited improvements in water quality and the recovery of Humpback Whales since hunting ceased in 1962. However, the significant loss of coral cover and species specifically associated with the GBR, including Dugongs, inshore dolphins, turtles, sharks, fish, birds and some invertebrates, has led IUCN to state 'Until a clear roadmap on how [the GBR] will be managed and concrete indicators of values demonstrating positive trends are in place, the conservation outlook … at this point in time is of Significant Concern' (IUCN 2014).

In response to a specific request from the World Heritage Committee, the Australian and Queensland governments have both undertaken comprehensive strategic assessments

Table 5.5. Changing management priorities in the Great Barrier Reef

(Adapted from Australian Government 2012)

1960–69	1970–79	1980–89	1990–99	2000–2009	2010–present
• Limestone mining • Proposals for oil drilling	• Proposals for oil drilling • Crown-of-thorns Starfish	• Crown-of-thorns Starfish • Increasing tourism	• Commercial fishing (including netting and trawling) • Shipping issues	• Water quality issues • Coastal development • Threats to inshore biodiversity (especially from netting)	• Climate change issues • Coastal development (especially ports) • Water quality • Unsustainable fishing

of the GBR (GBRMPA 2013) and the adjoining coastal area (DSDIP 2013). The GBR Strategic Assessment concluded that, without additional management intervention, there is a high risk of further declines in the condition of biodiversity and heritage values and the community benefits they support (GBRMPA 2013).

Today, many of the threats facing the future of the GBR are beyond the direct jurisdiction of the GBRMPA (adapted from GBRMPA 2014a; Day 2015):

- **Climate change** – the most serious threat facing the GBR, likely to have far reaching consequences for the future. Predictions of climate change impacts include continuing sea level and sea temperature rises and increased acidification of the ocean (Johnson and Marshall 2007). More extreme weather including more severe cyclones is also occurring and predicted to increase.
- **Declining water quality** – agricultural practices in the GBR catchment are slowly improving with reduced nutrient and sediment loads entering the GBR through decreased use of fertilisers and pesticides, and reductions in runoff (Reef Water Quality Protection Plan Secretariat 2013). However, a significant lag time is likely before measurable water quality improvements are achieved in the GBR. Emerging water quality issues including urban and industrial pollution are of increasing concern, along with incipient issues such as antifoulants, heavy metals, polychlorinated biphenyls, micro-plastics, pharmaceuticals and 'personal products' and nano-particles. The environmental impacts of pharmaceuticals (including synthetic oestrogen in contraceptive pills, antibiotics and anti-depressants) are largely unknown – but many are endocrine-disruptors and bio-accumulate in the food chain (Brodie and Waterhouse 2012).
- **Loss of coastal habitats due to development** – clearing and modifying coastal habitats that buffer and support the GBR are likely to continue to present a high risk, especially to inshore biodiversity (e.g. through loss of connectivity between coastal and marine environments). Coastal development, and particularly expanding port operations and further port developments, will continue to have impacts. These coastal developments are likely to include large amounts of capital dredging and maintenance dredging, both of which also affect water quality. Furthermore, activities such as marine reclamation and major infrastructure will further impact coastal processes.
- **Impacts from some fishing activities** – Although there have been some improvements in the management of fishing in the GBR, community and scientific concerns remain about the status of some targeted species, as well as concerns about the sustainability of fishing spawning aggregations, illegal fishing activity, discarded catch and impacts on other species of conservation concern (GBRMPA 2014a).

Other key issues facing GBR include:

- increasing population and urban pressures
- increasing numbers and size of ships, and the associated port activities
- declining resources for management, especially for compliance, monitoring and heritage protection.

Shipping use in the GBR has increased substantially since 2000, driven mainly by industrial and mining activity (GBRMPA 2014a). However, the number of shipping incidents has remained relatively stable in recent years due primarily to the introduction of additional management measures by the Australian Maritime Safety Authority (AMSA), the responsible federal agency. For example, in 2014 following approval from the International Maritime Organisation, AMSA designated a two-way shipping route for the entire length of the GBR.

Some concerns remain, given that ships are allowed to transit along the coast inside the outer reefs; however, there are many reasons why this is actually safer than requiring all ships to operate solely outside the GBR. Ships operating inside the GBR are continuously tracked and monitored and, if a breakdown does occur, it is easier for a ship to anchor until assistance arrives rather than being carried by wind or waves onto the reefs, given that it is too deep to anchor on the seaward side of the outer reefs.

There are also misguided public perceptions that some other pressures such as Crown-of-thorns Starfish, increasing tourism or Indigenous hunting are the major causes of the decline in the values of the GBR. In reality, these pressures are much less significant compared with the primary pressures mentioned above, and should not be regarded as major causes of the declining values.

Future directions?

Although the past management approaches in the GBR are considered by many to have been both effective and appropriate, clearly today's pressures and cumulative impacts mean that these past approaches are not enough to retain the values for which the GBR has been long been acclaimed. The Australian Government's *GBR Outlook Report 2014* (GBRMPA 2014a, p. v), concluded:

> *'Notwithstanding positive actions since 2009, the greatest risks ... have not changed. Climate change, poor water quality from land-based run-off, impacts from coastal development, and some remaining impacts of fishing remain the major threats to the future vitality of the Great Barrier Reef'.*

The above-mentioned major threats are only some of the 41 threats impacting the GBR that are listed in the 2014 Outlook Report (GBRMPA 2014a, p. 256). Of those 41 threats, all but three (sea level rise, ocean acidification and altered ocean currents) are happening now, with the remaining three threats all expected to happen in 10+ years. Of greatest concern are the cumulative impacts of all these threats, many of which are likely to act synergistically.

The 2014 Outlook Report (GBRMPA 2014a, p. vi) also concluded:

> *'Even with the recent management initiatives to reduce threats and improve resilience, the overall outlook for the Great Barrier Reef is poor, has worsened since 2009 and is expected to further deteriorate in the future. Greater reductions of threats at all levels, Reef-wide, regional and local, are required to prevent the projected declines in the Great Barrier Reef and to improve its capacity to recover'.*

In the period 2011–2014, the World Heritage Committee raised concerns about the GBR and questioned whether the Outstanding Universal Value was being maintained. Consequently, three of the Committee's decisions (in 2012, 2013 and 2014) warned of the '... possible inscription of the property on the List of World Heritage in Danger' (World Heritage Committee 2014).

This warning is not surprising given that the Australian Government's own Strategic Assessment (GBRMPA 2013) showed that 24 out of the 41 metrics or values that collectively comprised the Outstanding Universal Value of the GBR World Heritage Area had deteriorated since its 1981 inscription on the World Heritage List (GBRMPA 2014a; Hughes *et al.* 2015). Of those 24 values, 10 were assessed as 'poor' rather than 'good' or 'very good' and these 10 include significant attributes such as coral assemblages, breeding colonies of seabirds, seagrass, Dugongs, Indigenous connection to country and a component of the integrity of the GBR.

The *Reef 2050 Long-Term Sustainability Plan* (Commonwealth of Australia 2015), released in March 2015, was criticised by many scientists and NGOs who recommended additional actions needed to be included (Australian Academy of Science 2015; Hughes *et al.* 2015). The criticism included a greater focus on climate change and taking action on unsustainable activities, such as the disposal of maintenance dredge spoil and unsustainable fishing. Furthermore, concerns were raised that the available and proposed resourcing were insufficient to properly achieve the aims of the plan.

In July 2015, the World Heritage Committee decided not to officially list the GBR as World Heritage in Danger, but chose to closely observe further developments in the GBR. Australia was requested to submit a progress report on the planned protective measures by late 2016, and the Committee's expectations by 2020 were summarised by IUCN:

> '*.... it is ... fundamental that the plan delivers its anticipated results towards halting and reversing the decline of the property in order to confirm that the GBR does not face an ascertained or potential danger to its Outstanding Universal Value*'. (World Heritage Committee 2015)

Conclusions

Today, the GBR is under unprecedented pressures (both human-induced and natural) and these pressures are expected to increase. There are real concerns that many of the values for which the GBR is internationally recognised are deteriorating and may continue to decline. In 2014, IUCN rated the conservation outlook of the GBR 'of Significant Concern' (IUCN 2014).

Solutions to tackle these pressures are not simple, because they are the accumulation of many additive and synergistic activities rather than one or two identifiable pressures. However, if measures to adapt to pressures such as climate change are not identified and taken, and if certain development decisions continue to remove options for long-term conservation, the GBR's future is unlikely to be secure (Hughes *et al.* 2015).

In the past, several fundamental requisites have helped the GBR to become the success story that it has proven to be. For the GBR to reclaim the mantle of one of the world's best-managed MPAs, the following requisites must be maintained, and in many cases improved (recognising they need to occur within the over-arching context of global warming and climate change):

1. Accept that many of the GBR's values are deteriorating, and working collaboratively towards restoring those values so that their status is the same, or as close as is ecologically possible, to what existed when the area was declared a World Heritage Area in 1981.
2. Maintain the essential close working partnership between the Australian and Queensland governments and their agencies (fundamentally important for effective GBR management).
3. Maintain strong partnerships between the GBRMPA officers and other agencies, stakeholders, councils, Traditional Owners, other Indigenous groups, community members and researchers.

4 Ensure that genuine community education, information and engagement are always key components of management (given the size and complexity of the GBR, there is no other option).
5 Reinstate adequate resourcing to the managing agencies (and particularly the GBRMPA) to be more commensurate with what the GBR generates for the national economy.
6 Ensure effective compliance and enforcement to implement all the available legislative tools.
7 Develop realistic and achievable indicators of the key values to demonstrate the trends that are occurring.

The future of the GBR will depend on the Australian and Queensland governments taking their responsibilities more seriously than has been evident from many of the development and resourcing decisions made in the decade between 2005 and 2015. The Australian public and the global community also must make clear that they want the values of the GBR restored as suggested above and to ensure that those values remain protected in accordance with the clear objectives of the legislation and with global expectations.

Acknowledgements

Numerous people provided comments on earlier drafts of this chapter; thanks therefore to Geoff Wescott, James Fitzsimons, Helene Marsh, Richard Kenchington, Graeme Kelleher, Trevor Ward, John Tanzer, Di Tarte and Erik Olsen as well as editorial assistance from Liz Tynan and CSIRO. All maps courtesy of the Spatial Data Centre, the GBRMPA; the assistance of Rhonda Banks in preparing the maps and Sonia Batley and Joanna Ruxton for sourcing the GBRMPA photographs is much appreciated.

Biography

Jon Day spent 39 years as a professional protected area planner and manager, 28 years of which were in the Great Barrier Reef. From 1975 to 1997, Jon worked for protected area agencies in Victoria, the Northern Territory, Queensland and Canada. Following his appointment as a GBRMPA Director in 1998, Jon was variously responsible for biodiversity conservation, all aspects of heritage conservation, the Outlook Report and Indigenous Partnerships in the GBR. Jon's most notable achievement was commencing and coordinating the GBR Representative Areas Program for which he was awarded an Australian Public Service Medal and a Smithsonian-Queensland fellowship in 2004. He departed the GBRMPA in 2014 to undertake a PhD at James Cook University, Townsville.

Endnotes

1 These 13 port areas cover a very small area compared with the GBR (1533 km^2 = 0.45% of the Marine Park), but some port activities have the potential for significant impacts on the adjoining areas.

2 For the purposes of national and international law, any waters on the landward side of a defined baseline are usually considered to be 'internal waters'. Where the coastline is deeply indented (e.g. long narrow bays that are longer than they are wide, extending far inland), or where there is a fringe of islands along the coast, the baseline is usually defined

by a system of straight lines joining discrete points at low-water between 'straight baseline end points'. Similarly, long narrow channels, such as that between Hinchinbrook Island and the mainland, are also normally considered to be 'internal waters'.

3 The area of reefs depends on what is actually mapped as coral, with some inter-reefal areas leading to uncertainty; the best estimates today suggest that 10% of the entire GBRMP is the most appropriate figure.

4 Nine broad 'meso-scale' regions were previously described covering the GBR and became part of the Integrated Marine and Coastal Regionalisation of Australia (Commonwealth of Australia 2006); the scale of these bioregions, however, were considered by the GBRMPA to be too broad for planning an effective no-take network across the GBR. Consequently, a finer scale bioregionalisation for the GBR was developed in 1999–2000 as one of the fundamental components of the Representative Areas Program. This resulted in 70 ecologically and geographically different bioregions mapped at a scale of tens to hundreds of kilometres.

5 The term 'Outstanding Universal Value' is the central cornerstone for many aspects of World Heritage, including nominations and reporting. Outstanding Universal Value relates to the natural and/or cultural values of a site being universally recognised as globally significant, and being 'so exceptional to be of common importance for present and future generations of all humanity'. A Statement of Outstanding Universal Value is required for each World Heritage property and, once it is approved, that statement provides the basis for the future protection and management of the property.

6 This ban on oil drilling and mining was unprecedented in Australia; there have been periodic unsuccessful attempts to have this wound back in the GBR.

7 'Traditional Owner' is defined in the *GBRMP Act* as '…an Indigenous person:
  (a) …recognised in the Indigenous community or by a …representative Aboriginal or Torres Strait Islander body:
    (i) as having spiritual or cultural affiliations with a site or area in the Marine Park; or
    (ii) as holding native title in relation to that site or area; and
  (b) who is entitled to undertake activities under Aboriginal or Torres Strait Islander custom or tradition in that site or area'.

8 The arrangements for trawl included a cap on fishing effort, major closures of previously un-trawled fishing grounds (some 96 000 km^2), mandatory controls including a satellite-based vessel monitoring system and turtle excluder and by-catch reduction devices, and a A$20 million program to buy licences in an over-capitalised fishery.

9 Key international conventions include: World Heritage Convention 1972; Convention on the International Trade in Endangered Species 1973; International Convention for the Prevention of Pollution from Ships 1973; Convention on the Conservation of Migratory Species 1979; UN Convention on the Law of the Sea 1982; Convention on Biological Diversity 1992; UN Framework Convention on Climate Change 1992.

10 This is more an administrative arrangement than a legal requirement; the Commonwealth's responsibility to 'provide for the long term protection and conservation of the environment, biodiversity and heritage values of the GBR Region' could legally be argued to include the conservation of all natural resources in the Region including fish, and any activities that impact fish.

11 'Legacy impacts' are impacts from activities that occurred long ago but which are still having severe and long-lasting impacts on the values of the GBR.

12 'Cumulative impacts' are defined as 'the successive and combined effects of impacts on the environment, taking into account the direct, indirect and consequential impacts, and the incremental and compounding effects of these impacts over time' (GBRMPA 2013, p. 6–4).

References

Arkema KK, Abramson SC, Dewsbury BM (2006) Marine ecosystem-based management: from characterization to implementation. *Frontiers in Ecology and the Environment* **4**, 525–532. doi:10.1890/1540-9295(2006)4[525:MEMFCT]2.0.CO;2.

Attorney-General's Department (2014) *Offshore Constitutional Settlement.* Attorney-General's Department, Canberra, <https://www.ag.gov.au/Internationalrelations/InternationalLaw/Pages/TheOffshoreConstitutionalSettlement.aspx>.

Australian Academy of Science (2015) *Reef 2050 Long-term Sustainability Plan – Position Statement.* 1 April 2015 edition. Australian Academy of Science, Canberra, <https://www.science.org.au/reef-2050-long-term-sustainability-plan>.

Australian Government (2012) 'State Party Report on the state of conservation of the Great Barrier Reef World Heritage Area (Australia) – Property ID N154'. Australian Government, Canberra, <http://www.environment.gov.au/resource/state-party-report-state-conservation-great-barrier-reef-world-heritage-area-australia-2012>.

Australian National Audit Office (ANAO) (2015) *Regulation of Great Barrier Reef Marine Park Permits and Approvals.* Australian Government, Canberra, <http://www.anao.gov.au/~/media/Files/Audit%20Reports/2015-2016/03/ANAO_Report_2015-2016_3.pdf>.

Banks J (1770) The *Endeavour.* Journal of Joseph Banks, 1768–1771. Volume 2: 15 August 1769 – 12 July 1771. State Library of New South Wales, Sydney (transcribed by JC Beaglehole 1962), <http://acms.sl.nsw.gov.au/_transcript/2015/D33961/a1193b.html#a1193257>.

Bowen J, Bowen M (2011) *The Great Barrier Reef: History, Science, Heritage.* Cambridge University Press, Cambridge, UK.

Brodie J, Waterhouse J (2012) A critical review of environmental management of the 'not so Great' Barrier Reef. *Estuarine, Coastal and Shelf Science* **104–105**, 1–22. doi:10.1016/j.ecss.2012.03.012.

Commonwealth of Australia (2006) *A Guide to the Integrated Marine and Coastal Regionalisation of Australia.* Version 4.0. Department of the Environment and Heritage, Canberra.

Commonwealth of Australia (2015) *Reef 2050 Long-term Sustainability Plan.* Commonwealth of Australia, Canberra.

Condon BC, Erwin CA, Peck DR, Baker GB, Double MC, O'Neill P (2007) Vulnerability of seabirds on the Great Barrier Reef to climate change. In *Climate Change and the Great Barrier Reef: a Vulnerability Assessment* (Eds JE Johnson and PA Marshall) pp. 427–463. Great Barrier Reef Marine Park Authority and Australian Greenhouse Office, Townsville.

Day JC (2002a) Marine park management and monitoring – lessons for adaptive management from the Great Barrier Reef. In *Managing Protected Areas in a Changing World: Proceedings of the 4th International Conference on Science and Management of Protected Areas, May 2000, Waterloo, Canada* (Eds S Bondrup-Nielsen, NWP Munro, G Nelson, JH Martin Willison, TB Herman and P Eagles) pp. 1258–1282. Science and Management of Protected Areas Association, Canada.

Day JC (2002b) Zoning – lessons from the Great Barrier Reef Marine Park. *Ocean and Coastal Management* **45**, 139–156. doi:10.1016/S0964-5691(02)00052-2.

Day JC (2008) The need and practice of monitoring, evaluating and adapting marine planning and management – lessons from the Great Barrier Reef. *Marine Policy* **32**, 823–831. doi:10.1016/j.marpol.2008.03.023.

Day JC (2011) Protecting Australia's Great Barrier Reef. *Solutions* **2**, 56–66.

Day JC (2015) Marine spatial planning (MSP): one of the fundamental tools to help achieve effective marine conservation in the Great Barrier Reef. In *Transboundary Marine Spatial*

Planning and International Law. (Eds D Hassan, T Kuokkanen and N Soininen) pp. 103–131. Earthscan, Routledge, London.

Day JC, Dobbs K (2013) Effective governance of a large and complex cross-jurisdictional marine protected area: Australia's Great Barrier Reef. *Marine Policy* **41**, 14–24. doi:10.1016/j.marpol.2012.12.020.

Day JC, Fernandes L, Lewis A, De'ath G, Slegers S, Barnett B, *et al.* (2002) The Representative Areas Program for protecting biodiversity in the Great Barrier Reef World Heritage Area. In *Proceedings of the Ninth International Coral Reef Symposium, Bali, Indonesia, October 2000.* Vol, 2. (Eds K Moosa, S Soemodihardjo, A Soegiarto, K Romimohtarto, A Nontji, Soekarno and Suharsono) pp. 687–696. Ministry of Environment, Indonesian Institute of Sciences, International Society for Reef Studies, Indonesia.

Day JC, Fernandes L, Lewis A, Innes J (2004) RAP – An ecosystem level approach to biodiversity protection planning. In *Proceedings of the Second International Tropical Marine Ecosystems Management Symposium, Manila, Philippines, 24–28 March 2003.* International Coral Reef Action Network, Manila.

Day JC, Dudley N, Hockings M, Holmes G, Laffoley D, Stolton S, *et al.* (2012) *Guidelines for Applying the IUCN Protected Area Management Categories to Marine Protected Areas.* International Union for Conservation of Nature and Natural Resources, Gland, Switzerland.

De'ath G, Fabricius KE, Sweatman H, Puotinen M (2012) The 27-year decline of coral cover on the Great Barrier Reef and its causes. *Proceedings of the National Academy of Sciences of the United States of America* **109**, 17995–17999. doi:10.1073/pnas.1208909109.

Deloitte Access Economics (2013) *Economic Contribution of the Great Barrier Reef 2011–2012.* Great Barrier Reef Marine Park Authority, Townsville, <http://www.environment.gov.au/system/files/resources/a3ef2e3f-37fc-4c6f-ab1b-3b54ffc3f449/files/gbr-economic-contribution.pdf>.

Department of State Development Infrastructure and Planning (DSDIP) (2013) *Great Barrier Reef Coastal Zone Strategic Assessment 2013* (Draft report for consultation). Queensland Government, Brisbane, <http://www.statedevelopment.qld.gov.au/resources/great-barrier-reef-coastal-zone-strategic-assessment-final-reports.html>.

Emslie MJ, Logan M, Williamson DA, Ayling AM, MacNeil A, Ceccareli D, *et al.* (2015) Expectations and outcomes of reserve network performance following re-zoning of the Great Barrier Reef Marine Park. *Current Biology* **25**, 983–992. doi:10.1016/j.cub.2015.01.073.

Fernandes L, Day J, Lewis A, Slegers S, Kerrigan B, Breen D, *et al.* (2005) Establishing representative no-take areas in the Great Barrier Reef: large-scale implementation of theory on marine protected areas. *Conservation Biology* **19**, 1733–1744. doi:10.1111/j.1523-1739.2005.00302.x.

Fernandes L, Day J, Kerrigan B, Breen D, De'ath G, Mapstone B, *et al.* (2009) A process to design a network of marine no-take areas: lessons from the Great Barrier Reef. *Ocean and Coastal Management* **52**, 439–447. doi:10.1016/j.ocecoaman.2009.06.004.

Flinders M (1814) *A Voyage to Terra Australis.* Vol II (1966 facsimile edn). G and W Nicol, London.

GBRMPA (2003) *Great Barrier Reef Marine Park Draft Zoning Plan: Consultation Draft for Public Review.* Great Barrier Reef Marine Park Authority, Townsville.

GBRMPA (2004) *Great Barrier Reef Marine Park Zoning Plan 2003.* Great Barrier Reef Marine Park Authority, Townsville.

GBRMPA (2005) *Great Barrier Reef Marine Park Heritage Strategy.* Great Barrier Reef Marine Park Authority, Townsville.

GBRMPA (2010) *Overview of the Representative Areas Program*. Great Barrier Reef Marine Park Authority, Townsville, <http://www.gbrmpa.gov.au/zoning-permits-and-plans/rap>.

GBRMPA (2012) *Great Barrier Reef Biodiversity Conservation Strategy 2012*. Great Barrier Reef Marine Park Authority, Townsville.

GBRMPA (2013) *Great Barrier Reef Region Strategic Assessment Report*. Great Barrier Reef Marine Park Authority, Townsville.

GBRMPA (2014a) *Great Barrier Reef Outlook Report 2014*. Great Barrier Reef Marine Park Authority, Townsville.

GBRMPA (2014b) *Annual Report 2013–14*. Great Barrier Reef Marine Park Authority, Townsville.

Harrison HB, Williamson DH, Evans RD, Almany GR, Thorrold SR (2012) Larval export from marine reserves and the recruitment benefit for fish and fisheries. *Current Biology* **22**, 1023–1028. doi:10.1016/j.cub.2012.04.008.

Hughes TP, Day JC, Brodie J (2015) Securing the future of the Great Barrier Reef. *Nature Climate Change* **5**, 508–511. doi:10.1038/nclimate2604.

IUCN (1981) *World Heritage Nomination IUCN Technical Review: Great Barrier Reef*. IUCN, Gland, Switzerland, <http://whc.unesco.org/archive/advisory_body_evaluation/154.pdf>.

IUCN (2014) *World Heritage Outlook – Great Barrier Reef*. IUCN, Gland, Switzerland, <http://www.worldheritageoutlook.iucn.org/search-sites/-/wdpaid/en/2571

Johnson JE, Marshall PA (Eds) (2007) *Climate Change and the Great Barrier Reef: A Vulnerability Assessment*. Great Barrier Reef Marine Park Authority and Australian Greenhouse Office, Townsville.

Kenchington RA, Day JC (2011) Zoning, a fundamental cornerstone of effective Marine Spatial Planning: lessons learnt from the Great Barrier Reef, Australia. *Journal of Coastal Conservation* **15**, 271–278. doi:10.1007/s11852-011-0147-2.

Kerrigan B, Breen D, De'ath G, Day JC, Fernandes L, Tobin R, *et al.* (2010) *Classifying the Biodiversity of the Great Barrier Reef World Heritage Area*. Research Publication No. 104. Great Barrier Reef Marine Park Authority, Townsville.

Lawrence D, Kenchington RA, Woodley S (2002) *The Great Barrier Reef: Finding the Right Balance*. Melbourne University Press, Melbourne.

Lucas PHC, Webb T, Valentine PS, Marsh H (1997) *The Outstanding Universal Value of the Great Barrier Reef World Heritage Area*. Great Barrier Reef Marine Park Authority, Townsville.

Macintosh A, Bonyhady T, Wilkinson D (2010) Dealing with interests displaced by marine protected areas: a case study on the Great Barrier Reef Marine Park Structural Adjustment Package. *Ocean and Coastal Management* **53**, 581–588. doi:10.1016/j.ocecoaman.2010.06.012.

Marsh HD, Hodgson A, Lawler I, Grech A, Delean S (2007) 'Condition, status and trends and projected futures of the dugong in the Northern Great Barrier Reef and Torres Strait; including identification and evaluation of the key threats and evaluation of available management options to improve its status'. Marine and Tropical Sciences Research Facility Report Series. Reef and Rainforest Research Centre Limited, Cairns.

McCook LJ, Ayling T, Cappo M, Choat JH, Evans RD, De Freitas DM, *et al.* (2010) Adaptive management of the Great Barrier Reef: a globally significant demonstration of the benefits of networks of marine reserves. *Proceedings of the National Academy of Sciences of the United States of America* **107**, 18278–18285. doi:10.1073/pnas.0909335107.

Osmond M, Airame S, Day JC, Caldwell M (2010) Lessons for marine spatial planning: a comparison of three Marine Protected Area processes. *Ocean and Coastal Management* **53**, 41–51. doi:10.1016/j.ocecoaman.2010.01.002.

Reef Water Quality Protection Plan Secretariat (2013) *Reef Water Quality Protection Plan 2013.* Government of Australia, Canberra and Government of Queensland, Brisbane.

Royal Commission into Exploratory and Production Drilling for Petroleum in the Area of the Great Barrier Reef (1974) Queensland State Archives Agency ID1748, Brisbane, <http://www.archivessearch.qld.gov.au/Search/AgencyDetails.aspx?AgencyId=1748>.

Ruckelshaus M, Klinger T, Knowlton N, DeMaster DP (2008) Marine ecosystem-based management in practice: scientific and governance challenges. *BioScience* **58**, 53–63. doi:doi.org/10.1641/B580110.

Russ GR, Cheal AJ, Dolman AM, Emslie MJ, Evans RD, Miller I, *et al.* (2008) Rapid increase in fish numbers follows creation of world's largest marine reserve network. *Current Biology* **18**, 514–515. doi:10.1016/j.cub.2008.04.016.

Sweatman H (2008) No-take reserves protect coral reefs from predatory starfish. *Current Biology* **18**, 598–599. doi:10.1016/j.cub.2008.05.033.

Wescott G, Fitzsimons J (2011) Stakeholder involvement and interplay in coastal zone management and marine protected area planning. In *Marine Resources Management.* (Eds W Gullett, C Schofield and J Vince) pp. 225–238. LexisNexis Butterworths, Sydney.

World Heritage Committee (2014) *Decision: WHC 38 COM 7B.63, Great Barrier Reef (Australia) (N 154) (Doha, Qatar 2014).* UNESCO, Paris, <http://whc.unesco.org/en/decisions/6049>.

World Heritage Committee (2015) *Recording of Committee Discussion about the Great Barrier Reef, 1st July 2015. (Bonn, Germany).* UNESCO, Paris, <http://whc.unesco.org/en/sessions/39com/records/?day=2015-07-01#trv_TzLFMKfE12876>.

Chapter 6

Marine protected areas in the Antarctic and Sub-Antarctic region

Lyn Goldsworthy, Bob Zuur and Gilly Llewellyn

Introduction

This chapter focuses on marine protected areas (MPAs) in the exclusive economic zones (EEZs) around Australia's southern-most islands and in the waters of Eastern Antarctic where Australia's Antarctic operations are primarily focused. It compares the application of the process to the Heard Island and McDonald Islands Marine Reserve, within Australia's EEZ, to the proposed East Antarctic Representative System of Marine Protected Areas (EARSMPA), located in the Southern Ocean high seas and managed by the Commission for the Conservation of Antarctic Marine Living Resources (CCAMLR[1]).

The Southern Ocean covers 10% of the world's oceans and includes some of the most productive marine areas in the world. Specialised benthic and pelagic habitats support a diverse range of species and foraging grounds for seabirds and mammals (Griffiths 2010). Many species are endemic, often limited to very small and distinct areas, are slow growing, and have largely evolved without impact from human activity until recent times (Verde and di Prisco 2013). These factors make both species and habitats particularly vulnerable to changes to their environments.

There is currently no single mechanism for the establishment of MPAs in high seas areas beyond national jurisdiction. The Convention on Biological Diversity provides the global framework for biodiversity conservation and sustainable resource use, and, in 2008, adopted a set of criteria to identify 'ecologically and biologically significant areas (EBSAs) in high seas areas' (Convention on Biological Diversity 2008). Regional fisheries management arrangements, including CCAMLR, are generally responsible for specific designations of MPAs. Designation then depends on decisions of several or all members of the particular arrangement as well as the specific objectives of the arrangement.

The Australian Government has gazetted two MPAs in Southern Ocean seas in its EEZs around the Australian sub-Antarctic Heard Island and McDonald Islands (HIMI) and around Macquarie Island (Fig. 6.1).

Heard Island is the only sub-Antarctic island group with an intact terrestrial ecosystem, without plants or animals introduced directly by humans (Department of the Environment 2015a). It is the location of Big Ben: an active volcano and Australia's largest

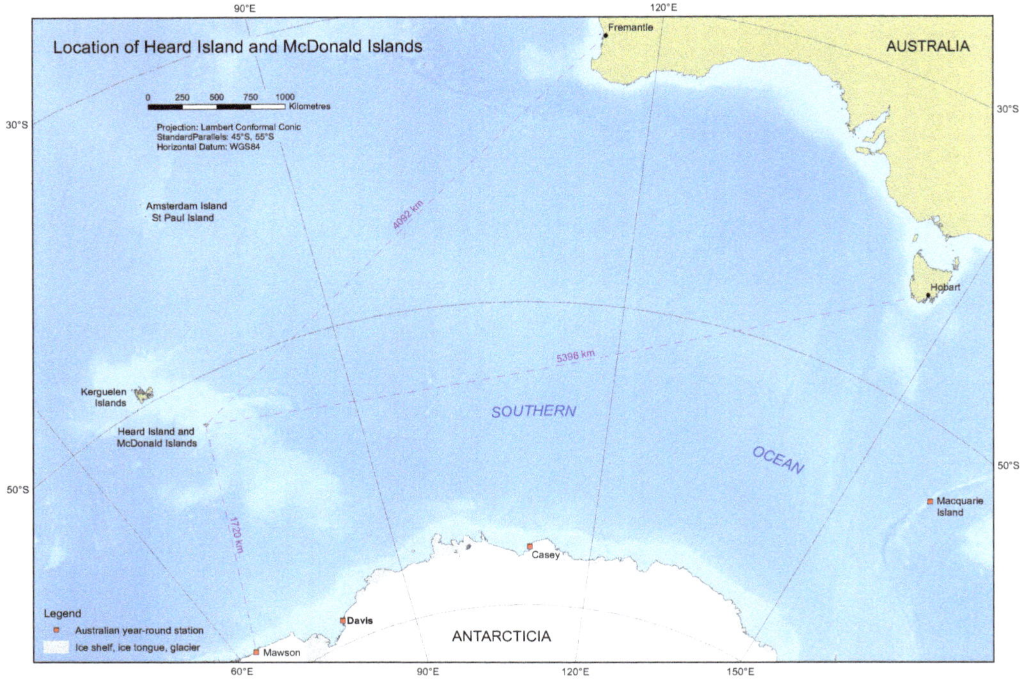

Fig. 6.1. Location of Macquarie Island and Heard Island and McDonald Islands in relation to Australia and Antarctica. (Map courtesy of the Australian Antarctic Division © Commonwealth of Australia 2015. The coastline of Antarctica is from the Antarctic Digital Database © 1993–2015 Scientific Committee on Antarctic Research. The bathymetry is from the GEBCO_08 Grid, version 20091120.)

mountain, and is home to vast numbers of penguins and seals, and to threatened great albatrosses. HIMI is on the Kerguelen Plateau and abuts the French EEZ around the Kerguelen Islands.

Australia's World Heritage-listed Macquarie Island is located 1500 km south-east of Tasmania and contains significant populations of sub-Antarctic wildlife (Figs 6.2, 6.3), including nearly 4 million seabirds. Not only does it provide nesting habitat for four threatened albatross species but it is the only breeding ground for Royal Penguins (with an estimated population of ~850 000 pairs) (Department of the Environment 2015b).

The Australian Government is also currently championing a network of MPAs on the high seas of the Southern Ocean, focusing particularly within the East Antarctic where it has significant research and operational experience. While this includes waters within the Australian Antarctic Territorial claim as well as waters off Adélie Land claimed by France (with which Australia is partnering in this proposal, along with the European Union), the area is managed by CCAMLR not by any individual country (Box 6.1).

This latter ambition is an articulation of Australia's commitment to both sustainable global oceans management and its particular desire to ensure the stable management of the Antarctic and Southern Ocean through participation in the Antarctic Treaty System. The principal national interest of Australia in Antarctica is stated as maintaining and strengthening the Antarctic Treaty System (DSEWPC and AAD 2011).

Fig. 6.2. Southern Elephant Seals tussling at Macquarie Island Marine Reserve. (Photo: Bob Zuur)

Fig. 6.3. Southern Elephant Seals tussling in front of King Penguins, Macquarie Island Marine Reserve. (Photo: Bob Zuur)

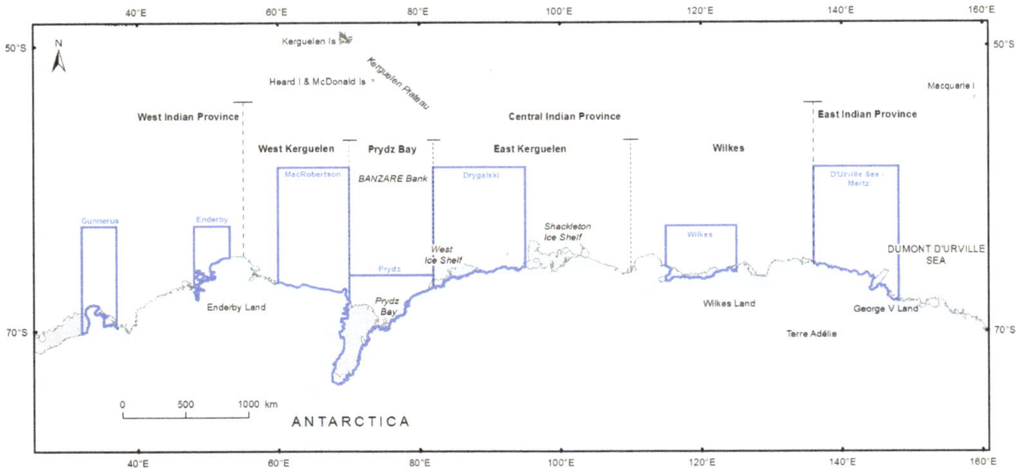

Fig. 6.4. Proposed MPAs as part of East Antarctic Representative System of Marine Protected Areas 2011. (Map courtesy of the Australian Antarctic Division © Commonwealth of Australia 2011)

The proposed East Antarctic Representative System of MPAs (EARSMPA) originally included seven areas within CCAMLR's East Antarctic planning domain that encompassed representative examples of the diverse biodiversity and geological features in the domain (AAD 2013) (Fig. 6.4). The proposed system included unique continental ridge, canyon and seamount features, with associated distinctive deep water flora and fauna, and important feeding areas for marine mammals, penguins and other seabirds. It would also have provided for the establishment of ecosystem-sized reference areas to allow for the study of the effects of climate change on Antarctic and Southern Ocean ecosystems, as well as fishing in the region (CCAMLR 2011a; Antarctic Ocean Alliance 2012). In the process of negotiating the consensus required within CCAMLR, this system

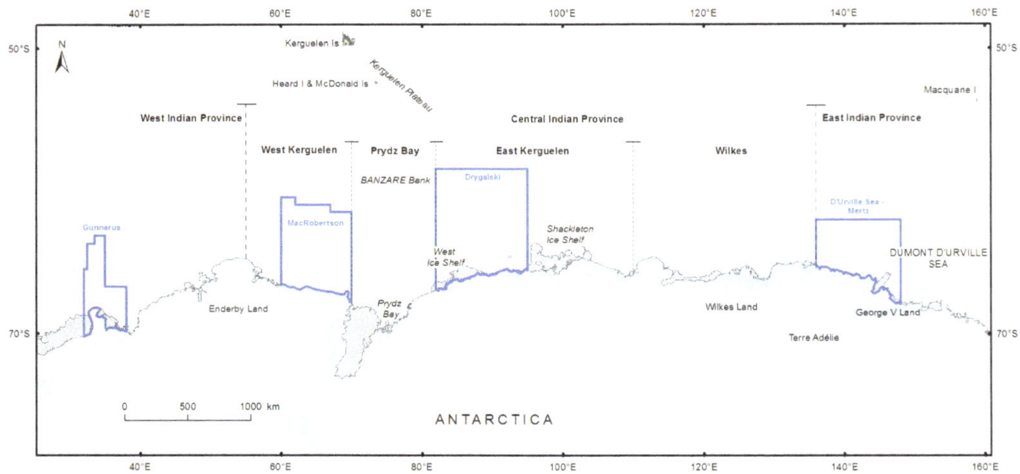

Fig. 6.5. Amended proposal for four MPAs as part of East Antarctic Representative System of Marine Protected Areas as of early 2015. (Map courtesy of the Australian Antarctic Division © Commonwealth of Australia 2014)

Box 6.1. Convention on the Conservation of Antarctic Marine Living Resources

The objective of the Convention on the Conservation of Antarctic Marine Living Resources (which established CCAMLR) is the 'conservation of Antarctic marine living resources' where conservation includes 'rational use'. Unlike fisheries agreements that focus only on the status of the commercial target species, the Convention requires consideration is given to all species in the ecosystem and to conserving ecological relationships.

The Convention defines three principles of conservation with which any harvesting activities must comply; these are:

- the prevention of the decrease in the size of any harvested population to levels below those that ensure its stable recruitment
- the maintenance of the ecological relationships between harvested, dependent and related populations, and the restoration of depleted populations
- the prevention of changes or the minimisation of the risk of changes in the marine ecosystem, which are not potentially reversible over two or three decades.

Article 1 of the Convention states that it applies 'to the Antarctic marine living resources of the area south of 60° South latitude and to the Antarctic marine living resources of the area between that latitude and the Antarctic Convergence which form part of the Antarctic marine ecosystem'. The Antarctic Convergence follows a line around the continent where cold, northward-flowing Antarctic waters meet the relatively warmer waters of the sub-Antarctic, forming a natural northern boundary for the Southern Ocean's marine life.

as of early 2015 had been reduced to four areas – Gunnerus, MacRoberston, D'Urville Sea-Mertz and Drygalski – each extending from the Antarctic coastline to a northern boundary corresponding to the Southern Antarctic Circumpolar Current Front (CCAMLR 2014)[2] (Fig. 6.5).

History and development

Heard Island and McDonald Islands Marine Reserve

The initial proposal for a marine reserve covering Australia's territorial islands of Heard Island and McDonald Islands and surrounding waters was announced in January 2001. The design approach was based on three elements: (i) consideration of all available scientific data; (ii) application of the principles of comprehensiveness, adequacy and representativeness; and (iii) inclusion of reference areas to assess impacts of fishing as well as the effectiveness of the reserve in achieving its objectives (Welsford *et al.* 2011). The Australian Antarctic Division and the Commonwealth environment department (then Environment Australia) embarked on an 18-month consultation process including the establishment of a Stakeholder Reference Group involving conservation groups, commercial fishing operators (including their skippers), scientists and other government agencies (M. Moore and M. Exel *pers. comm.* 2014). At the time there were two active fishing permits for the area. The government agencies encouraged an open and honest process where there was scope for everyone to share their points of view and for differences to be resolved (M. Exel *pers. comm.* 2014).

The reserve was announced in May 2002, encompassing 6.5 million ha of highly protected marine reserve, and an additional four areas adjoining the reserve (totalling 1.2 million ha) were set aside as a Conservation Zone under section 390D of the *Environment Protection and Biodiversity Conservation Act 1999* (EPBC Act) to allow for additional assessment for conservation and fishing resource values for future inclusion in the reserve. The remaining area of the 42 million ha EEZ is available for fishing. The fishery is managed by the Australian Fisheries Management Authority, in compliance with standards of CCAMLR as the area falls within the area of application of CCAMLR.

East Antarctica

In 1980, the Commission for the Conservation of Antarctic Marine Living Resources was established to manage living marine resources of the Southern Ocean (CCAMLR 1980) (see Box 6.1) and forms an important part of the Antarctic Treaty system. Australia's stated goals with respect to CCAMLR are to 'ensure the conservation of the marine living resources within the CCAMLR area ... to enhance Australia's influence in the Antarctic Treaty system ... and to maintain Australia's reputation as a responsible manager of marine resources' (AAD 2011).

In 1991, the Australian Government initiated a long-term marine conservation program to ensure the conservation and sustainable use of its marine and estuarine environments. A key part of this initiative was the commitment to the establishment of a National Representative System of Marine Protected Areas to protect the biotic diversity of Australia's marine ecosystems including seamounts and habitats in southern areas (ANZECC TFMPA 1998) (see Chapter 4).

In 2002, the World Summit on Sustainable Development (WSSD) called for the establishment of representative networks of MPAs consistent with international law and based on scientific information by 2012, as a relevant tool for ocean management (United Nations 2002). In the same year, CCAMLR discussed the WSSD proposal (CCAMLR 2002) and Australia played a leading role in promoting actions that led to the commitment to implementing a representative system of MPAs in CCAMLR waters, including hosting expert consultations to consider the application of broad-scale bioregionalisation for identification of MPAs (CCAMLR 2006). The Commission established a Scientific Advisory Subgroup on Protected Areas (2003), supported a workshop to advise on bioregionalisation to inform the placement of future Southern Ocean MPAs (2005), initiated discussions on principles for the establishment of the system of MPAs, and endorsed a work plan for its Scientific Committee to progress the development of MPAs (2006).

In 2003, a workshop hosted by WWF-Australia and the Antarctic Climate and Ecosystems Cooperative Research Centre brought together scientists of key CCAMLR countries to develop a biogeographic classification system of both pelagic and benthic biomes (Grant 2006). The outcomes of the workshop led to the adoption of network of nine biogeographic planning domains across the Southern Ocean by CCAMLR. Australia was an active participant in the scientific assessments upon which spatial management measures to support marine biodiversity conservation could then be applied. Analyses of the areas where benthic marine assemblages are likely to differ from each other in the Southern Ocean identified 23 ecoregions and nine bathomes (depth-related zones) (Douglass *et al.* 2014). Seven of these were located within the East Antarctica planning domain.

In 2010, Australia proposed the endorsement of a general conservation measure (CM) to support the establishment of a representative system of MPAs across the Convention Area, which later became CM 91–04 (CCAMLR 2010) (see Box 6.2). One fundamental element of the need for CM 91–04 expressed by Australia was that it allowed for different

Box 6.2. Conservation Measure 91–04 (2011) General framework for the establishment of CCAMLR MPAs

Conservation Measure 91–04 sets the framework for establishing MPAs under CCAMLR's jurisdiction. This provides that CCAMLR MPAs shall be established on the basis of the best available scientific evidence and shall contribute to the achievement of the following objectives:

1 protection of representative examples of marine ecosystems, biodiversity and habitats
2 protection of key ecosystem processes and habitats important to the life-history stages
3 establishment of scientific reference areas for monitoring natural variability and long-term change
4 protection of areas vulnerable to impact by human activities
5 protection of features critical for ecosystem functioning
6 protection of areas to maintain resilience in the face of change.

approaches within each planning domain, rather than a single prescribed format, to take into account capacity for large-scale MPAs, variability of available knowledge and the number of members interested in the other activities of the area.

In 2011, Australia, France and the European Union presented their proposal for a system of seven MPAs within the East Antarctica Planning Domain (see Fig. 6.4). The proposal was based on a comprehensive and inclusive scientific assessment, and thus represented the best scientific evidence available (CCAMLR 2011b), was in accordance with CM 91–04 and would assist CCAMLR to meet its commitment for a system of MPAs by 2012 (see Box 6.3). Each of the seven MPAs contained representative areas of biodiversity of the region and three were suggested as reference areas to allow for the study of the impacts of fishing activities and climate change on the ecosystem, which will assist in the development of effective management measures (CCAMLR 2011a). The system was framed as a multiple-use system where activities, including fishing, could be undertaken when those activities did not undermine the objectives of the individual MPA or the representative system as a whole.

Despite several revisions to take account of concerns expressed by other CCAMLR members, the proposal is yet to be accepted. The adoption of conservation measures requires consensus agreement by all 25 CCAMLR members including those with potential fishing interests in the area. Even with these revisions the proposal remains consistent with the principles of the National Representative System of Marine Protected Areas (ANZECC TFMPA 1998), agreed to by all Australian governments.

Current MPA estate in Australia's Antarctic and Sub-Antarctic waters

Heard Island and McDonald Islands

The Australian HIMI Territory was inscribed on the UNESCO World Heritage List in 1997 and included in the Australian National Heritage List in 2007. The inscribed area is 658 900 ha, of which 37 000 ha is terrestrial. The marine area extends 12 nautical miles (22 km) from the coastline.

Box 6.3. Key conservation values identified for protection in the 2011 proposals for East Antarctic Representative System of Marine Protected Areas[3]

- **Gunnerus MPA:** unique continental ridge and seamount features; representative biodiversity related to shelf, canyon and slope sea floor ecosystems.
- **Enderby MPA:** only representation of unique endemic mollusc in the East Antarctic Enderby Land province; important sea floor features likely to support distinct sea floor ecosystems.
- **MacRoberson MPA:** highly productive coastal and oceanic food webs providing summer foraging grounds for marine mammals and penguins; diverse sea floor ecosystems representing Central Indian Province; reference area for measuring natural variability and long-term changes in ecosystems and species.
- **Prydz MPA:** unique features such as southern-most embayment in the East Antarctic region and sea floor; important nursery area for Antarctic Krill and toothfish.
- **Drygalski MPA:** diverse sea floor environment on shelf and slope canyons and ice shelves; representation of greater oceanic food web and coastal food web adjacent to the ice shelves, which are important foraging areas for penguins, marine mammals and other seabirds; reference area for studying natural variability and long-term changes.
- **Wilkes MPA:** represents the unique sea floor of the Wilkes Sub-province; only area representative of biodiversity of shelf, canyon and slope ecosystems in region; reference area for evaluating the effects of bottom fishing in adjacent areas.
- **D'Urville Sea-Mertz MPA:** site of Antarctic Bottom Water formation, important for understanding climate change, and generating a range of habitats not found elsewhere; coastal and oceanic food webs; nursery area for Antarctic Silverfish; foraging area for marine mammals and penguins; registered CCAMLR vulnerable marine ecosystems.

In 2002, 6.5 million ha of waters surrounding Heard Island and McDonald Islands were declared a marine reserve under the *EPBC Act 1999* and designated IUCN Category Ia (Strict Nature Reserve) to protect habitats, ecosystems and native species in an as undisturbed state as possible. Public access is primarily limited to scientific research and environmental monitoring, and fishing and mineral access is excluded from the reserve (fishing is sustainably managed in the waters of HIMI outside of the reserve). The Reserve has thus been largely successful in striking a delicate balance of conservation goals and fishing ambition. At the time it was the largest marine area in the world to exclude commercial harvesting of marine life.

Four additional areas adjoining the marine reserve, totalling 1.2 million ha, were included as a conservation zone under the *EPBC Act 1999* while further assessments were undertaken. In March 2014, following a comprehensive scientific assessment of the region's conservation values and extensive consultation with key stakeholders, a further 6.2 million ha were added to the Marine Reserve on the basis of high conservation value (Fig. 6.6). The remaining conservation zone area was revoked (Fleming 2014). It is Australia's second largest IUCN Ia Strict Nature Reserve (after the Great Barrier Reef Marine Park).

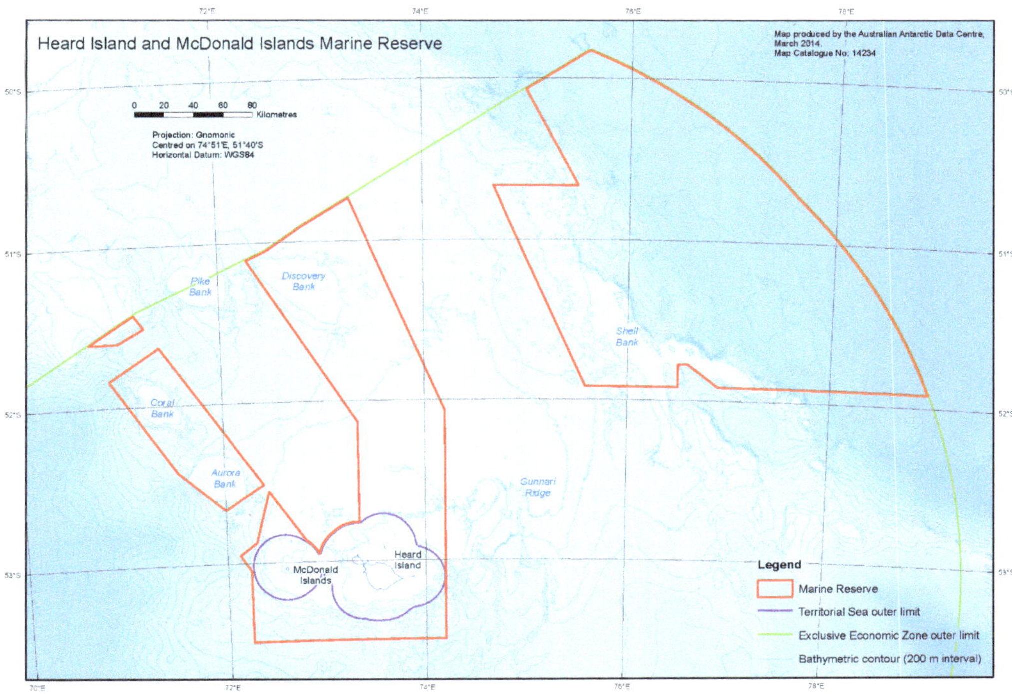

Fig. 6.6. Heard Island and McDonald Islands Marine Reserve. (Map courtesy of the Australian Antarctic Division © Commonwealth of Australia 2014)

The Reserve is managed under the *Heard Island and McDonald Islands Marine Reserve Management Plan 2014–2024* (Commonwealth of Australia 2014), which sets out the agreed management objectives and research and monitoring plans.

Macquarie Island

Macquarie Island and surrounding waters have a long and complex history of protection mechanisms. The Island was declared a Tasmanian Wildlife Sanctuary in 1933, a Tasmanian State Reserve in 1972, and an international UNESCO Biosphere Reserve in 1977. Commonwealth waters to the east and south of the Island were declared a Marine Park in 1999 and the waters surrounding the islands in Tasmanian jurisdiction proclaimed the Macquarie Island Marine Reserve in 2000 (Figs 6.7, 6.8).

In 1999, the Australian Government declared the 16.2 million ha Macquarie Island Marine Park in Commonwealth waters (3–200 nautical miles from the coast) to the east and south of Macquarie Island under the *National Parks and Wildlife Conservation Act 1974* (replaced a year later by the *EPBC Act 1999*). In 2007, this reserve was renamed the Macquarie Island Commonwealth Marine Reserve as part of the declaration of the South-east Commonwealth Marine Reserve Network (see Chapter 4). The reserve consists of a no-take sanctuary zone (58 000 km^2) and two habitat protection zones that were established to protect the habitat of seabirds and seals, and the benthic and pelagic fauna that depend on the area (104 000 km^2 combined) (Kriwoken and Holmes 2007). It is managed as part of the *South-east Commonwealth Marine Reserve Network Management Plan 2013–2023* (Director of National Parks 2013) (Fig. 6.9).

Fig. 6.7. King Penguins in front of rusting boilers that rendered down thousands of the birds into oil a century ago, Lusitania Bay, Macquarie Island. (Photo: Bob Zuur)

The Macquarie Island Marine Reserve, an extension of Tasmania's existing terrestrial Macquarie Island Nature Reserve, is completely no-take (fishing not permitted). These waters are both a nature reserve under the *Nature Conservation Act 2002* and a marine nature reserve under *Fisheries Rules 1999*. The 75 000 ha area encompassed by the reserve

Fig. 6.8. Antarctic Fur Seal, Macquarie Island. (Photo: Bob Zuur)

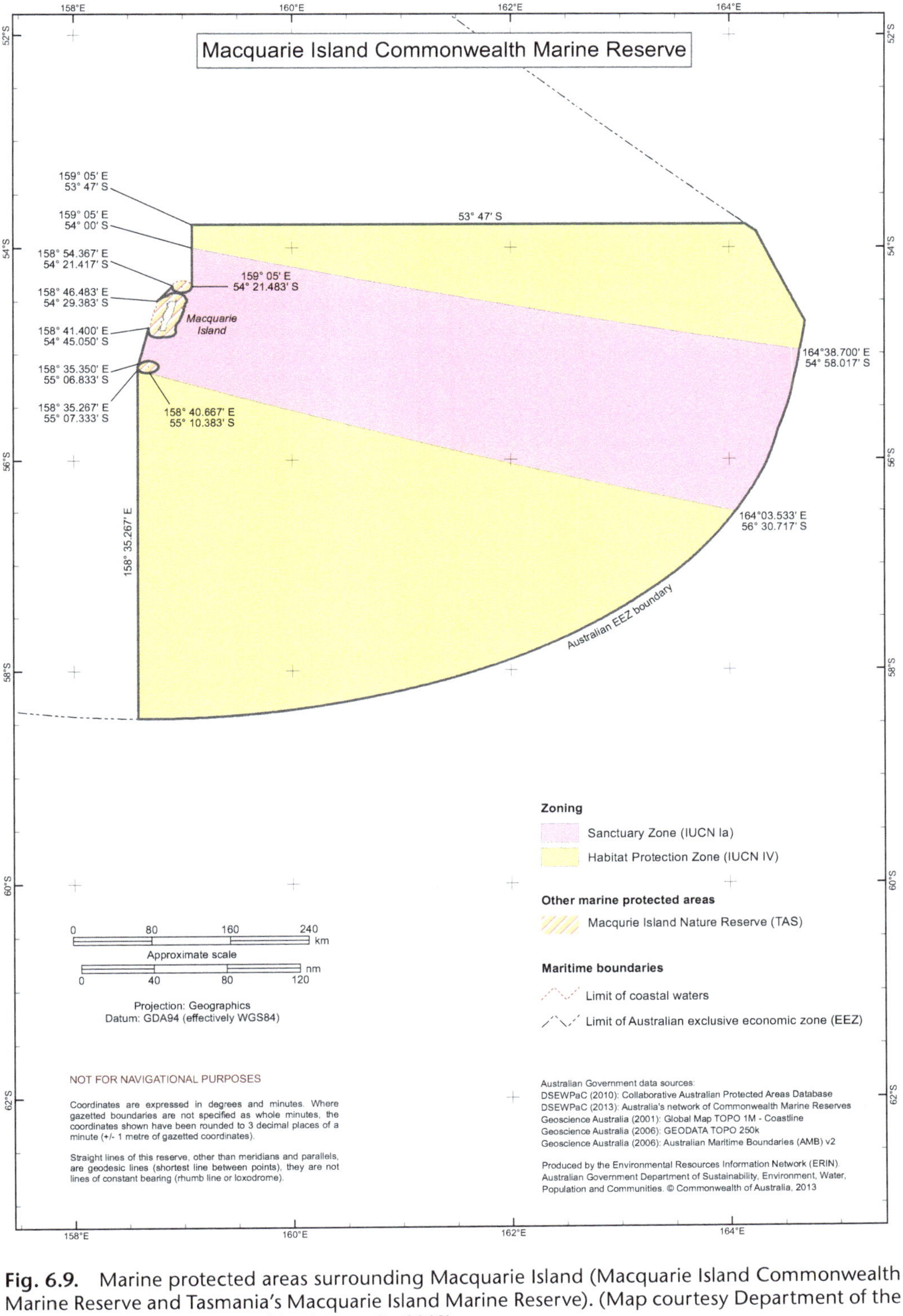

Fig. 6.9. Marine protected areas surrounding Macquarie Island (Macquarie Island Commonwealth Marine Reserve and Tasmania's Macquarie Island Marine Reserve). (Map courtesy Department of the Environment © Commonwealth of Australia 2013)

includes Macquarie Island's state waters from 0–3 nautical miles outwards (almost 75 000 ha of marine waters, see Chapter 10).

East Antarctic Representative System of Marine Protected Areas

Australia, France and the European Union first formally presented their proposal for the EARSMPA to the Scientific Committee and Commission meetings of CCAMLR in 2011 (CCAMLR 2011a,b). The original proposal comprised one area in each of the seven biogeographic regions identified within the East Antarctic Planning Domain, located to 'conserve representative areas of marine biodiversity in the region based on analyses of biology, ecology and biogeography of biota in the region ... [and to] provide reference areas for measuring the effects of climate change independent from the effects of human activities, and provide the reference areas necessary for managing ecosystem effects of fishing (CCAMLR 2011a,b) (Fig. 6.4 and Box 6.3). The total area encompassed in these seven proposed MPAs was 160 million ha (European Commission 2013).

The proposal is based on a 'multiple use' approach, where research and exploratory fishing activities would be allowed when they are judged consistent with the maintenance of the objectives of the MPAs and considered sustainable under existing conservation measures. The management plan for the EARSMPA would outline the criteria for assessing consistency. New conservation measures may be developed as necessary to guide that assessment (CCAMLR 2013, 2014).

The proposal was resubmitted for consideration in the following CCAMLR meeting in 2012 and for discussion at the Special Meetings of the Scientific Committee and Commission in July 2013. A revision was then presented to the 2013 CCAMLR Commission meeting where further changes, including the removal of three of the originally proposed areas (Enderby, Prydz and Wilkes), were made, bringing the total area to 109 million ha (Fig. 6.5). As of May 2015, the authors understand that Australia together with its co-proponents are preparing further revisions, still based on the multiple use concept, to meet the concerns of specific members.

If adopted, the EARSMPA would complement other protection measures for Australian Antarctic waters such as the Southern Ocean Whale Sanctuary (Box 6.4).

Successes and challenges

Heard Island and McDonald Islands

The success of the timely establishment of the Heard Island and McDonald Islands Marine Reserve can be partly contributed to the commitment to high-level stakeholder engagement by the Department of the Environment (then Environment Australia) and then the Australian Antarctic Division (AAD) as the responsible government agencies. These agencies went beyond the required public consultation procedures of passively seeking comments via written submissions, by establishing and then hosting regular meetings of the HIMI Stakeholder Reference Group. This Group was intimately involved in the development of the proposal. The strong support of the Environment Minister and his Parliamentary Secretaries at a time when the government was not generally focused on environmental initiatives, was also key.

A further reason for success was the support of the fishing industry for the establishment of MPAs. The MPAs gave the industry clarity regarding areas available for managed fishing in waters outside the MPA and direct involvement in the research fishing (under AAD's management) to assist the review of the conservation zone.

The structure of the AAD also contributed to the success of the designation, review and amendment of the HIMI reserve. This agency has responsibility for Southern Ocean

Box 6.4. Australia and the Southern Ocean Whale Sanctuary

Australia is party to the Southern Ocean Whale Sanctuary established in 1994 by the International Whaling Commission. Despite its name, this is not a marine protected area as defined by the IUCN but all commercial whaling is banned in this 50 million km^2 (5000 million ha) area.

Japan unsuccessfully attempted to remove the sanctuary in 2004 and continued to undertake whaling, rebadging their program as 'scientific whaling'. Australia successfully challenged this activity in the international Court of Justice (ICJ), which ordered in March 2014 that 'Japan revoke any extant authorisation, permit or licence to kill, take or treat whales in relation to JARPA II, and refrain from granting any further permits under Article VIII, paragraph 1, of the Convention, in pursuance of that programme'. The decision of the ICJ is final and the right of appeal does not apply in this context; Japan said it would abide by the decision. However, in November 2014 Japan announced a research plan for its New Scientific Whale Research Program in the Antarctic Ocean, to replace previous programs.

Within Australia, all Commonwealth waters are part of the Australian Whale Sanctuary. Under the *Environment Protection and Biodiversity Conservation Act 1999*, it is illegal to kill, injure or interfere with cetaceans (whales, dolphins and porpoises) in Australian waters.

policy, research and monitoring. This vertical integration has allowed codification of a conservation outcome directly integrated with the conduct of science. Although several other federal agencies, as well as the Tasmanian Government (which has experience in managing Macquarie Island Marine Reserve), have played an important role, the overall design and outcome has been shaped within one agency.

In the early 2000s, illegal fishing around HIMI was a major threat to the sustainability of targeted Toothfish stocks. However, since 2004 no illegal fishing has been reported in this area. This is largely the result of a concerted surveillance effort by Australia and France and high-profile prosecutions of those found (AAD 2010).

East Antarctica

CCAMLR makes decisions on the basis of consensus by its 25 members, each of whom have differing values and interests. Any MPA proposal in CCAMLR waters must ensure protection of the environment while taking into account various interpretations of 'rational use' and freedom of scientific research. In addition, there are varying interpretations of how the precautionary principle should be applied (e.g. some members prefer the 'prove the threat' approach) and there are many views on what 'best available science' and 'representative' mean.

A large majority of members support the EARSMPA proposal. However, objection to the proposal has arisen both because of outright opposition by a few members to large-scale MPAs (in spite of CCAMLR's commitment to establish them) and opposition to specific matters. Outright opposition is often driven by the desire to protect national fishery interests but also seems to be underpinned by geopolitical interests and a resistance to what is perceived by some members as MPA proposals reinforcing claims to sovereignty. A strong desire by some members to ensure open access across the entire CCAMLR zone for commercial fishing into the future remains the most challenging issue to be overcome for creating MPAs on the high seas, although it is worth noting that some of the MPA proponents (i.e. Australia, France, New Zealand and the UK) also undertake fishing in these waters.

Some CCAMLR members raised concerns around the size, activity restrictions and permanency for MPA proposals, with some suggesting MPAs should only be established for short periods of time (ASOC 2013). This is inconsistent with all widely accepted international definitions of MPAs (Day *et al.* 2012) and would not support effective use of the areas for research requiring long-term datasets or for reference areas monitoring complex ecological processes.

The EARSMPA proposal raises particular challenges in itself. It is a partnership of three Members, one of which is a Union of several individual members (i.e. the European Union). Furthermore, the EARSMPA proposal is based on multiple-use management with the expectation that existing and new conservation measures would be applied to ensure activities are consistent with the maintenance of the MPA's conservation objectives. Some members appear to have difficulty understanding this concept compared with the simpler demarcation of no-take zones (as with the Ross Sea MPA proposal (MFAT 2014)). This may be because it is easier to agree on smaller no-take restrictions on single species or habitats, than large-scale areas that aim to protect large integrated habitats and ecosystems and permit activities consistent with the objectives of the protected areas.

In addition, several CCAMLR members were involved in discussions concerning marine protected zones or areas in other parts of the world and appear concerned that decisions made in CCAMLR do not create precedents for those other discussions. Furthermore, diplomatic differences between nations on matters that have nothing to do with CCAMLR can have a significant impact on the capacity of those nations to negotiate in any formal negotiating arena. For example, the Ukraine conflict may have exacerbated previous policy differences and restricted negotiations.

Progress is slower and less congenial on the high seas because of the much wider range of state, commercial and civil society actors, the range of their expectations and (especially) differing legal understandings regarding the right to fish on the high seas, and consensus decision-making processes. The impact of government positions relating more to their geopolitical positions in other parts of the world as noted above add another level of challenge. CCAMLR members also articulate differing interpretations of the United Nations Convention on the Law of the Sea conditions on the 'right to fish', which is increasingly common in other international fora (UNESCO Division for Ocean Affairs and the Law of the Sea 2015).

The strong support of the lead government agency, the Australian Antarctic Division, has again been key in leading the coalition of EARSMPA proponents and providing significant research and policy work that formed the base of this proposal. The ongoing support of the Australian Environment Ministers has ensured sufficient resources (at the time of publishing) to ensure Australia could play this leading role.

Despite these decision-making difficulties, EARSMPA has been progressing steadily, albeit along a sluggish path, to contribute to CCAMLR's commitment to a representative system of MPAs in Antarctic waters.

Conclusion

Substantial progress in achieving marine protection was possible in the Southern Ocean within Australian jurisdiction (Macquarie Island, and Heard Island and McDonald Islands) because the lead agency (Australian Antarctic Division), operating through national legislation, was dedicated to delivering the World Summit on Sustainable

Development's commitment to marine protection, because they used an open and inclusive process, and because of the commitment to collaboration from the commercial fishers licenced to fish in Australian waters. Striking a research-driven balance between maintaining fishing opportunities and effectively protecting a representative network of MPAs has been congenially achieved.

The development of a representative system in Southern Ocean waters beyond national jurisdiction has been somewhat slower. It is more difficult to designate MPAs on the high seas where the agreement of all signatories to the relevant convention is required.

Australia and several other CCAMLR members have long been committed to the 'values' based approach to identifying and delineating MPAs based on the intrinsic and scientific values of an area. However others, especially those with strong fishing priorities, tend to approach the issue from a perspective of requiring that 'threats' from destructive fishing practices be proven before marine protection is designated. The IUCN approach to MPAs allows several approaches, including multiple use. This allows for these two perspectives to be reconciled, by having both values-based designation and threats-based management. Although CCAMLR recognises this sophisticated approach to marine conservation through its Conservation Measure 91–04, it has yet to be implemented through a consensus of CCAMLR members designating the EARSMPA.

Whether CCAMLR continues to be the leading regional forum where global commitments to biodiversity conservation and sustainable fishing are resolved in a cooperative environment remains to be seen. Whether Australia can maintain a leading role in that direction given changing government priorities also remains to be seen.

Acknowledgements

We thank Margaret Moore, Alistair Graham and Paul Gamblin for helpful suggestions and for filling in some details, and, in particular, Wendy Howe for editing and keeping the drafting on track.

Biographies

Lyn Goldsworthy AM has undertaken extensive work in global conservation advocacy and non-government governance. A primary focus of her work over the past 30 years has been on strengthening the conservation objectives of the Antarctic Treaty system and was awarded an Order of Australia (Member) for this work. She served for 8 years on the Antarctic Science Advisory Committee and has attended more than 30 Antarctic Treaty and CCAMLR meetings as environmental NGO advisor on the Australian delegation. She is currently Executive Director of the Frank Fenner Foundation and also provides strategic advice on Antarctic and deep ocean issues. She worked with the WWF Antarctic and Southern Ocean Initiative between 2011 and 2015.

Bob Zuur is Manager of WWF's Antarctic and Southern Ocean Initiative. A marine biologist who studied Antarctic fish growth and who has a Masters degree in public policy, Bob had a broad career in research and environmental management before joining WWF.

Gilly Llewellyn has an undergraduate degree in Natural Sciences from Cambridge University, a PhD in Earth Sciences from Harvard University, and more than 10 years' experience in scientific research and on-ground conservation. Gilly is Conservation Director for WWF-Australia.

Endnotes

1 For further information about CCAMLR see https://www.ccamlr.org/

2 At the XXXIV CCAMLR Meeting in October 2015, the Gunnerus area was removed, leaving just three areas from the original seven. Some small changes were also made to the text to meet specific member suggestions. Despite these changes, the proposal was not accepted and will be considered at the next CCAMLR Meeting in October 2016 (CCAMLR 2015). As the proposal is still in negotiation, details of specific boundaries are not yet in the public arena. Permission to view details may be requested from the CCAMLR Secretariat.

3 http://www.antarctica.gov.au/law-and-treaty/ccamlr/marine-protected-areas

References

AAD (2010) *Australia's Fisheries in the Heard Island and MacDonald Islands Region*. Australian Antarctic Division, Hobart, <http://www.antarctica.gov.au/law-and-treaty/ccamlr/australias-fisheries-in-the-heard-island-and-mcdonald-islands-region

AAD (2011) *Australia's Primary Goal Within CCAMLR*. Australian Antarctic Division, Hobart, <http://www.antarctica.gov.au/law-and-treaty/ccamlr/australias-primary-goal-within-ccamlr>.

AAD (2013) *A Proposal for a Representative System of Marine Protected Areas in the East Antarctic Planning Domain*. Australian Antarctic Division, Hobart, <http://www.antarctica.gov.au/law-and-treaty/ccamlr/marine-protected-areas>.

Antarctic Ocean Alliance (2012) *Antarctic Ocean Legacy: Protection for the East Antarctic Coastal Region, 18 October 2012*. Antarctic Ocean Alliance, <http://awsassets.wwf.org.au/downloads/mo029_antarctic_ocean_legacy_protection_for_the_east_antarctic_coastal_region_18oct12.pdf>.

ANZECC TFMPA (1998) *Guidelines for Establishing the National Representative System of Marine Protected Areas*. Australia and New Zealand Environment and Conservation Council Task Force on Marine Protected Areas, Environment Australia, Canberra.

ASOC (2013) *AOA Briefing 1: Duration of MPAs*. CCAMLR paper SM-II/BG/05, 12 June 2013, Antarctic and Southern Ocean Coalition, Washington DC, <http://www.asoc.org/storage/documents/CCAMLR-SM-IIBG05_AOA_Briefing_1_Duration_of_MPAs.pdf>.

Commission for the Conservation of Antarctic Marine Living Resources (CCAMLR) (1980) *The Convention on the Conservation of Antarctic Marine Living Resources*. CCAMLR, Hobart, <https://www.ccamlr.org/en/organisation/camlr- convention-text>.

CCAMLR (2002) *Report of the Twenty-first Meeting of the Commission of CCAMLR* (*CCAMLR XXI*). Hobart Australia, 21 October–1 November 2002. Commission for the Conservation of Antarctic Marine Living Resources, Hobart, <https://www.ccamlr.org/en/ccamlr-xxi>.

CCAMLR (2006) *Report of the Twenty-fifth Meeting of the Scientific Committee of CCAMLR (SC-CCAMLR XXV)*. Hobart, Australia, 23–27 October 2006. Commission for the Conservation of Antarctic Marine Living Resources, Hobart, <https://www.ccamlr.org/en/sc-camlr-xxv>.

CCAMLR (2010) *Report of the Twenty-ninth Meeting of the Commission (CCAMLR-XXIX)*. Hobart, Australia, 25 October–5 November 2010. Commission for the Conservation of Antarctic Marine Living Resources, Hobart, <https://www.ccamlr.org/en/ccamlr-xxix>.

CCAMLR (2011a) *Report of the Thirtieth Meeting of the Commission (CCAMLR-XXX)*. Hobart, Australia, 24 October–4 November 2011. Commission for the Conservation of Antarctic Marine Living Resources, Hobart, <http://www.ccamlr.org/en/system/files/e-cc-xxx.pdf

CCAMLR (2011b) *Report of the Thirtieth Meeting of the Scientific Committee of CCAMLR (SC-CCAMLR XXX).* Hobart, Australia, 24 October–4 November 2011. Commission for the Conservation of Antarctic Marine Living Resources, Hobart, <https://www.ccamlr.org/en/sc-camlr-xxx>.

CCAMLR (2013) *Report of First Intersessional Meeting of the Scientific Committee Commission (CCAMLR-SM-II).* Bremerhaven, Germany, 11–13 July 2013. Commission for the Conservation of Antarctic Marine Living Resources, Hobart, <http://www.ccamlr.org/en/system/files/e-sc-im-i.pdf>.

CCAMLR (2014) *Report of Thirty-third Meeting of the Scientific Commission (CCAMLR XXXIII).* Hobart, Australia, 20–31 October 2014. Commission for the Conservation of Antarctic Marine Living Resources, Hobart, <https://www.ccamlr.org/en/sc-camlr-xxxiii>.

CCAMLR (2015) *Report of the Thirty-fourth Meeting of the Commission (CCAMLR-XXXIV).* Hobart, Australia, 19–30 October 2015. Commission for the Conservation of Antarctic Marine Living Resources, Hobart, <https://www.ccamlr.org/en/ccamlr-xxxiv>.

Commonwealth of Australia (2014) *Heard Island and McDonald Islands Marine Reserve Management Plan 2014–2024.* Department of the Environment, Canberra.

Convention on Biological Diversity (2008) *COP9 Decision IX/20, Marine and Coastal Biodiversity.* 19–30 May 2008. COBD, Bonn, Germany, <http://www.cbd.int/decision/cop/default.shtml?id=11663>.

Day J, Dudley N, Hockings M, Holmes G, Laffoley D, Stolton S, *et al.* (2012) *Guidelines for Applying the IUCN Protected Area Management Categories to Marine Protected Areas.* IUCN, Gland, Switzerland.

Department of the Environment (2015a) *World Heritage Places-Heard and McDonald Islands.* Australian Government, Canberra, <http://www.environment.gov.au/heritage/places/world/heard-mcdonald>.

Department of the Environment (2015b) *World Heritage Places-Macquarie Island.* Australian Government, Canberra, <http://www.environment.gov.au/heritage/places/world/macquarie-island>.

Director of National Parks (2013) *South-east Commonwealth Marine Reserve Network Management Plan 2013–2023.* Australian Government, Canberra.

Douglass LL, Turner J, Grantham HS, Kaiser S, Constable A, Nicoll R, *et al.* (2014) A hierarchical classification of benthic biodiversity and assessment of protected areas in the Southern Ocean. *PLoS ONE* **9**, e100551. doi:10.1371/journal.pone.0100551.

DSEWPC and AAD (2011) *Australian Antarctic Science Strategic Plan 2011–12 to 2020–21.* Department of Sustainability, Environment, Water, Population and Communities and Australian Antarctic Division, Australian Government, Canberra.

European Commission (2013) *Questions and Answers-East Antarctic Representative System of Marine Protected Areas (EARSMPA).* 12 July 2013, European Commission, Brussels, <http://ec.europa.eu/information_society/newsroom/cf/mare/itemdetail.cfm?item_id=11370>.

Fleming J (2014) Expansion of the Heard Island and McDonald Islands Marine Reserve. *Australian Antarctic Magazine* **26**, 34.

Grant S (2006) *Bioregionalisation of the Southern Ocean: report of the Experts Workshop.* Hobart, September 2006. Antarctic Climate and Ecosystems Cooperative Research Centre and WWF-Australia, Hobart, <https://data.aad.gov.au/analysis/crc/acebio/>.

Griffiths HJ (2010) Antarctic marine biodiversity – what do we know about the distribution of life in the Southern Ocean? *PLoS ONE* **5**, e11683. doi:10.1371/journal.pone.0011683.

Kriwoken LK, Holmes ND (2007) Emerging issues of Australia's Sub-Antarctic Islands: Macquarie Island and Heard Island and McDonald Islands. In *Looking South: Australia's*

Antarctic Agenda. (Eds L Kriwoken, J Jabour and AD Hemmings) pp. 149–164. Federation Press, Leichhardt, NSW.
MFAT (2014) *Ross Sea Region Marine Protected Area – updated proposal*. New Zealand Ministry of Foreign Affairs and Trade, Wellington, New Zealand, <http://www.mfat.govt.nz/ross-sea-mpa/tabs/proposal.php>.
United Nations (2002) *United Nations World Summit on Sustainable Development (WSSD) Johannesburg Summit – Key Outcomes of the Summit*. UNESCO, Paris, <http://www.unesco.org/education/tlsf/mods/theme_a/img/02_WSSDOutcomes.pdf
UNESCO Division for Ocean Affairs and the Law of the Sea (2015) *Ad Hoc Open-ended Informal Working Group to study issues relating to the conservation and sustainable use of marine biological diversity beyond areas of national jurisdiction*. 13 March 2015, UNESCO, Paris, <http://www.un.org/depts/los/biodiversityworkinggroup/biodiversityworkinggroup.htm>.
Verde C, di Prisco G (Eds) (2013) *Adaptation and Evolution in Marine Environments, Volume 2: The Impacts of Global Change on Biodiversity*. Springer-Verlag, Berlin, Germany.
Welsford D, Constable A, Nowara G (2011) The Heard Island and McDonald Islands Marine Reserve and Conservation Zone. A model for Southern Ocean marine reserves? In *The Kerguelen Plateau: Marine Ecosystem and Fisheries*. (Eds G Duhamel and D Welsford) pp. 297–304. Societe Francaise d'Ichtyologie, Paris.

Chapter 7

The Western Australian marine conservation reserves system

Barry Wilson

Introduction

Western Australia has by far the longest coastline of any of the Australian states, spanning more than 21 degrees of latitude, embracing tropical waters of the Kimberley in the north, temperate waters of the south coast and a wide zone of biogeographic overlap on the west coast. The coastline is extremely varied, including rocky ria shores with inundated gulfs and a mountainous hinterland, long stretches of straight or curving beach, rocky shores of Proterozoic granite headlands and other igneous rocks, and long tracts of Quaternary and Tertiary limestone cliffs. It is swept by two major oceanic current systems. Western Pacific water is driven into the north-eastern Indian Ocean by the Indo-Pacific through flow and mixes with Indian Ocean water. From this source, the Western Australian Holloway and Leeuwin Currents are derived and flow southwards along the North West Shelf and the west coast, delivering propagules of tropical marine organisms far down into the temperate zone. The south coast is serviced by a different ocean current system. It is swept by the cooler westerly, circumtropical South Indian Current. The fauna of this temperate south coast is primarily endemic but the South Indian Current delivers propagules of some cosmopolitan Southern Ocean species. These two oceanic current systems and their biogeographic services interact at the south-west corner of the continent.

Biogeographically the waters of Western Australian cover two of the world's primary marine biogeographic realms (Wilson 2013). The tropical north is a province of the vast Indo-West Pacific realm while the south coast is part of the Southern Australian realm of the Southern Ocean and the two support marine faunas that are distinctly different. For example, the tropical northern coast has diverse and species-rich coral reefs and mangrove habitats, Dugongs, Saltwater Crocodiles and sea turtles. In contrast, nearshore waters of the south coast have vast seagrass meadows and kelp beds, seal colonies and Little Penguins. The distinctly different character of northern and southern marine faunas of Western Australia has its origin in the ancient tectonic and biogeographic history of the world, sustained by contemporary oceanic current and climatic conditions.

Superimposed over these primary historical and global climatic factors there are significant regional changes in climatic conditions along the vast length of the Western

Australian coast. Much of it has an arid or semi-arid hinterland with few rivers and estuaries. However, the Kimberley and the south-west of Western Australia are relatively wet with major rivers and extensive development of estuarine habitats.

Biodiversity and biogeomorphic diversity on this scale is a marvellous asset but with the geographic scale involved, developing a network of marine protected areas (MPAs) that is representative and adequately protects this biodiversity and the region's coastal ecological integrity is a major challenge. This is especially so because the ecological, taxonomic and distributional data needed for well-informed reserve selection and management decision making remain inadequate for the purpose and must be gathered as the marine conservation program is put in place. It is necessary to make provisional decisions using available information and to be willing to amend them as the knowledge base grows. This is a difficult way to proceed in a social and political environment that demands 'certainty' and justification of interference with people's established activities. Nevertheless, substantial progress has been made and the Western Australian marine conservation reserve system now includes a large number of important reserves that represent much of the state's marine biodiversity. Adequate management of this system is another matter.

History and development of the Western Australian MPA network

Early development of the idea

Reservation of marine areas for environmental protection and resource management purposes was an idea that arose in Western Australia in the 1960s, as it did elsewhere. The concept was linked with another: that is, that conservation reserves should be selected to be representative of regional habitats and biodiversity. Conservation reserves were thought of as places where ecosystems could be sustained and biodiversity protected. This remains a keystone of conservation management, modified by contemporary concepts of dispersal and connectivity.

In 1962, a Western Australian Sub-Committee of the Australian Academy of Science (AAS) noted the 'need for a pattern of reserves to be representative of all major communities of natural wildlife and scenery types that would preserve biological communities for all time' (AAS 1965, p. 74). This concept became known as the principle of 'Conservation through Reserves'. In 1972, the Western Australian Environmental Protection Authority (EPA) established the 'Conservation Through Reserves Committee' (CTRC) comprising Dr David Ride (zoologist and Director of the Western Australian Museum), Mr John Morgan (Western Australian Surveyor General), Dr Basil Balme (Senior Lecturer in Geology, University of Western Australia) and Professor Reg Appleyard (Professor of Economic History, University of Western Australia).

Initially, the scope of the Western Australian CTRC was concerned with terrestrial biota and landscapes but marine reserves were also considered. The introduction of the CTRC's first report to the EPA (CTRC 1974, p. 0–23) included a long discussion of the legislative issues involved in establishing marine conservation reserves, concluding that there was 'a need for new legislation to permit aquatic reserves to be set aside' because the Western Australian *Land Act 1933*, which provided for reservation of terrestrial reserves, could not be used to designate reserves below low-water mark.

In the body of its 1974 report to the EPA, the CTRC identified several coastal marine and estuarine areas as worthy MPAs (referred to then as 'aquatic reserves'). The CTRC recommended that aquatic reserves should be controlled under the fisheries legislation. This was suggested as an interim measure, because there was 'no other statutory provision which can be used for this purpose despite the fact that the *Fisheries Act [1905]* was intended to provide for the conservation and exploitation of commercially important marine life' (CTRC 1974, p. 0–23).

Subsequently the EPA referred the CTRC recommendations, with amendments and additions, to the Western Australian Government (EPA 1975, 1976, 1978, 1983 – referred to as the 'EPA Systems proposals'). The aquatic reserve recommendations were endorsed in principle by government and many were later implemented.

The Western Australian Government also implemented the EPA recommendations for new legislation. Legislation was introduced that provided:

- for aquatic reserves managed by government agencies in both the conservation jurisdiction (*Conservation and Land Management Act 1984*) and fisheries jurisdiction (*Fish Resources Management Act 1994*)
- that Ningaloo Marine Park and future aquatic areas reserved for conservation and recreational purposes should be managed by the state conservation agency, not by the Fisheries Department
- that fishing activities (where permitted) within marine parks should be regulated and administered by the Fisheries Department under the provisions of the *Fisheries Act 1905*, in collaboration with the then Department of Conservation and Land Management.

These decisions proved to be controversial and resulted in confusion about the respective roles of the two departments.

Establishment of the Western Australian MPA program

The EPA (1975) recommendation for the reservation of Ningaloo Reef Tract received early attention. It was proposed as an aquatic reserve vested in the National Parks Authority (NPA) managed together with the adjacent Cape Range National Park and with advice from the Director of Fisheries and Wildlife in regard to the aquatic habitats and biota. In 1978 the NPA established a Marine Park Working Group to develop a management plan for the proposed Ningaloo Marine Park. The Working Group comprised Mr Richard May (NPA Planning Officer – Convenor), Dr Rod Lenanton (fisheries scientist, WA Fisheries and Wildlife Department), and Dr Barry Wilson (this author – marine zoologist, WA Museum; later replaced in 1979 by Dr Paddy Berry, also marine zoologist, WA Museum).

However, before the NPA plans for the Ningaloo Marine Park were completed there was a radical re-arrangement of administration of environmental management agencies within the Western Australian Government. The NPA and the Wildlife Section of the Department of Fisheries and Wildlife were dismantled and their staff and resources regrouped within a new agency called the Department of Conservation and Land Management (CALM) controlled under a new *Conservation and Land Management Act 1984*. The new department was charged with development of the marine reserves program and designation of many of the EPA recommendations followed, the first, in 1987, being the Ningaloo Marine Park and the Marmion Marine Park, north of Perth.

In 1986, an inter-agency Marine Parks and Reserves Selection Working Group (MPRSWG) was established by CALM to review the representativeness of the EPA recommendations and recommend further development of a representative marine reserve system. In 1994, the Working Group produced the report *A Representative Marine Reserve System for Western Australia* (CALM 1994). This comprehensive study followed the principle of representativeness required by the CTRC and was generally consistent with other international selection criteria promoted by the IUCN. It included an account of Western Australian coastal and estuarine habitats and biota, as then known, arranged in four regional sections – Kimberley, Pilbara-Canning, West Coast and South Coast. Within these broad biogeographical units, the Working Group recognised 'major distinctive coastal types', based on biogeomorphic features, which proved to be a very useful selection tool for new marine reserves.

Introduction of new fishery legislation in Western Australia in 1994 provided for Fish Habitat Protection Areas, which are MPAs set aside for a variety of purposes relating to fish and fish habitat conservation and management. There was duplication of intent between reserve provisions of the *Fish Resources Management Act 1994* and the *Conservation and Land Management Act 1984*. In effect, Western Australia set up two MPA systems designated and administered by two different agencies. It remains for the two departments to ensure their efforts are collaborative and complementary.

MPAs reserved under the *Conservation and Land Management Act 1984*

The CALM Act sets out the categories of marine reserve, the procedures to be followed in their declaration and management, the requirements for reserve management plans, and provides for establishment of a 'controlling body' in which marine reserves are vested (presently the Marine Parks and Reserves Authority). Marine reserves are established by order of the Governor published in the *Western Australian Government Gazette* (the *Gazette*) and may be classified as Class A, meaning that their purpose and boundaries may not be altered except by Parliament. Marine reserves may not be classified at all if the Minister for Mines objects to a Class A classification, but that situation has not arisen. Reserves designated as Classes B and C may be altered by the Minister by notice in the *Gazette* but these Classes have not been used for marine reserves. At the time of this publication these sections of the Act are under review.

Categories of reserve

Initially, the CALM Act provided for two reserve categories, 'Marine Park' and 'Marine Nature Reserve'. An amendment in 1997 added a third category, 'Marine Management Area'. The purposes of these reservations are as follows:

Marine Nature Reserve: 'the conservation and restoration of the natural environment; the protection, care and study of indigenous flora and fauna; and the preservation of any feature of archaeological, historic or scientific interest'.

Recreation is not included in the purposes of the Marine Nature Reserve category but the word 'study' is interpreted to include education and observation by visitors as well as scientific study.

Marine Park: '... that level of recreational and commercial activity which is consistent with the proper conservation and restoration of the natural environment, the protection of indigenous flora and fauna and the preservation of any feature of archaeological historic or scientific interest'.

It is important to note that recreational and commercial activities are determined as possible purposes of reservation to the extent that they consistent with the primary purposes. Specific purposes are stated in the Management Plan for each park and are normally manifested in a zoning scheme (see below).

Marine Management Area: '... managing and protecting the marine environment so that it may be used for conservation, recreational, scientific and commercial purposes'.

This definition of purpose is similar to that for Marine Park but allows for compromise and a best result from integrated management of diverse purposes that are specified in individual reserve management plans. Importantly, it also specifically identifies aquaculture, commercial fishing, pearling, mining, seismic surveys, exploratory drilling and production of petroleum as permitted commercial purposes.

Zoning

As implementation of the representative MPA plan proceeded it was found necessary to devise and add provisions for a zoning arrangement for Marine Parks. Amendments to the CALM Act were introduced in 1997 that achieved these objectives, allowing for the establishment of 'general use areas', 'sanctuary areas', 'recreation areas' and 'special purpose areas':

- General use areas allow for diverse activities that are consistent with the park's conservation objectives as given in the management plan.
- A sanctuary area is essentially a 'no-take' zone with the same purposes as a Marine Nature Reserve except that its boundaries may be amended or cancelled through the management review process and does not need Parliamentary approval. For this reason, it is a more flexible management tool. (To date there is only one Marine Nature Reserve in Western Australia – Hamelin Pool). Sanctuary areas may serve several functions, including undisturbed research, monitoring and reference areas, and observation areas where swimmers may observe fish and other communities undisturbed by fishing.
- Recreation areas may also be observation areas but may include recreational fishing free from commercial fishing activities.
- Special purpose areas allow for specific management objectives ranging from strictly conservation (e.g. protection of whale breeding), to restricted commercial activities (such as anchorages or fish farms). Such purposes could be accommodated in a general use area but specifying them as a special purpose provides certainty to the users.

The 1997 amendments contain no specific provisions for zoning Marine Nature Reserves or Marine Management Areas but Section 62 of the Act allows the Minister to classify marine reserves into zones if s/he sees fit. Alternatively, the requirement may be dealt with by insertion of other reserve categories as enclaves.

MPAs reserved under the *Fish Resources Management Act 1994*

The *Fish Resources Management Act 1994* allows the Minister (for Fisheries) by order published in the *Gazette*, to set aside areas as Fish Habitat Protection Areas (FHPAs) for purposes including 'the conservation and protection of fish, fish breeding areas, fish fossils or the aquatic eco-system' and 'the management of fish and activities relating to the appreciation or observation of fish'. Before doing so, the Minister is to 'approve a draft management plan for management of the area' and publish the draft for public comment. There is no requirement for the Minister to publish a final plan that incorporates changes made following the public comment phase. FHPAs may be no-take areas intended for strict conservation, research or observation purposes or they may be areas where fishing is restricted to recreational activities or specific fisheries.

The Minister may, by order in the *Gazette*, revoke or amend the orders setting the reserve aside. In this regard, the level of statutory protection of FHPAs is significantly less than that provided to Class A MPAs declared under the CALM Act.

The current MPA estate in Western Australia

Number and extent of existing MPAs

As at December 2014, Western Australia's MPA system is not yet adequately representative of the conservation and recreation values of the Western Australian coast, but it is extensive and includes many areas that are very large and include important marine ecosystems.

The dual MPA system in Western Australia, at the time of writing, comprises 16 Class A CALM Act marine reserves and 7 Fish Resources and Management Act FHPAs. The former (Table 7.1, Fig. 7.1) are managed by the Department of Parks and Wildlife (DPAW – successor of the Department of CALM) and the latter (Table 7.2) by the Department of Fisheries.

There is also one reserve (Scott Reef Nature Reserve) established under the Land Act that is managed by DPAW and includes an extensive area of intertidal coral reef habitat, and a MPA surrounding Rottnest Island, established under the *Rottnest Island Authority Act 1987* and managed by the Authority. Neither of these MPAs is included in Table 7.1 or Fig. 7.1.

The EPA Systems reports (EPA 1975, 1976, 1983) recommended that the boundaries of many existing and proposed coastal national parks and nature reserves include important intertidal marine or estuarine habitats and they qualify as MPAs but they are not included in Table 7.1.

Figure 7.1 also shows several MPAs off the Western Australian coast that lie within Commonwealth jurisdiction that have been established under Commonwealth legislation. It is important to note that where they lie adjacent or proximal to state MPAs they are managed together as ecological entities by Commonwealth and state agreement. In 2013, further MPAs were designated in Commonwealth waters off the Western Australian coast (see Chapter 4); these are currently under review and are not shown in Fig. 7.1.

Seven of the original eleven EPA Systems Report recommendations have been implemented, some with extended boundaries (Table 7.1). Of the 42 additional reserves recommended by the MPSWG, seven have been implemented (and one implemented as a FHPA) and 11 are under active consideration. Only one of the 14 recommendations for the south coast has been implemented.

Two large MPAs have been established that were not originally included, in their present form, in either the EPA Systems or MPRSWG recommendations. Both embrace several of the original areas within a larger boundary than was originally conceived in the MPRWSG report.

The first, the Lalang-garram/Camden Sound Marine Park in the Kimberley bioregion is a large area that includes the original recommendations for Prince Frederick Harbour and St George Basin, representing Kimberley gulf and estuarine habitats, and an additional large area of adjacent coastal waters (Fig. 7.1). Inclusion of the intertidal parts of Montgomery Reef, one of the earlier recommendations, has been deferred pending resolution of negotiations with Indigenous people. The impetus for designation of this large MPA was not its representativeness of Kimberley bioregion habitats and biodiversity. It followed the discovery that Camden Sound is a principal breeding area for the Humpback Whale. The selection process was based more on 'ecological importance' than representation. However, the large area includes important habitats representative of the region, such as the extensive fringing and patch coral reefs and some very large gulf and mangrove ecosystems.

The second, the Montebello/Barrow Islands marine reserves (of different categories), includes original recommendations for reservation of several representative ecosystems in

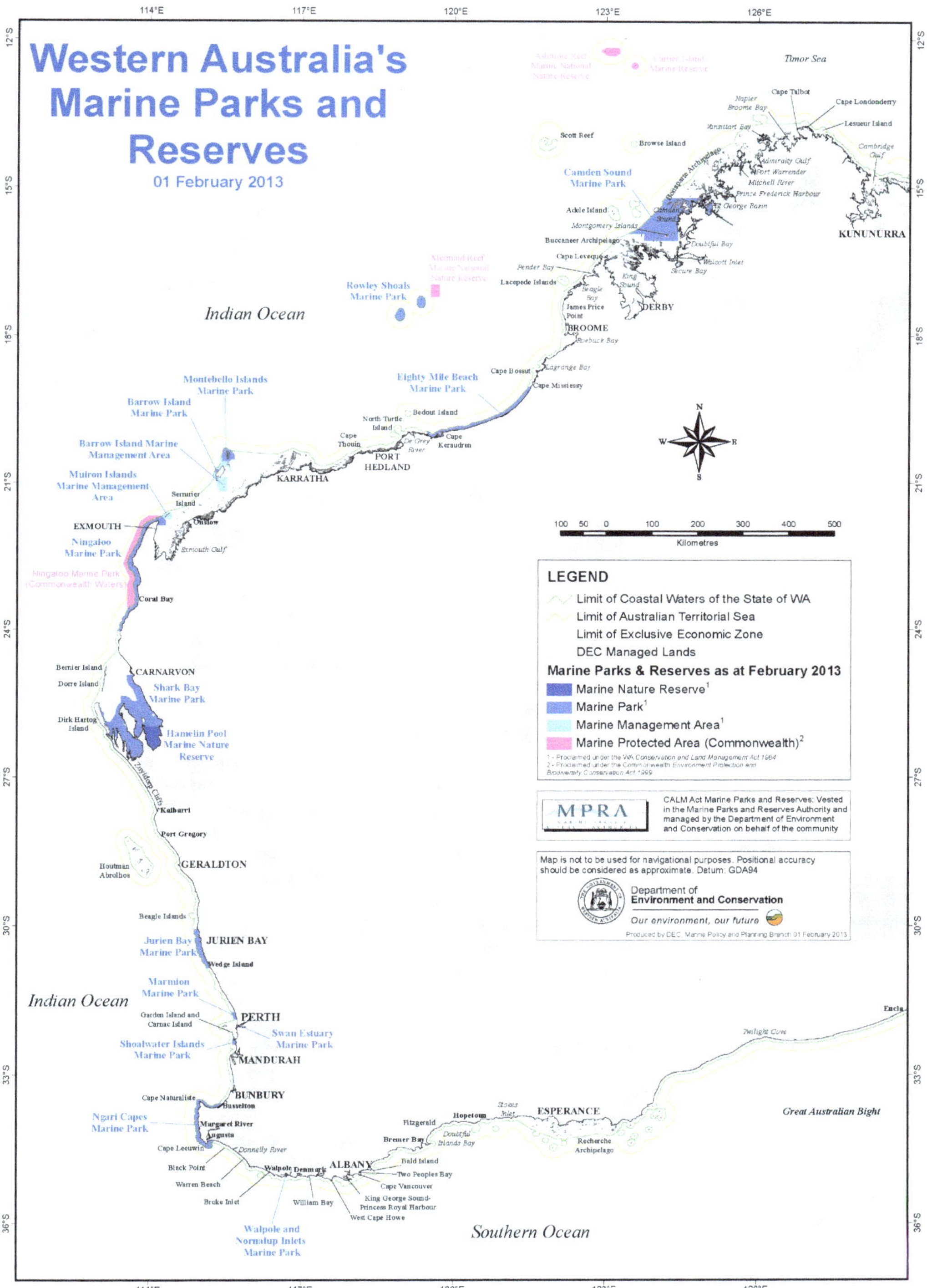

Fig. 7.1. Existing marine protected areas (Marine Parks, Marine Nature Reserves, Marine Management Areas) in Western Australian waters at August 2014. Note proposed marine parks not shown. (Source: Department of Parks and Wildlife)

Table 7.1. Marine protected areas in Western Australian waters vested in the Marine Parks and Reserves Authority or Conservation Commission, indicating implementation progress since initiation of the MPA program

(The term 'no further action' in part B of this table means that development of an original proposal has not proceeded but the area remains on the list of areas worthy of consideration as a MPA. Details of recommendations (column 2) may be found in publications cited in the Reference list *viz* – CALM Kimberley Report (Burbidge *et al.* 1991); EPA System Reports 2, 6, 7, 9 (EPA 1975, 1976, 1978, 1983); MPRSWG (CALM 1994).)

A. Existing MPA areas (at June 2015)					
Candidate areas	**Recommended by**	**Characteristics**	**Current status**	**Management Plan?**	**Zoning?**
KIMBERLEY					
Lalang-garram/Camden Sound Marine Park	WA Government, subsequent to previous recommendations	Inner shelf Humpback Whale breeding area and fringing coral reefs	Gazetted 19 June 2012	Yes	Yes
Prince Frederick Harbour – St George Basin	MPRSWG	Ria rocky coast and embayments with mangals; whale breeding area	Gazetted 19 June 2012	Yes	Yes
Montgomery Reef	CALM Kimberley Report, MPRSWG	Inner shelf platform reef	Part Lalang-garram/Camden Sound Marine Park – proposal in preparation	Yes	Yes
PILBARA and CANNING COASTS and ROWLEY SHOALS					
Rowley Shoals (Clerke Reef, Imperieuse Reef)		Oceanic slope atoll	Gazetted 25 May 1990 with later extensions	Yes	Yes
Pender Bay (West Dampierland)	MPRSWG	Wide embayment with mangals; whale breeding area	No further action		
Roebuck Bay–Lagrange Bay	CALM Kimberley Report, MPRSWG	Open sea embayment with mangals and mudflats; Ramsar-listed	In preparation		
Eighty Mile Beach Marine Park	MPRSWG	Open sea beach with sandy mudflats; Ramsar-listed	Gazetted 29 January 2013	Yes	Yes
Montebello Islands Marine Park	MPRSWG	Outer islands with diverse habitats including fringing and patch coral reefs	Gazetted 10 December 2004	Yes	Yes

Barrow Island Marine Park	MPRSWG	Outer islands with fringing coral reef and turtle nesting	Gazetted 10 April 2004	Yes	Yes
Barrow Island Marine Management Area	MPRSWG	Outer islands with diverse habitats including fringing and patch coral reefs	Gazetted 10 April 2004	Yes	Yes
Muiron Islands Marine Management Area	MPRSWG	Outer rocky and sandy islands with fringing coral reefs	Gazetted 14 January 2005	Yes	Yes
WEST COAST					
Ningaloo Marine Park	EPA System 9, MPRSWG	Fringing-barrier coral reef; recreational activities	Gazetted 3 April 1987 with later extensions	Yes	Yes
Shark Bay Marine Park	EPA System 9	Gulf with shallows and seagrass meadows; recreational activities	Gazetted 30 November 1990	Yes	Yes
Hamelin Pool Marine Nature Reserve	EPA System 9	Saline gulf with stromatolites and coquina	Gazetted 25 May 1990	Yes	No
Jurien Bay Marine Park	MPRSWG	Limestone reefs and lagoons; major rock lobster fisher; recreational activities	Gazetted 26 June 2005	Yes	Yes
Marmion Marine Park	EPA System 6, MPRSWG	Limestone reefs and lagoons; recreational activities	Gazetted 13 March 1987	Yes	Yes
Shoalwater Islands Marine Park	EPA System	Limestone reefs and lagoons; sea bird rookeries; recreational activities	Gazetted 25 May 1990	Yes	Yes
Swan Estuary Marine Park	EPA System	Estuary with migratory bird feeding areas; recreational activities	Gazetted 25 May 1990	Yes	Yes
Ngari Capes Marine Park	MPRSWG	Rocky shore and bay habitat with sea grass meadows; recreational activities	Gazetted 12 May 2012	Yes	Yes
SOUTH COAST					
Walpole and Nornalip Inlets National Park	MPRSWG	Estuary	Gazetted 8 May 2009	Yes	Yes

B. Proposed MPA areas					
Candidate Areas	**Recommended by**	**Characteristics**	**Current status**	**Management Plan?**	**Zoning?**
KIMBERLEY					
North Kimberley Marine Park	New proposal	Varied coastal types including rocky archipelago, coral reefs and mangals	Proposal in preparation		
Cambridge Gulf	MPRSWG	Estuarine gulf with mangals	Part of proposed North Kimberley Marine Park		
Cape Londonderry	MPRSWG	Rocky coast with fringing coral reefs	Part of proposed North Kimberley Marine Park		
Vansittart Bay-Admiralty Gulf	MPRSWG	Ria rocky coast and embayments with mangals	Part of proposed North Kimberley Marine Park		
Lalang-garram/ Horizontal Falls Marine Park	New proposal	Macro-tidal gulfs with mangals and mud-dominated coral reefs	Proposal in preparation		
North Lalang-garram Marine Park	New proposal	Rocky coastline and a number of islands fringed with *wooddooroo* (coral reefs)	Proposal in preparation		
Walcott Inlet and Secure Bay	CALM Kimberley Report, MPRSWG	Ria rocky coast with enclosed estuarine gulfs and mangals	Proposal in preparation – part of Horizontal Falls Marine Park		
Buccaneer Archipelago	CALM Kimberley Report, MPRSWG	Archipelago with diverse habitats including fringing coral reefs	No further action		
Adele Reef	CALM Kimberley Report	Mid shelf platform coral reef	No further action		
Browse Reef	CALM Kimberley Report, MPRSWG	Oceanic platform coral reef	No further action		
PILBARA and CANNING COASTS and ROWLEY SHOALS					
Pender Bay (West Dampierland)	MPRSWG	Wide embayment with mangals; whale breeding area	No further action		

Lacapede Islands	EPA System, CALM Kimberley Report, MPRSWG	Coastal islands with surrounding reef platform; seabird rookeries	No further action		
Roebuck Bay–Lagrange Bay	CALM Kimberley Report, MPRSWG	Open sea embayment with mangals and mudflats; Ramsar-listed	In preparation		
North Turtle and Bedout Islands	MPRSWG	Outer sandy islands with sea bird nesting	No further action		
Cowrie Beach	MPRSWG	Mainland beach with turtle nesting	No further action		
Dampier Archipelago Marine Park	MPRSWG	Archipelago with diverse habitats incl. Fringing coral reefs and mangals	Proposal prepared – pending	Indicative	Indicative
Cape Preston Marine Management Area	MPRSWG	Rocky headland with mangals and nearshore islands	Proposal prepared – pending	Indicative	Indicative
Robe (Fortesque–Ashburton deltas)	MPRSWG	Mangal delta habitats and nearshore islands – boundaries not suggested	No further action		
Serrurier Island	MPRSWG	Outer sandy island, fringing coral reef	No further action		
Exmouth Gulf (southern and eastern shores)	MPRSWG	Mangal habitats	No further action		
WEST COAST					
Red Bluff to Point Quobba	MPRSWG	Rocky shore	No further action		
Kalbarri	MPRSWG	Rocky shore	No further action		
Port Gregory	MPRSWG	Mainland lagoon with coral communities	No further action		
Beagle Islands	MPRSWG	Outer islands with limestone reefs and coral communities	No further action		
Peel-Harvey Inlet	MPRSWG	Estuary; recreational activities	No further action		
Leschenault Inlet	MPRSWG	Estuary; recreational activities	No further action		

Candidate Areas	Recommended by	Characteristics	Current status	Management Plan?	Zoning?
SOUTH COAST					
Hardy Inlet	MPRSWG	Estuary			
Black Point	MPRSWG	Open rock coast	No further action		
Warren Beach	MPRSWG	Open beach coast	No further action		
Broke Inlet	MPRSWG	Estuary	No further action		
Donelly and Gardner Rivers	MPRSWG	Estuary	No further action		
William Bay	MPRSWG	Open rocky coast	No further action		
West Cape Howe	MPRSWG	Open rocky coast	No further action		
King George Sound – Princess Royal Harbour	MPRSWG	Embayments	No further action		
Cape Vacouver	MPRSWG	Open rocky coast	No further action		
Fitzgerald Biosphere Reserve	MPRSWG	Open rocky coast	No further action		
Stokes Inlet	MPRSWG	Estuary	No further action		
Recherche Archipelago	MPRSWG	Open rocky coast	No further action		
Twilight Cove	MPRSWG	Open coast	No further action		

Table 7.2. Existing Fish Habitat Protection Areas in Western Australian waters, vested in the Minister for Fisheries at August 2014

Existing areas	Characteristics	Gazettal	Management plan?	Zoning?
Abrolhos Islands Fish Habitat Protection Area	Shelf-edge coral banks and islands in a high latitude location, large area with very high conservation and scientific values; centre of a major rock lobster and fin fish industry; undeveloped tourist potential	16 February 1999	Yes	Yes – four Reef Observation Areas set aside for conservation and observation of marine life and habitats
Cottesloe Reef Fish Habitat Protection Area	Small mainland area set aside for recreation but not commercial fishing	21 September 2001	Yes	No
Lancelin Island Lagoon Fish Habitat Protection Area	Small mainland area set aside for recreation (swimming and snorkelling); fishing prohibited	20 March 2001	Yes	No
Kalbarri Blue Holes Fish Habitat Protection Area	Small mainland area set aside for recreation (swimming and snorkelling); fishing prohibited	9 November 2007	Yes	No
Point Quobba Fish Habitat Protection Area	Small mainland area with significant coral communities, set aside for recreation including restricted fishing (rock oysters and squid)	26 October 2004	Yes	No
Miaboolya Beach Fish Habitat Protection Area	Small mainland area set aside for restricted recreation but excluding net fishing	27 June 2003	Yes	No

the West Pilbara – the coral reefs and lagoons of the Montebellos, the fringing reefs of the Barrow west coast, and the patch reefs and shallow platform habitats of the Lowendal Islands and the Barrow Shoals. These areas lie within a major precinct of the offshore oil and gas industry, presently undergoing intensive development. Industry itself, through the agency of the Australian Petroleum Production and Exploration Association, set up a study of how the proposed marine reserves might proceed and be sustainably managed within an industrial area like this, recognising the legitimacy of both high conservation values and high economic values (APPEA 1997). The key to success of this approach would be a management plan that identified the conservation values and the proposed industrial activities and formalised their co-existence, subject to agreed objectives and constraints, with appropriate management and monitoring.

An outcome of the discussion of the Montebello/Barrow complex was the realisation within the management agency that the existing legislation did not provide for multiple-use on this wide scope. Legislation for a new reserve category called Marine Management Area followed in 1997. Subsequently, two areas, the Montebellos and Biggada Reef on the west coast of Barrow, were reserved as Marine Park and a large surrounding area was reserved as the Barrow Island Marine Management Area (Fig. 7.1). The port area on the central part of the east side of Barrow and around the Lowendal Islands, where most of the industrial infrastructure is located, was not included. This Montebello/Barrow complex of MPAs is now in place and development is subject to environmental protection prescriptions, such as water quality, that are contained within its management plan.

MPAs 'in the pipe-line'

Proposals for additional marine reserves are currently being prepared for presentation to the Western Australian Parliament for consideration.

Roebuck Bay Marine Park

A proposal for the reservation of Roebuck Bay Marine Park in the southern Kimberley is currently being prepared. This was a recommendation of the MPRSWG report that stressed the importance of the intertidal mud flats of the bay as a Ramsar wetland and feeding area for migratory birds. An indicative management plan has been released for public comment that proposes that the reserve should be called the Yawuru Nagulagun / Roebuck Bay Marine Park.

Dampier marine protected areas

In the Pilbara, proposals for reservation of MPAs in the Dampier Archipelago are well advanced. A draft management plan was published for public comment in 2005. The proposals encountered heavy opposition from sections of the local community and industry. Boundary issues with the Dampier Port Authority were resolved. The amended proposals have been deferred but are likely to be taken up again in the near future.

Horizontal Falls Marine Park

This new proposal, arising subsequent to the MPRSWG report, lies in the southern part of the Kimberley bioregion immediately south and adjacent to the Lalang-garram/Camden Sound Marine Park and on the eastern side of the existing Port of Yampi Sound. It will include the waters of Collier Bay, Walcott Inlet, Secure Bay and Talbot Bay and embrace the scenically spectacular coastline of the Yampi Peninsula with its macro-tidal gulfs and tidal waterfalls and unique mud-dominated coral reefs.

North Kimberley Marine Park

This very large area includes the northern part of the Kimberley bioregion with its newly discovered extensive, species-rich coral reefs in the Bonaparte Archipelago (Fig. 7.2), and the coastal waters of Joseph Bonaparte Gulf bioregion (excluding the port area of Cambridge Gulf).

The scenic coast of the Kimberley bioregion, with its extremely rich natural heritage and cultural heritage of rock art and sacred sites, has recently been placed on the National Heritage List. It has become a region with a high tourism profile. Recognition of this, and a clear public expectation that its wilderness and heritage values will be protected, has led the Western Australian Government to develop a concept of a 'Great Kimberley Marine Park' as a series of multiple-use MPAs stretching from the Yampi Peninsula to the Northern Territory border. At present, it is not proposed that this should be a declared marine

Fig. 7.2. The fringing reef on the north-west shore of South Maret Island, Bonaparte Archipelago, Kimberley bioregion, exposed at low spring tide. This extremely species-rich intertidal reef is typical of those of turbid coastal waters of the north Kimberley. The reef front is dominated by robust domal corals, mainly faviids. It is part of the proposed North Kimberley Marine Park. (Photo: Barry Wilson)

park in its own right; it is seen as a network of contiguous marine parks that are established separately. The 'great' park would include the newly established Lalang-garram/Camden Sound Marine Park and the proposed Horizontal Falls and North Kimberley Marine Parks. Much of this area is covered by Indigenous Land Rights and these reserve proposals are being discussed as joint management enterprises. When in place, the network of Kimberley marine parks will be one of the largest and most spectacular and heritage rich of all the Australian MPAs.

The fact that most coastal land and waters of the Kimberley are now subject to Indigenous Land Rights determinations adds a dimension not experienced previously. There will be extensive use of Indigenous Land Use Agreements (ILUAs) and joint management agreements and the process will embrace a new concept that deals with Aboriginal Customary Activities (DPAW Policy Number 86 September 2013). The coastal lands adjoining existing and proposed Kimberley MPAs are now mostly Indigenous Protected Areas (IPAs).

Gaps in the system

When the reserve proposals presently being processed are implemented the Bonaparte Gulf, Kimberley and Canning bioregions will be well represented by MPAs (see also Chapter 17). Of the original EPA Systems and MPRSWG recommendations for the Kimberley, the most conspicuous omissions will be the Buccaneer Archipelago and the mid-shelf platform reefs of the Adele complex that represent types of coral reef that are not included elsewhere.

In the Pilbara, the coral reef ecosystems around the nearshore and offshore islands will be adequately represented but consideration still needs to be given to mainland coastal habitats, especially mangroves and estuaries. In particular, the habitats and biota of the highly productive and biogeomorphically distinctive deltas of the Fortesque and Ashburton Rivers are presently conspicuous gaps in the representative MPA system (Fig. 7.3).

All the major recommendations for the West Coast have been implemented, with Ningaloo and Shark Bay Marine Parks two of the largest nearshore marine reserves in the country, both World Heritage Sites.

The South Coast, between Augusta and the South Australian border is the area that remains deficient in MPAs. The only one of the MPRSWG recommendations (Table 7.1) to have been implemented is the small area of waters enclosed by the Walpole and Nornalup Inlets National Park, representing the estuarine habitats of the western South Coast high rainfall zone.

Two of the MPRSWG recommendations for the South Coast – Fitzgerald and Recherche Archipelago – are large areas with very high conservation, scientific and recreational values. When the original MPRSWG report was presented to the South Coast communities it was poorly received, largely because of misunderstanding about the purpose and management of the CALM Act marine reserve categories. In response to this,

Fig. 7.3. The Ashburton Delta, West Pilbara. Like most estuaries of the arid West Pilbara, the sharply zoned mangrove forest is backed by wide supra-littoral mud flats. Only seven mangrove species are present with *Avicennia marina* (light green) and *Rhizophora stylosa* (dark green) dominant but there is a diverse invertebrate fauna. Successive Holocene dunal ridges between the sand bar and the Pleistocene hinterland indicate phases of inundation and delta development. This site is a recommended MPA. (Photo: Barry Wilson)

the Western Australian Government undertook an integrated regional planning study that considered all current and potential uses of coastal resources of the region (South Coast Regional Marine Planning Working Group 2010). It made a wide range of coastal planning recommendations for the South Coast including that, in this broader context, the MPA system proposed by the MPRSWG should be reviewed. This has not yet been done, but it is understood that the matter will be reviewed when the Kimberley MPA program has been completed.

Current policy

The initial policy of the Western Australian MPA program was to establish a conservation reserve system representative of major communities of natural wildlife and scenery types of the state's coastal waters (EPA 1975, 1976, 1978, 1983; Government of Western Australia 1994). However, because of the high level of connectivity between marine communities, reservation and management of representative samples, by themselves, cannot adequately protect and conserve marine biodiversity and habitats. To achieve their purpose, marine conservation reserves must be set in a regional context of ecological management. This realisation led to adoption of two concepts:

- multiple-use marine reserves within the MPA system
- the MPA program becoming embedded in a broader policy of 'integrated coastal management'.

The Western Australian MPA program has evolved to embrace the concept of multiple-use Marine Parks. Conservation objectives may remain primary but other activities are permitted subject to defined management processes, embedded within approved management plans. Introduction of the reserve category Marine Management Area developed this concept, enabling defined conservation objectives and industry objectives to be managed together to the benefit of both. A more recent policy development seeks to take this a further step by encouraging activities that are positive for people's health and wellbeing while maintaining conservation values.

When the MPA program stalled on the south coast, the potential for an integrated coastal management approach was examined and a recommendation that the South Coast MPA program should be reconsidered in that context. The principle of integrated coastal management has been widely accepted and it is likely that the existing MPA systems in the Canning-Pilbara and West Coast regions may be 'retro-fitted', within this broad regional planning policy context.

An outcome of community consultations relating to the new Kimberley proposals has been adoption of policies that facilitate Aboriginal involvement in planning and management of conservation reserves, including MPAs.

Challenges

1. Identification using best available information

The first challenge in establishing a MPA system is the identification of suitable candidate areas. Many attempts to introduce scientific rigour into the process assume that biodiversity conservation is the sole or main purpose of reservation and neglect to consider other statutory functions of diverse zone categories. In Western Australia, selection criteria relate directly to legislated definitions of the state's reserve categories.

The Western Australian Marine Parks and Reserves Selection Working Group began its task with the objective to establish a pattern of representative reserves. The concept of 'major distinctive coastal type' was adopted and proved to be a useful first-cut tool in selecting representative areas. However, while analysis of biological and geological data relating to the IUCN categories is where the selection process begins, sociological, economic and political issues are brought in and the process is never entirely scientific.

2. Establishment with public input

It is a well-known dictum that conservation parks cannot succeed without community support and the expansive provisions for consultation in the Western Australian legislation are intended to achieve that. But the challenges at the public consultation stage of MPA establishment become political, with little to do with science or design principles. Whatever the technical merits of a proposal might be, it will fail at this point if there is significant community objection to it.

The challenge is to present a technically sound proposal supporting the area's worthiness as a candidate for reservation for the intended purposes, and to anticipate and respond to likely local objections to it. The MPRSWG plan for MPAs on the south coast failed because it was premature on both counts and the whole process must begin again. The Dampier Archipelago marine reserves proposal in 2005 also failed to proceed beyond the initial community consultation stage because of objections from several stakeholder groups. Since then, significant boundary changes have been negotiated but other issues remain. Notably, the no-take principle and the effectiveness of sanctuary zones as a management tool remains an unresolved state-wide political issue. Nevertheless, reservation of significant MPAs in the Dampier area may be regarded as postponed, not rejected, and an amended proposal is expected to proceed in the near future.

The Department did a better job with the Ningaloo Marine Park public consultation process. When Richard May and I first presented the idea of a Ningaloo Marine Park to a community meeting in Exmouth in 1986, we received a very hostile reception. Yet two decades later the community presented a petition to the Western Australian Government complaining about a decision to allow an offshore oil well to be drilled in the vicinity of 'their' marine park. Submissions to the release of the *Ningaloo Marine Park Draft Management Plan* (CALM 2004) were four to one in favour (see also Chapter 21). It takes time to achieve public acceptance at this level but that is where you have to be and you have to work hard to get there.

3. Sustained funding for management

As it is everywhere, the most difficult challenge confronting the Western Australian marine reserves program is restricted funding for ongoing management. In dealing with this problem, it is vitally important that management objectives are identified, prioritised, implemented and regularly assessed so that expenditure can be defended. In 2004, CALM and the MPRA introduced an innovative Performance Assessment Report for Marine Parks to tackle this problem as a basis for the 10-year reviews of reserve management plans. The reports comprise an assemblage of 'report cards' for each of the management objectives of each reserve with 'input measures' and 'outcomes'.

4. Limited knowledge base

Periodic review of management outcomes is always challenged by a limited knowledge base upon which decisions may be made. Developing the knowledge base through ongoing experience complemented by research and monitoring is an essential component of MPA management. However, this process needs close direction towards specific management objectives that are defined in individual reserve management plans.

The original Management Plan and zoning scheme for Ningaloo Marine Park was produced in 1989 when very little environmental information was available. The park zoning scheme contained a series of Sanctuary Areas, selected somewhat subjectively to represent reef and lagoon habitat types and areas used for coral viewing. It assumed, on ecological principles, that a linear series of Sanctuary Zones along the length of the reef would provide the best probability of effective connectivity between fully protected areas but site-specific information base was so poor that the credibility of that original management plan was in question.

Without the benefit of a research and monitoring program, the knowledge base available for the 2004 management plan review was little better and the 2004 draft plan was heavily criticised by its opponents for its inadequate science base. Approval of the plan review was controversial and part of the government's political response was to announce allocation of a significant sum of money to support a 3-year, multidisciplinary research plan that would fill some of the knowledge gaps. This study has produced vital information and the next management plan review, due in 2015, will be far better informed.

The matter of most community concern during the 2004 debate on the Draft Ningaloo Management Plan Review related to the subject of no-take zones. The purposes of these areas as education and scientific research zones where the fauna remains undisturbed by human activities tended to be overlooked and the debate focused on whether they were effective in protecting fish stocks. (Two Western Australian views of the no-take argument may be found in Woodley *et al.* 2010 and Penn and Fletcher 2010.) Research and monitoring on ecological issues such as this, in a context of ongoing climate change and human disturbance, is a challenge to marine reserve management.

Conclusions

The Western Australian MPA program comprises a series of separate reserves, large and small, set within a matrix of state territorial waters. The notion that selected marine areas may be permanently reserved for specific purposes remains controversial. Nevertheless, it is gathering support as evidenced by acceptance of the concept of the Great Kimberley Marine Park, which proposes that the whole coast is to be reserved and zoned for multiple-uses, integrated with overall regional coastal planning.

The effect of reservation within an integrated management context is that an obligation is introduced for planning and management by diverse agencies, with approval processes and procedures that have statutory authority. This is increasingly seen as a better option for coastal management than the current situation where agencies independently apply their own objectives and planning procedures, with or without community discussion, and often conflicting with each other.

The key to achieving effective incorporation of MPAs into regional integrated marine environmental management lies in collaborative preparation of management plans that lay out specific objectives, and a sense of common purpose among the agencies responsible for their implementation and ongoing management. Though difficult to achieve administratively, management plans should have single budgets and performance assessments that are performed jointly.

The Western Australian MPA system is well advanced and has evolved and matured into a concept where conservation of marine biodiversity and habitats has become accepted as an essential objective and MPAs, with diverse purposes, have become an important element in marine environmental management.

One important lesson learned is that marine reserves for protection of biodiversity cannot be managed on a stand-alone basis. They may be small single-purpose no-take areas surrounded by areas managed for other purposes, or large multiple-use reserves (such as Shark Bay Marine Park) where biodiversity management is one of several objectives. In either case, several management agencies are likely to be involved and inter-agency collaboration is essential. Some people find inter-agency administrative systems inhibiting, even threatening. But they should be enabling, achieving better biodiversity and educational outcomes on a wider front. Devising inter-agency administrative systems that put integrated, multiple-use management into effect, while preserving biodiversity conservation objectives, is a challenge for marine reserve management authorities.

Acknowledgements

The author thanks Roxane Shadbolt, Mark Sheridan, Liesl Ludgerus, Michael Higgins and Ian Herford of the WA Department of Parks and Wildlife for updates of current circumstances and assistance with the tables and figure, and Eve Bunbury and Nikki Sarginson of the Department of Fisheries for information on current Fish Habitat Protection Areas.

Biography

Barry Wilson is an invertebrate zoologist and biogeographer, specialising in the biology and distribution of marine molluscs and coral reefs. He graduated from the University of Western Australia followed by a stint (1964–65) as Fulbright Scholar and Research Fellow at the School of Evolutionary Biology, Harvard University, USA. He has held positions as Curator of Molluscs and Head of the Natural Science Division at the Western Australian Museum, Director of the Museum of Victoria, WA Director of Nature Conservation, Managing Director of the Australian Wildlife Conservancy, and Chairman of the WA Marine Parks and Reserves Authority. He has published seven books on Australian marine fauna and biogeography. He is currently a Research Associate with the Western Australian Museum.

References

AAS (1965) *National Parks and Nature Reserves in Western Australia.* Report of the Western Australian Sub-Committee of the Australian Academy of Science Committee on National Parks, Canberra.

APPEA (1997) *Potential Arrangements for Multiple Use Management in the Montebello Island/Barrow Islands Region: A Petroleum Industry Perspective.* Australian Petroleum Production and Exploration Association Ltd., Perth.

Burbidge AA, McKenzie NL, Kenneally KF (1991) *Nature Reserves in the Kimberley, Western Australia.* Department of Conservation and Land Management, Perth.

CALM (1994) *A Representative Marine Reserve System for Western Australia.* Report of the Marine Parks and Reserves Selection Working Group. Department of Conservation and Land Management, Perth.

CALM (2004) *Ningaloo Marine Park Draft Management Plan: and Indicative Management Plans for the Extension to the Existing Park and the Muiron Islands Marine Management Area.* Department of Conservation and Land Management, Marine Parks and Reserves Authority, Perth.

CTRC (Conservation Through Reserves Committee) (1974) *Conservation Reserves in Western Australia.* Report to the Environmental Protection Authority, Perth.

EPA (1975) *Conservation Reserves for Western Australia. Systems 4,8,9,10,11,12.* Environmental Protection Authority, Perth.

EPA (1976) *Conservation Reserves for Western Australia. Systems 1,2,3,5.* Environmental Protection Authority, Perth.

EPA (1978) *Conservation Reserves for Western Australia. System 7.* Environmental Protection Authority, Perth.

EPA (1983) *Conservation Reserves for Western Australia. The Darling System – System 6.* Environmental Protection Authority, Perth.

Government of Western Australia (1994) *New Horizons: The Way Ahead in Marine Conservation and Management.* Department of Conservation and Land Management, Perth.

Penn JW, Fletcher WJ (2010) *The efficacy of sanctuary areas for the management of fish stocks and biodiversity in WA waters.* Fisheries Research Report No. 169. Department of Fisheries, Perth.

South Coast Regional Marine Planning Working Group (2010) *Oceans of Opportunity: Proposed Strategic Framework for Marine Waters of Western Australia's South Coast.* Report to the Minister for the Environment, Western Australia.

Wilson B (2013) *The Biogeography of the Australian North West Shelf.* Elsevier, Burlington, MA, USA.

Woodley S, Loneragan N, Babcock R (2010) 'Report on the scientific basis for and the role of Marine Sanctuaries in marine planning'. Prepared for the Inter Departmental Committee for the Management of the State's Marine Protected Areas, Perth.

Chapter 8

South Australia's experience: establishing a network of nineteen marine parks

Chris Thomas and Vera Hughes

Introduction

South Australia is at the centre of one of the world's largest marine ecosystems. The state's jurisdiction includes 6000 km of coastline and 60 000 km^2 of coastal waters, ranging from two large gulfs with some relatively sheltered shallow waters, to high-energy open coastlines with deep waters.

Southern Australia is also home to ecologically significant marine species, 85% of which are found nowhere else on Earth (Edyvane 1999). The significant species include Southern Right Whales (*Eubalaena australis*), Australian Sea Lions (*Neophoca cinerea*), Giant Australian Cuttlefish (*Sepia apama*) (Fig. 8.1), Leafy Seadragons (*Phycodurus eques*) and Great White Sharks (*Carcharodon carcharias*). Effective planning and management are crucial for the protection and conservation of our marine species and the waters they inhabit (IUCN-WCPA 2008).

Historically, South Australia has had several marine and coastal areas protected by aquatic reserves and Rock Lobster sanctuaries under the *Fisheries Act 1982* (now superseded by the *Fisheries Management Act 2007*), along with the Great Australian Bight National Park and areas incorporating islands and estuaries in conservation parks declared under the *National Parks and Wildlife Act 1972*. These areas were established to meet particular needs, for example to support sustainable fishing or to protect species habitat in a specific location, and were not implemented as components of a comprehensive plan to protect marine biodiversity and ecosystems.

As a result, although these areas provide some important conservation benefits, they did not in themselves deliver a robust network of marine protected areas that complies with international best practice or national commitments to develop a comprehensive, adequate and representative system. Many significant areas for conservation were under-represented, or not represented at all, in these earlier forms of protected areas.

In recognition of the unique nature of South Australia's marine environment and mounting evidence of change caused through human activity, the South Australian Government committed to conserve marine and estuarine biodiversity through the establishment of a system of marine parks to include comprehensive, adequate and representative

Fig. 8.1. Giant Australian Cuttlefish, Upper Spencer Gulf Marine Park. (Photo: David Pearce)

examples of marine biodiversity selected from the entire state's jurisdiction, while also providing for the ecologically sustainable use of marine resources.

The establishment of South Australia's network of 19 marine parks is one of the most influential conservation initiatives ever undertaken in the state because it directly affects existing and future uses of the marine environment, such as commercial and recreational fishing, shipping, mining, and commercial developments, and indirectly affects most South Australians who rely on the marine environment for social and economic benefits. Consequently, a comprehensive community engagement process supported the development of the network to help ensure the views of all interested community members were considered (Weatherill and Caica 2012).

Adjustment at this scale does not come easily. This work required the collaboration of interested South Australians, key stakeholders and the whole of government. The establishment of these marine parks zoned for multiple-uses was a major social change process, which was demanding and challenging for everyone involved.

History of South Australia's marine parks network

Before the systematic development of the South Australian marine parks network, the *National Parks and Wildlife Act 1972* and the former *Fisheries Act 1982* were used to establish various forms of protected areas over marine waters for purposes relating to these Acts. However, it was recognised that purpose specific legislation would be better suited to achieve the desired state wide conservation outcomes.

Through the *Blueprint for the South Australian Representative System of Marine Protected Areas*, which was released in November 2004, the South Australian Government committed to the development of purpose-specific legislation to establish a comprehensive, adequate and representative network of marine parks spread across state waters to protect marine biodiversity and ecosystems (DEH 2004a).

Marine parks were a component of a broad marine conservation strategy endorsed by the government in 2004 in the Living Coast Strategy (DEH 2004b). Around this time, *Towards a System of Ecologically Representative Marine Protected Areas in South Australian Marine Bioregions* (Baker 2004) was released to support development of an ecologically representative system of marine protected areas in South Australia. This report explained the criteria for recommending 19 broad locations that could feature in this system. The significance of this commitment was recognised in *South Australia's Strategic Plan 2004*, which included the target to create 19 marine parks by 2010 (Government of South Australia 2004).

In 2005, a trial marine park process was undertaken in the area between the Fleurieu Peninsula and Kangaroo Island, now known as the Encounter Marine Park (Fig. 8.2). This trial uncovered a wealth of information and learnings that later informed the development of the South Australian *Marine Parks Bill 2006*. Subsequently, the *Marine Parks Act 2007* was approved by the Parliament in late 2007 and began operation in 2008.

The *Marine Parks Act 2007* prescribes a staged process for establishing marine parks and separates the proclamation of park outer boundaries from the development of park management plans and zoning and also from the development of regulations to prescribe the activities prohibited in each type of zone (see also Chapter 14).

The first stage of this process was completed when the outer boundaries of a network of 19 marine parks across South Australian state waters, based on the locations identified in

Fig. 8.2. Saltmarsh at Kangaroo Island, Encounter Marine Park. (Photo: Kris Carlyon)

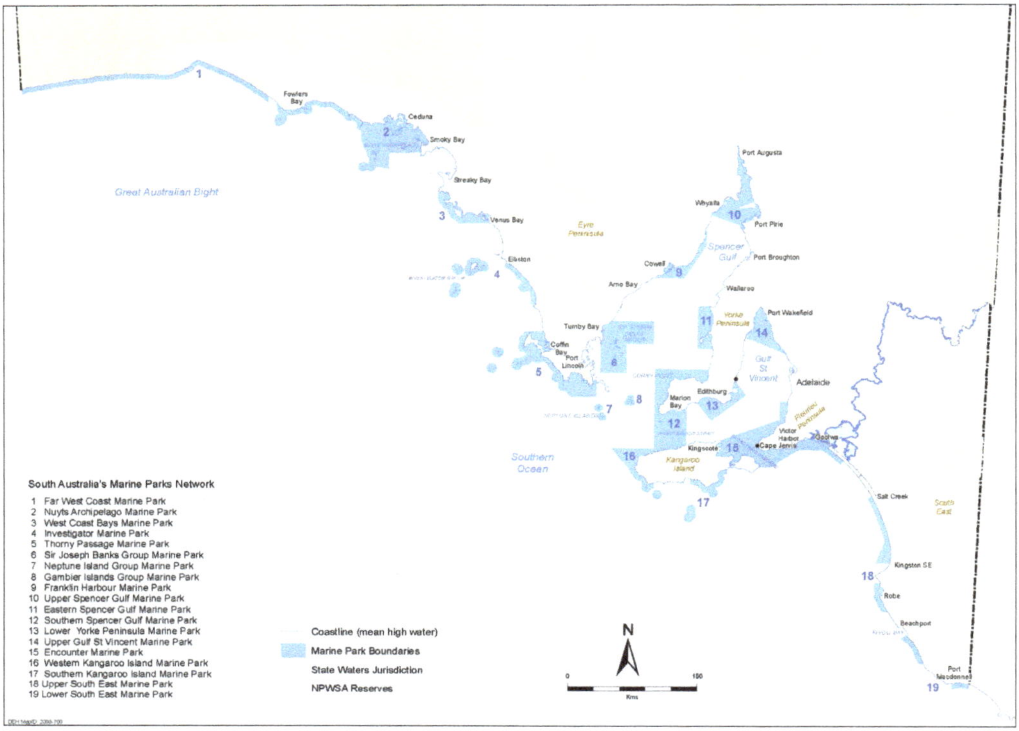

Fig. 8.3. South Australian marine park boundaries.

Baker (2004), were proclaimed in January 2009. The outer boundaries were available for 8 weeks of public consultation, which included 57 information days held across the state to provide members of the community the opportunity to seek information and express their views about the locations of the outer boundaries. Around 2360 written submissions were submitted and, in response, the boundaries of seven parks were amended in July 2009 (Fig. 8.3).

The next step was to develop draft management plans and zoning for each of the 19 marine parks. During 2010 and 2011, extensive work was undertaken to ensure advice on possible zone locations was received directly from local communities, key stakeholders and relevant government agencies.

To facilitate local community involvement, the Department of Environment, Water and Natural Resources (DEWNR) established 14 Local Advisory Groups across the state, providing the opportunity for around 180 volunteer members to develop their own proposals for zoning within each of the 19 marine parks. Each group held up to five meetings, which were open to the public. Many additional people attended and provided contributions.

To help guide this work, 14 design principles (DEH 2008) and a set of government policy commitments (Government of South Australia 2009) were released. The commitments, in particular, provided existing and future users, such as fishing, mining, aquaculture, shipping and developers, greater certainty and clarity about how their activities might be affected across the marine parks network.

The South Australian Government did not adopt an arbitrary percentage target for the extent of no-take sanctuary zones. Instead, it remained open to the advice of the community as to how to best design the marine parks network to protect and conserve marine biodiversity and marine habitats in a manner that accords with the objects of the *Marine Parks Act 2007*.

To assist community zoning discussions, some indicative zoning guidelines were provided. These guidelines were developed with due regard to advice from the Marine Parks Scientific Working Group and Marine Parks Council of South Australia, which provided independent advice on marine park matters. One of these zoning guidelines is that sanctuary zones should cover ~20–25% of each marine park, which is ~10% of state's waters.

Many discussions were held with these sectors to seek advice about zoning proposals. It is estimated that, during this time, Department officers had over 25 000 face to face conversations with community members.

A government steering committee held monthly meetings to provide input as proposals were developed. This committee was chaired by DEWNR and included representatives from Primary Industries and Regions SA, Department of Planning, Transport and Infrastructure, the Environment Protection Authority, the South Australian Tourism Commission, Department of Manufacturing, Innovation, Trade, Resources and Energy (now Department of State Development), Department of Treasury and Finance, and Department of the Premier and Cabinet. Many additional meetings were held with agencies to consider possible impacts on their client groups. This work helped expedite government approval processes and ensured potential conflicts were identified and resolved early in the process.

The zoning proposals from the Local Advisory Groups were taken to the broader community through state-wide information days and major events, such as the Adelaide Boat Show. A Key Stakeholder Forum was also convened in early 2012 to refine the zoning proposals further. This Forum included commercial and recreational fishers, conservationists and scientists who considered the network as a whole and provided advice about the zoning proposed by the local groups. This advice identified priority areas for locating sanctuary zones, which helped consolidate the zones proposed by the Local Advisory Groups and also helped support the achievement of broad ecological, social and economic outcomes.

These priority areas were subsequently posted on the South Australian Government website for public preview and comment with full draft management plans including draft zoning released for 8 weeks of public review in August 2012. The marine park management plans and zoning were finalised in November 2012, following over 8000 public submissions, 50 public information days and 50 amendments to the draft zoning (DEWNR 2012). The final zoning across the 19 marine parks contains 83 highly protected zones, which in total take up 6% of South Australian waters. Some 80% of these zones were derived from Local Advisory Group advice (Fig. 8.4).

This resulted in zoning that formed the basis of a comprehensive, adequate and representative network of marine parks, while still recognising the current and future socio-economic requirements of local communities and others who use the parks.

The current MPA estate in South Australia

South Australia has 19 marine parks proclaimed under the *Marine Parks Act 2007*, which collectively cover 26 937 km^2 or around 44% of South Australian waters (Fig. 8.3).

Marine parks are divided into different zones, which provide various levels of protection to habitats and species (Table 8.1). The types of zones are:

- **restricted access zone (RAZ)** – primarily established so that an area may be managed by limiting access to the area

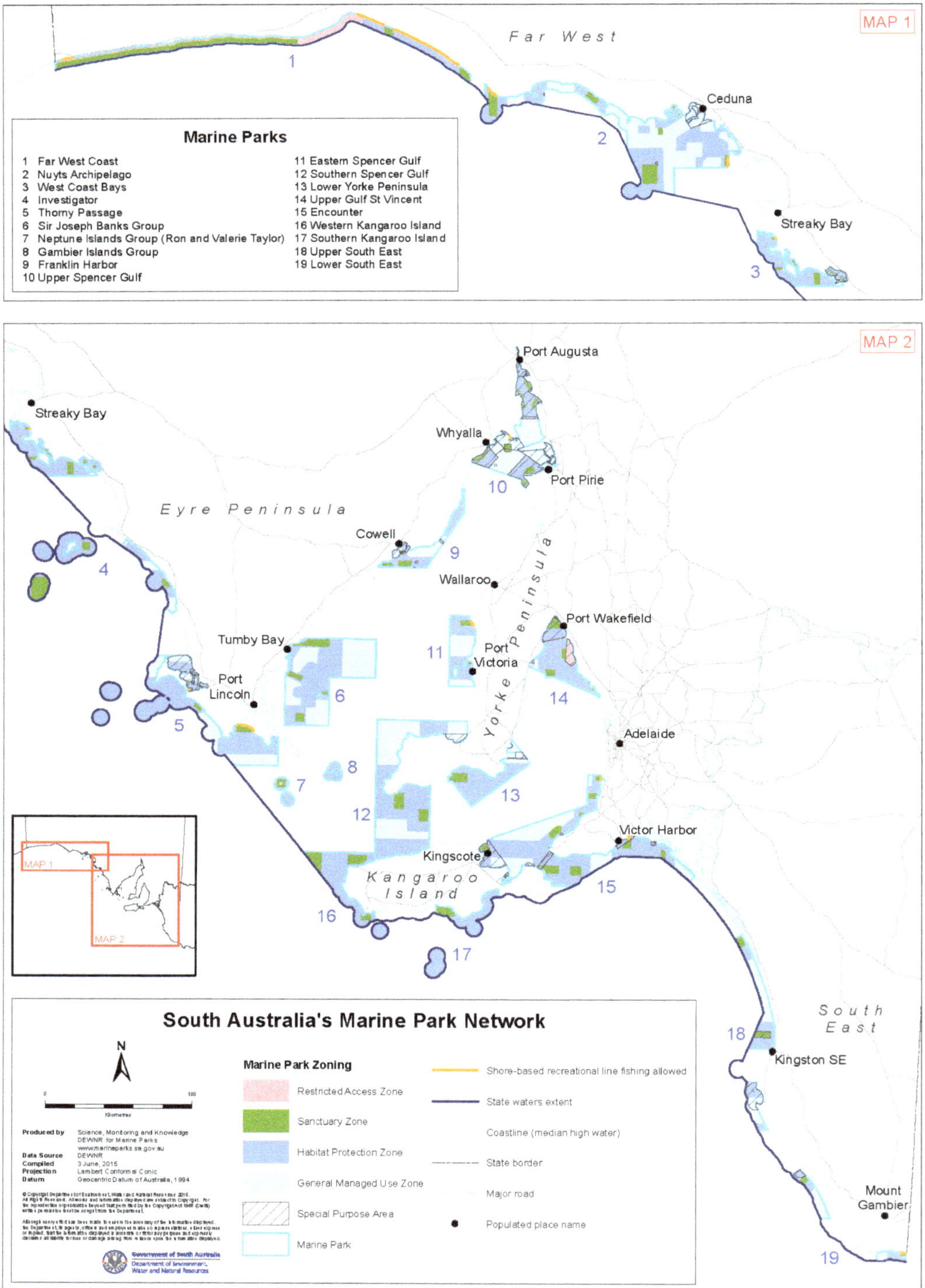

Fig. 8.4. Zones within South Australian marine parks.

Table 8.1. Area of zones within South Australian marine parks and their IUCN protected area management category (see Day *et al.* 2012)

Zone	IUCN category	Area (km^2)	Proportion of state waters
Restricted access zone	Ia	693	1%
Sanctuary zone	II	3014	5%
Habitat protection zone	IV	15 029	25%
General managed use zone	VI	8201	13%
Total		**26 937**	**44%**

The restrictions on activities in the various zones are set out in the *Marine Parks (Zoning) Regulations 2012*, finalised at the same time the 19 management plans were authorised. A summary of these arrangements is provided in Table 8.2.

- **sanctuary zone (SZ)** – primarily established so that an area may be managed to provide protection and conservation for habitats and biodiversity within a marine park, especially by prohibiting the removal or harm of plants, animals or marine products
- **habitat protection zone (HPZ)** – primarily established so that an area may be managed to provide protection for habitats and biodiversity within a marine park, while allowing activities and uses that do not harm habitats or the functioning of ecosystems
- **general managed use zone (GMUZ)** – primarily established so that an area may be managed to provide protection for habitats and biodiversity within a marine park, while allowing ecologically sustainable development and use.

Special purpose areas have also been established in some specific locations to allow activities that would not otherwise be permitted. These special purpose areas overlay the zoning and provide for activities such as shore-based recreational fishing, transhipment and anchoring of large vessels, harbour maintenance, cables/pipelines, Department of Defence activities and aquaculture. There is also a special purpose area over the whole of the Upper Spencer Gulf Marine Park to provide for some developments determined to be

Table 8.2. Aims of zones as established by the *Marine Parks Act 2007* and the *Marine Parks (Zoning) Regulations 2012*

Zone	Marine Parks Act requirements	Examples of restricted activities
Restricted access zone	Manage by restricting access	All entry prohibited. Permits may be issued to allow entry, expected primarily for research.
Sanctuary zone	Prohibits removal or harm of plants, animals or marine products to protect and conserve biodiversity and habitats	Fishing, aquaculture, establishing structures of any kind, discharge of waste waters are all prohibited.
Habitat protection zone	Manages harm to habitats and the functioning of ecosystems by prohibiting activities that may harm habitats and ecosystem functions	Dredging (except for port operators), trawling, anchoring of vessels over 80 m, removal or harm to habitat forming plants, algae or sessile animals are all prohibited.
General managed use zone	Managed to allow ecologically sustainable development and use	General duty of care applies. Littering prohibited.

in South Australia's interests such as public infrastructure or major developments (Government of South Australia 2009).

The 19 marine parks established under the *Marine Parks Act 2007* deliberately overlaid several protected areas previously established under the *National Parks and Wildlife Act 1972* and the *Fisheries Management Act 2007*. However, some of these earlier declarations are not within the new marine parks (e.g. some aquatic reserves). In addition, a separate piece of legislation created the Adelaide Dolphin Sanctuary – the *Adelaide Dolphin Sanctuary Act 2005*: (a) to protect the dolphin population of the Port Adelaide River estuary and Barker Inlet; and (b) to protect the natural habitat of that population. Finally, several terrestrial national parks, conservation parks and other reserves along the coast are reserved down to the low water mark and have some marine components.

Current policy

The 19 marine park management plans establish the zones in each park and also set out strategies to achieve the objects of the *Marine Parks Act 2007*. These broad strategies are common to all management plans and specific management priorities for each park are identified from them. The strategies are:

1 **Protection** – Development and administration of appropriate legislation and regulations, management actions, government partnerships, permits and statutory referrals.
2 **Stewardship through community involvement** – Provision of opportunities for public appreciation, education and enjoyment by promoting activities such as school education, citizen science, sustainable nature-based tourism, and cooperation with Aboriginal communities to conserve country, plants, animals and culture.
3 **Performance, assessment, knowledge and review** – Development and implementation of a monitoring, evaluation and reporting program to inform on-going management and the statutory review of management plans (required at least every 10 years). Partnerships with universities, researchers and community organisations will be formed to do this work.
4 **Compliance** – Compliance will be supported by education, across government collaboration with already existing programs, effective deterrence programs and adaptive management through ongoing review of compliance activity.

Commercial fishing in marine parks

One of the marine park policy commitments made in 2009 (Government of South Australia 2009) was that the implementation of the final marine parks zoning would have no more than a 5% economic impact on the commercial fishing industry's state-wide gross value of production (GVP) (as per the 2007 EconSearch independent fishing assessment report) and this commitment was met with the estimated impact of the final zoning at 1.7% of GVP.

The South Australian Government's approach to the management of displaced commercial fishing effort used the following steps:

- pragmatic zoning to avoid displacement
- redistribution of effort (where possible) without impacting ecological or economic sustainability of the fishery

- market-based buyback of fishing licences or adjustments of quotas or other mechanisms of sufficient effort to avoid negative impacts on the fishery
- compulsory acquisition of fishing authorities (as a last resort option) or adjustments of quotas or other mechanisms (Government of South Australia 2011).

Although the 19 management plans were authorised in November 2012 and took effect immediately, the *Marine Parks (Zoning) Regulations 2012* prescribe that restrictions on fishing in sanctuary zones, the key change for most users, commenced on 1 October 2014. This provided both commercial and recreational fishers the opportunity to adjust to the new marine parks requirements.

In addition, this delay also provided the opportunity for the South Australian Government to focus on the residual impact of displaced commercial fishing by implementing the South Australian Marine Parks Voluntary Catch/Effort Reduction Program (Government of South Australia 2013).

Guided by fisheries managers and scientists, the commercial fishing effort expected to be displaced by the full implementation of sanctuary zones was calculated and applications were sought for the voluntary surrender of licences and quota in the relevant fishing sectors (Ward *et al.* 2012). Considerably more applications were received than required for five of the six fisheries involved. The sixth fishery had only a small number of affected fishers and separate negotiations with them to obtain the required affected quota were also successfully concluded.

Key factors for success

There were many factors that both contributed to and challenged the successful delivery of South Australia's marine parks, and the following were especially significant.

Government support

The scale of the marine parks program and its influence across the state on very diverse activities meant that its success was dependent on full support from relevant government ministers and agencies. It was critically important that this support was led at the highest level of government.

The involvement of all relevant agencies from the beginning ensured that informed advice was provided to ministers as government decisions were made. This helped facilitate a quick timeline and also encouraged other agencies to allocate resources to this work, because it was clear that the program was a government priority. As potentially conflicting issues arose, the imperative to complete the project drove through obstacles and ultimately ensured program targets were achieved.

The network approach – economies of scale

The decision to develop a complete network of marine parks across state waters was key to the program's success. Although the 3-year process to proclaim the 19 outer boundaries and develop management plans for each park was long, momentum was maintained.

While demands on key stakeholders, community members and government officers were intense, the network approach provided economies of scale for everyone. One process served all users and effort was not replicated sequentially from one region to another across the state over a longer time period.

The state-wide approach provided consistency and certainty for everyone. All users were treated equitably and parks were established across all regions, ensuring any impacts and benefits were similar. Planning for the complete network also provided greater flexibility when considering zoning, because a range of options was usually available to protect particular species and/or habitats.

Throughout the program, there was significant pressure to achieve a high volume of work in a short amount of time. This included briefings for ministers, Parliament and Cabinet, media interviews, freedom of information requests, a Parliamentary inquiry and public requests for information. For example, from 2009 to 2012, around 800 letters seeking information about the program were written to the South Australian Government. The Marine Parks Team – comprising up to 40 people with skills in project management, community engagement, planning, research, GIS mapping, policy, administration and media – was responsible for facilitating quick responses to all of these requests.

Managing misinformation

From 2008 to 2012, there were thousands of media stories on the marine parks program largely instigated by non-government sources such as conservationists, local government, commercial fishers, aquaculture operators and recreational fishers.

The majority of articles were on the critical side, with regional media having proportionately more coverage than metropolitan media. Several stories were founded on misunderstandings or deliberate misinformation. For example, the media frequently ran articles that challenged the need for marine parks, given that the state's fisheries were already being effectively managed. Challenges were also extended to the science behind marine parks, in an attempt to redirect government effort towards restoring degraded areas, rather than conserving healthy ones for future benefit.

This misinformation campaign was foreshadowed in the research undertaken in the lead up to this work from 1999 to 2006 (Compas *et al.* 2007). This research found that newspaper reports over this time were largely ineffective in conveying the significance of the local ecology, resulting in a poorly informed public.

Through the development of the draft management plans and zoning to the implementation of the final plans, the South Australian Government was left to reconcile these divergent stakeholder views and attempt to correct the ongoing misinformation about the purpose and merits of South Australia's marine parks program.

Working with recreational fishers

The South Australian Government recognised that the recreational fishing sector includes the largest number of people potentially affected by the implementation of marine parks. Several concerns were raised publicly by fishers throughout the process about the potential negative consequences that could result from the parks. Significant efforts were made to resolve these concerns and provide support to fishers to adapt to change.

For example, extensive advice was sought from local communities to ensure popular fishing areas were identified early, and that zoning resulted in as little displacement impact as possible. Furthermore, almost 2 years 'grace period' was provided before fishing restrictions came into effect in marine parks sanctuary zones on 1 October 2014. Fishers across the state were also offered a state-wide *Recreational Fishing Guide* (DEWNR 2014), a CD of GPS coordinates of the zones and a free smartphone app, to help locate the sanctuary zones on the water. Signs were also placed at key boat ramps and other access points across the state displaying the new sanctuary zones and fishing restrictions.

Fig. 8.5. Western Blue Groper, a species that will benefit from sanctuary zones within South Australia's marine parks. (Photo: DEWNR)

Future directions

South Australia now has a network of 19 marine parks, established in accordance with the *Marine Parks Act 2007*. The network forms the foundation of a comprehensive, adequate and representative system of marine parks for South Australia (Marine Parks Council of South Australia 2012) (Fig. 8.5).

The implementation of the 19 management plans remains a South Australian Government priority, with the focus on implementing strong and effective management of each marine park. A 4-year implementation plan was developed in 2014 to deliver on the strategic priorities identified in each management plan.

An ongoing budget of approximately $2 million per annum has been allocated to support the implementation plan. In addition, the government has provided an additional $4 million over 4 years (from 2014/15) to boost education, monitoring and compliance, as well as an additional $3.25 million over 3 years (from 2014/15) to enhance recreational fishing opportunities in and around marine parks, by improving access, participation and minor infrastructure at key locations. The aim of this is to increase community use of parks and to sustain local jobs that support these activities.

The South Australian Government intends to continue to emphasise community stewardship programs to increase understanding of the value of the marine environment and to support voluntary compliance with fishing restrictions. Compliance challenges were acknowledged, especially since parks are spread across the entire South Australian coastline. Compliance will include both voluntary and enforcement measures, recognising that these will necessarily be strategically deployed at priority locations. A risk assessment has been undertaken across the state to assess the likelihood of potential breaches at key locations, such as sanctuary zones, to inform the development of a compliance work plan.

The development and implementation of the monitoring, evaluation and reporting program to consider the ecological, social and economic effects of the management plans and to support the mandatory review of the plans is also a priority. As with marine parks everywhere, on-water activities are expensive and resource hungry and any protected area program could readily spend as much as was available.

In addition, the South Australian Government recognised that the initial implementation of the management plans caused concern in some local communities and independent early regional impact assessments were undertaken in three locations where particular concerns were raised and also agreed to start a program for the review of management plans within the first 4 years of full implementation of the management plans.

Lessons and conclusions

South Australia's 19 marine parks form the basis of a comprehensive, adequate and representative system of MPAs. Some environmental, social and economic benefits of these parks may start to become evident in several years and others may take decades to be fully understood (Ballantine 2014; Edgar *et al.* 2014). Some management initiatives will be successful and others may need to be improved.

The South Australian Government's marine parks project was lengthy and complicated to develop and implement, and many lessons were learnt along the way. Some are specific to the South Australian experience and some are more broadly applicable to other marine protected areas and change initiatives. Important concepts include:

1 Have a clear vision of the final outcome that is founded on good policy.
2 Define the milestones required to achieve your vision and do not become sidetracked.
3 Expect setbacks and look for opportunities for increased success when these occur.
4 Involve local communities in clearly defined roles with tangible outcomes. This will provide invaluable local knowledge and the base for ongoing stewardship.
5 Pay close attention to the balance between ecological imperatives and the community's capacity to change.
6 Involve all relevant government agencies from the beginning.
7 Find partners from community, government and industry.
8 Provide as much certainty as possible as early as possible to help reduce fear of the unknown.
9 Recognise that the fear of change causes distress and that distress can be compounded by the general lack of trust people have in the integrity of government initiatives.
10 Assemble a well-qualified and dedicated team.
11 Provide good support materials – maps are particularly important.
12 Keep it simple. Although detailed technical information must be available, it should not drive public communications.
13 Have strong, high-level government support.
14 Deal with ambiguity and misinformation clearly, calmly and consistently.
15 Provide effective communication about the purpose of the initiative.
16 Provide effective communication about the science supporting the initiative.
17 There is no silver bullet. Not everyone will be happy at the end because it is not possible to deliver everything that everyone wants. Expect basic acceptance by most, acknowledging compromise along the way, with some minority opposition persisting.

Acknowledgements

There were many contributors to the marine parks program. The authors acknowledge: the Department of Environment, Water and Natural Resources Marine Parks Team; the Marine Parks Local Advisory Groups; the Marine Parks Council of South Australia; the Scientific Working Group; the across government steering committee; the ministers responsible; the hundreds of stakeholder contributors; and the thousands of individuals who provided their views.

Biography

Chris Thomas is responsible for managing South Australia's Marine Parks Program. Chris came to South Australia in 2004 from Queensland, where he had been the Director of Tourism and Recreation for the Great Barrier Reef Marine Park Authority. Chris has a passion for the outdoors and regularly enjoys life in South Australia's marine parks.

Vera Hughes is responsible for marine parks legislative and government matters and fondly remembers when a briefing request was not urgent unless it was due on the same day it was requested. She lives in a house by the sea where she can see Encounter Marine Park from her window.

References

Baker JL (2004) *Towards a System of Ecologically Representative Marine Protected Areas in South Australian Marine Bioregions – Technical Report*. Prepared for Coast and Marine Conservation Branch, Department for Environment and Heritage, Adelaide.

Ballantine B (2014) Fifty years on: Lessons from marine reserves in New Zealand and principles for a worldwide network. *Biological Conservation* **176**, 297–307. doi:10.1016/j.biocon.2014.01.014.

Compas E, Clarke B, Cutler C, Daish K (2007) Murky waters: media reporting of marine protected areas in South Australia. *Marine Policy* **31**, 691–697. doi:10.1016/j.marpol.2007.03.001.

Day J, Dudley N, Hockings M, Holmes G, Laffoley D, Stolton S, *et al.* (2012) *Guidelines for Applying the IUCN Protected Area Management Categories to Marine Protected Areas*. IUCN, Gland, Switzerland.

DEH (2004a) *Blueprint for the South Australian Representative System of Marine Protected Areas*. Department for Environment and Heritage, Adelaide, <http://www.environment.sa.gov.au/files/bb30ede9-6916-4e0d-8ba9-9e4d00e15ca3/mpa_blueprint.pdf>.

DEH (2004b) *Living Coast Strategy for South Australia*. Department for Environment and Heritage, Adelaide.

DEH (2008) *Design Principles Guiding the Development of South Australia's Marine Park Boundaries*. Department for Environment and Heritage, Adelaide, <http://www.environment.sa.gov.au/files/43f2b2d6-e479-4f9c-a247-9e2500c88449/boundary_principles.pdf>.

DEWNR (2012) *Summary of Submissions Report, Draft Management Plans for South Australia's Marine Parks Network*. Department of Environment, Water and Natural Resources, Adelaide, <http://www.environment.sa.gov.au/files/8e5752fb-97f4-4ced-a398-a12e00e5def1/mp-gen-submissionsummary-nov12.pdf>.

DEWNR (2014) *Recreational Fishing in SA Marine Parks.* Department of Environment, Water and Natural Resources, Adelaide, <http://www.environment.sa.gov.au/marineparks/enjoy/fishing>.

Edgar GJ, Stuart-Smith RD, Willis TJ, Kininmonth S, Baker SC, Banks S, *et al.* (2014) Global conservation outcomes depend on marine protected areas with five key features. *Nature* **506**, 216–220. doi:10.1038/nature13022.

Edyvane KS (1999) *Conserving Marine Biodiversity in South Australia. Part 1. Background, Status and Review of Approach to Marine Biodiversity Conservation in South Australia.* South Australian Research and Development Institute, Department of Primary Industries and Resources, Adelaide, <http://www.sardi.sa.gov.au/__data/assets/pdf_file/0004/94594/marine_biodiversity_part1_full_version.pdf>.

Government of South Australia (2004) *South Australia's Strategic Plan 2004.* South Australian Government, Adelaide.

Government of South Australia (2009) *Whole of Government Commitments for Marine Parks.* South Australian Government, Adelaide, <http://www.environment.sa.gov.au/files/4ae2a839-218e-4dec-b53a-9e7e00f5176b/mp-gen-policycommitt09.pdf>.

Government of South Australia (2011) *Marine Parks Displaced Commercial Fishing Policy Framework, South Australian Government. April 2011.* South Australian Government, Adelaide, <http://www.environment.sa.gov.au/files/dd771b4d-581c-422d-abda-a14100b3a0fa/mp-gen-displaced-commercial-fishing-policy-framework.pdf>.

Government of South Australia (2013) *SA Marine Parks Commercial Fisheries Voluntary Catch/Effort Reduction Program, July 2013.* South Australian Government, Adelaide, <http://www.pir.sa.gov.au/__data/assets/pdf_file/0007/188809/SA_Marine_Parks_-_Commercial_Fisheries_Voluntary_Catch_Effort_Reduction_Program_Plan_-_15_July_2013.pdf>.

IUCN-WCPA (2008) *Establishing Resilient Marine Protected Area Networks – Making it Happen.* IUCN World Commission on Protected Areas, National Oceanic and Atmospheric Administration and The Nature Conservancy, Washington, DC.

Marine Parks Council of South Australia (2012) *Advice to the Minister for Sustainability, Environment and Conservation on the Draft Marine Parks Management Plans Public Consultation.* 16 November 2012. Marine Parks Council of South Australia, Adelaide, <http://www.environment.sa.gov.au/files/2c6baa32-1c79-4771-8b1f-a12e00e54aec/mp-gen-mpcouncil-advicetominister-Nov12.pdf>.

Ward TM, Burch P, Gorman D, Begg GA (2012) *Estimates of Historical Commercial Fishery Catches/Effort in Final Sanctuary and Habitat Protection Zones in South Australia's Marine Parks.* SARDI Publication No. F2011/000307-8. SARDI Research Series No. 670. Report to PIRSA Fisheries and Aquaculture. South Australia Research and Development Institute (Aquatic Sciences), Adelaide, <http://www.sardi.sa.gov.au/__data/assets/pdf_file/0009/179451/Habitat_Protection_Zones_in_SA_Marine_Parks_2.pdf>.

Weatherill J, Caica P (2012) *South Australian Marine Parks Take Effect Today.* Media release from Premier Jay Weatherill, Minister for State Development and Paul Caica, Minister for Sustainability, Environment and Conservation, 29 November 2012, Government of South Australia, Adelaide, <http://www.premier.sa.gov.au/images/news_releases/12_11Nov/marineparks.pdf>.

Chapter 9

Victoria's chequered history in the development and implementation of marine protected areas

Geoff Wescott

Introduction

Victoria has a comparative long (1600 km) east–west running coastline, which means most of it is south facing with predominant swells from the south-west on the western coast and from the south-east on the eastern coast. The combination of these swells blocked centrally by Tasmania and the shallowness of Bass Strait means a varied and diverse suite of coastal waters occurs.

Victoria has a proud history of the development of national parks and protected areas on land, which saw an increase of over 14-fold in national park area between 1975 and the turn of the century (Wescott 1995; Coffey *et al.* 2011). This was largely a result of the methodical region-by-region work of the Land Conservation Council (LCC, and its successor the Environment Conservation Council). The LCC followed a statutory process by which it published descriptive resource documents for public land in a region, called for public input, constructed a suite of proposed recommendations for that public land (e.g. national park, timber production) called for a second round of public comment and then published a set of final recommendations delivered to the government. The government then made the final decision, generally accepting and implementing most of the major recommendations (Coffey *et al.* 2011).

For the majority of the marine environment, the coastal waters did not go through this process (although the intertidal zone was tackled in study areas abutting the coast) until a descriptive report dedicated to the entire marine and coastal public estate of Victoria was published in 1993 (LCC 1993), some 20 years after the LCC had been established. Nevertheless, when this process was started it was methodical and considered and resulted in a suite of new marine protected areas (MPAs) – all of which were highly protected ('no-take').

In November 2002, when Victoria declared a suite of 13 marine national parks and 11 marine sanctuaries, the Victorian Government claimed this as a world first (Fyfe 2002). All of these additional reserves were single no-take reserves (a unique position in Australia), which were added to the existing network of multiple use marine and coastal parks and reserves declared previously.

As outlined below, there has not been a single addition to this combined suite of reserves since 2002.

History and development of the MPA network in Victoria

The 2002 network of MPAs summarised above had their genesis in an Australian Labor Party electoral commitment (when in opposition) exactly 20 years earlier (Wescott 2006).

Following the declaration of a small marine reserve in 1978, the next substantive MPAs were not declared until 1983, in South Gippsland (these declarations distinguished by a Victorian Supreme Court challenge to restrictions placed on commercial fishing; Wescott 1990), with a further isolated declaration in the same region in 1988 (see below).

Fourteen years elapsed before the next MPAs were declared – the network of no-take marine national parks and marine sanctuaries declared in 2002 that covered 5.6% of Victorian coastal waters.

This 'fits and starts' record of declarations is the hallmark of Victoria's MPA development over the past 30 years and continues through to today. There have been four distinct phases in the development of MPAs in Victoria.

The first phase occurred up until to 1982. Prior to the first political commitment to a system of MPAs in 1982 (see Wescott 1990), Victoria had followed the globally familiar approach of proposing one reserve and seeing that proposal through to conclusion. The sole proposal promoted by the Victorian National Parks Association and the Australian Marine Sciences Association was the Harold Holt Reserve established in 1978 under fisheries legislation. This was composed of five small distinct areas at the southern end of Port Phillip Bay (VEAC 2014). There was no further distinct proposal by the Victorian Government or NGOs until the election of the Labor Government in 1982, marking the commencement of the second phase (1982–1991) although another very small reserve was declared, again under fisheries legislation, at Point Cooke in 1982 (VEAC 2014)

In this second phase, the LCC hastily inserted in the draft recommendations for the South Gippsland area investigation a reference to considering proposals for MPAs in the last public consultation phase (LCC 1982a). The LCC investigations until that point across all of Victoria had only been considering public land in terrestrial systems (and associated inland waterways) down to low-water mark.

When the final recommendations were published late in 1982 (6 months after the draft recommendations), there were several MPAs recommended to the Victorian Government (LCC 1982b). The recommendations took the local community by surprise, but not the conservation groups, because they responded to the reference to considering MPAs in the final phase of the LCC process by including recommendations for MPAs in their submissions and reminding the LCC of the new government's electoral commitment to new MPAs. The subsequent declaration of a series of marine and coastal parks, marine reserves and marine parks in 1983 (Nooramunga Marine and Coastal Park, Corner Inlet Marine and Coastal Park, Shallow Inlet Marine and Coastal Park, Wilsons Promontory Marine Park, and Wilsons Promontory Marine Reserve) generated controversy, especially locally where they were unexpected and probably contributed to the delay until the next proposed MPAs (Bunurong Marine Park) in the late 1980s.

Bunurong Marine Park (Fig. 9.1) was declared in a return to the more traditional approach – it was proposed by the South Gippsland Conservation Society and taken on board by the Victorian Government in a modified form.

The Victorian Government's approach of attaching consideration of MPA proposals to the land-based investigations of the LCC and appearing to have suddenly decided to do so (by having the LCC consider it half way through the South Gippsland process) was probably both unnecessarily provocative to regional communities and was a haphazardly

Fig. 9.1. Temperate rocky shores of the Bunurong Marine National Park in South Gippsland, formerly reserved as the Bunurong Marine Park. (Photo: James Fitzsimons)

random process that, although replicating the approach, for example, towards forests on land (Coffey *et al.* 2011), meant that only small sections of coastal waters were considered each time a terrestrial region with some coastal frontage came up for review.

Between the late 1980s and early 1990s, the Victorian Government made frequent promises for a full marine and coastal LCC investigation (i.e. across the whole state), and in 1991 the government finally instructed the LCC to commence the study just before then Labor Government's defeat at an election (see Wescott 1988 for these commitments).

The third phase was the drawn-out LCC Marine, Coastal and Estuarine Investigation between 1992 and 2000. Most LCC studies up until that point were completed in approximately 2 years from initial publication of a descriptive report through two phases of public consultation to final recommendations to government. The marine, coastal and estuarine study was to last 8 years after seeing the descriptive report published in 1993 (LCC 1993).

The first set of final recommendations was about to be published in 1997 when suddenly the LCC was disbanded. The LCC was replaced by the Environment Conservation Council (ECC), which had a similar role of investigating and recommending on public land use, but with a greater emphasis on considering economic and social impacts of public land-use decisions (see Coffey *et al.* 2011).

The ECC continued with the marine, coastal and estuarine investigation and maintained some momentum, producing its final recommendations in 2000 (ECC 2000), by this time under a Labor Government, which had been re-elected in 1999 after 7 years in opposition. The primary ecological purpose of the 13 marine national parks and 11 marine sanctuaries recommended by the ECC was to maintain examples of Victoria's biodiversity and associated ecological processes, including their variation in space and time, in a relatively natural condition for their intrinsic value to future generations. They were also intended to serve as benchmarks against which other marine areas could be compared. The areas were not designed or intended to rehabilitate fish stocks. The ECC recommendations took into account a nationally agreed regional ecosystem-based classification for the Australian marine environment that recognises five biophysical regions in Victoria. The classification was known then as the Interim Marine and Coastal Regionalisation of Australia (IMCRA version 3.3).

The fourth and final phase of establishment of Victoria's current MPA system occurred between 2000 and 2002. There had been, by 2002, six formal public submission periods and 4500 written submissions received (see Wescott and Fitzsimons 2011), and yet some groups still claimed a lack of public consultation.

The minority Labor Government initially attempted to introduce to Parliament a suite of MPAs slightly different to those in the final recommendations of the ECC – less one proposed marine sanctuary (Ricketts Point Marine Sanctuary at Beaumaris in Port Phillip Bay) – in an attempt to pacify vocal opposition from the recreational and commercial fishing groups and in order to gain the support of the three independent cross benchers (two of whom held coastal seats with recommended MPAs in them). The government withdrew the proposed legislation in 2001 when it realised it did not have enough votes to pass it in either of the houses of parliament.

This marked the first occasion in the 26 years since the passing of the Victorian *National Parks Act 1975* that the two major parties – the Australian Labour Party and the Liberal Party – had not mutually supported a proposed addition of a national park to the Act.

The legislation was re-introduced in 2002 with the Ricketts Point Marine Sanctuary re-instated but more significantly including a comprehensive compensation package of measures to assist those who claimed to be disadvantaged by the new MPAs and a package of resources to ensure enforcement of the MPA's boundaries and restrictions. This time the legislation was supported and the suite of marine national parks and marine sanctuaries were declared in November 2002.

The compensation package, the declaration of all ECC recommended MPAs and the fact that no piers or jetties, and no very popular beach fishing areas, were included in the marine national parks and marine sanctuaries (some boundaries looked very peculiar because they omitted a beach from a very close offshore national park) were key to their final passage through parliament and implementation on the ground (Wescott 2006).

Soon after, Parks Victoria (2003) produced a major marine parks implementation strategy to cover the period 2003–2010 with considerable expert input as a model for delivery of the marine national parks and marine sanctuaries. Despite the detail in this document, its implementation was poor, as discussed below.

Since the declaration of the suite of MPAs in 2002, there have been no further additions to the MPA estate in Victoria, despite the addition of several terrestrial national parks during this period. Both major political parties went to the extent of promising at the 2010 state election that they would not be declaring any additional MPAs in their next term of office – a promise the narrowly elected Coalition Government kept. Neither party made any commitment either for or against MPAs in the 2014 election campaign.

The 'hangover' from the long and bitter campaign described above can be identified almost as a fifth phase in Victoria's attempts to establish a MPA system that even goes close to its extensive land-based system, which covers nearly 20% of land by area in highly protected parks and reserves.

In 2010, the Liberal-National Coalition was elected with a policy commitment to review the current MPAs in Victoria, with the review to be undertaken by the Victorian Environmental Assessment Council (VEAC), the successor to the ECC. The VEAC study was commissioned to examine the performance and management of Victoria's MPAs and involved two public consultation phases, in 2012 and 2013. It is interesting to note that VEAC has never been asked to examine the performance and management of any single terrestrial land use designation, whether it was reserved for productive use (e.g. timber production) or conservation purposes.

VEAC's final report to government focused on the ecological and biodiversity outcomes of the MPA system, identification of ongoing threats to effective management, and made 38 recommendations on how management, planning and resourcing could be improved (VEAC 2014).

In 2010, a scheduled audit on the management of the MPAs was carried out by the Victorian Auditor General's Office (VAGO). VAGO (2011) reported on how effectively the Victorian MPAs had been managed to protect biodiversity. The focus was primarily on the manager of the MPAs, Parks Victoria, but also covered the then Department of Sustainability and Environment, which is responsible for policy and creation of public protected areas. The overall conclusion of the report was that Parks Victoria could not 'demonstrate' (author's emphasis) that marine biodiversity was being protected (although it was stated that this did not necessarily mean that it was not being protected) due to systemic weaknesses with planning, program management and resource allocation. It stated that Parks Victoria could demonstrate little actual environmental activity within the MPAs and had used funds allocated to the marine parks for other activities.

Although Parks Victoria had developed management plans for all 24 marine national parks and marine sanctuaries, no final management plans existed for marine and coastal parks and other MPAs that also have significant environmental values.

A series of recommendations were made (VAGO 2011, p. ix) to overcome these issues and also to establish a better understanding of the service agreement between the Department of Sustainability and Environment and Parks Victoria. The former expired Marine Strategy (Parks Victoria 2003) was in part replaced, presumably in response to the VAGO recommendations, by a program plan in 2010 (Parks Victoria 2010). These developments were severely impacted upon by very substantial budget cuts and staff cuts in each Budget of the Victorian Government between 2010 and 2014.

Victoria's current MPA estate

Currently, Victoria possess 24 high protection, no-take (IUCN Category II and III) MPAs – 13 Marine National Parks and 11 smaller Marine Sanctuaries, covering 5.1% (52 573 ha) of Victoria's coastal waters (VEAC 2014). These are reserved under Schedules 7 and 8 of the *National Parks Act 1975*.

In addition, there are six other MPAs of various designations – marine reserves, marine parks and marine and coastal parks – totalling 66 524 ha or 6.6% of Victoria's coastal waters. These are multiple-use MPAs (but with no legislated internal zoning) and some of these areas contain coastal land and islands. The primary purpose for which the multiple-use areas were established was to protect biodiversity to the extent possible while accommodating a variety of uses, including fishing. Their primary ecological purpose is to protect areas

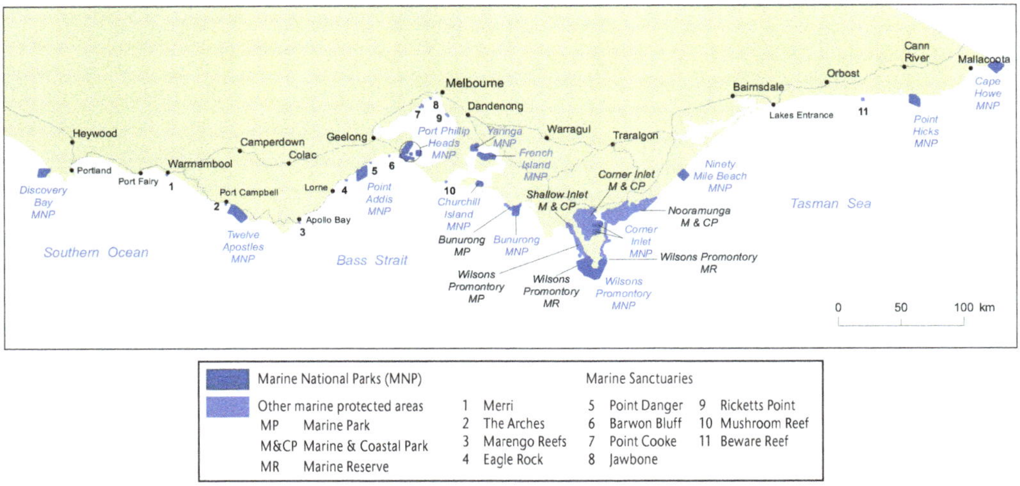

Fig. 9.2. Location of marine protected areas in Victoria (from VEAC 2014).

containing significant natural ecosystems. Like the no-take areas, these areas have a social purpose of providing opportunities for enjoyment, appreciation and understanding of the natural environment (VEAC 2014). These are reserved under Schedule 4 of the *National Parks Act* and are classified as IUCN Category VI protected areas.

Overall, Victoria has 11.7% of its 1 017 400 ha of marine environment (coastal waters) in five different types of MPAs (VAGO 2011; VEAC 2014). This estate is shown in Fig. 9.2.

Some primarily terrestrial national parks and coastal parks located on the coast have reservation down to the low-tide area and thus incorporate elements of the intertidal zone and/or estuarine systems. Although the section of these protected areas covering the intertidal zone could technically be considered as MPAs (and thus be considered for accounting purposes, and have been in the past), a reduced level or lack of enforcement may render their effectiveness for biodiversity conservation to be less than for true MPAs that extend to high-water mark (Fitzsimons and Wescott 2008). Although the Victorian *Fisheries Act 1995* has the provision to create Fisheries Reserves, which could include provisions for habitat protection or nursery, spawning protection, and so on, no existing fisheries reserves would qualify as MPAs in Victoria at present.

Special management areas (SMAs) were recommended by the ECC in Victoria to protect specific features in the marine environment while allowing other extractive resource uses (ECC 2000). In certain instances, the boundary of these SMAs were left undefined owing to the spatially dynamic nature of the special value, such as seagrass beds or Blue Whale calving areas. This spatial and temporal flexibility poses challenges for the establishment of these SMAs via traditional declaration and protection mechanisms, such as through legislation and regulations, which are not particularly suited to rapid changes. These would not be considered as MPAs under present definitions (Fitzsimons and Wescott 2008).

Victoria's current MPA policy

In short, there is no specific Victorian Government (or Opposition) MPA policy at present committed to any review or consideration of extending Victoria's MPA system. At the 2014 Victorian election, neither party commented specifically on developing further MPAs or specifically reviewing current MPAs. But both parties were committed to continuing a state-wide assessment of public land use categories by VEAC (which includes the sea bed and waters above it) and both supported the *Victorian Coastal Strategy 2014* (Victorian Government 2014), which proposed an investigation into marine planning in Victorian coastal waters. As well, the new Victorian Labor Government committed itself to a 'new Marine and Coastal Act' (ALP 2014, p. 4), the development of which commenced on 3 June 2015 by announcing an Expert Panel to oversee the process.

Parks Victoria is still responsible for managing these MPAs and ~70% of the Victorian coastline (Fig. 9.3) and has produced a study of marine natural values (Barton *et al.* 2012).

More broadly, Victoria, as part of Australia's commitment to the Convention on Biological Diversity Aichi Target 11, has committed to 10% reservation of the marine environment in representative, well-connected manner that particularly represents areas of high biodiversity value and ecosystem services by 2020. The extent to which Victoria's systems does this needs review in order to ensure its national and international commitments are being met.

There have been some advances from outside government, though. The VNPA has carried out four nature conservation reviews across Victoria since 1970. These have historically been very significant: the first review contributed to the approach adopted by the LCC to identifying national parks; the second caused a state-wide review to be performed in response by the LCC; and the third concentrated on marine areas and the box-ironbark forests of central Victoria and was a major contributor to the eventual successful declaration of MPAs in Victoria (Traill and Porter 2001).

Fig. 9.3. Point Cooke Marine Sanctuary in Port Phillip Bay protects narrow sandy beaches, rocky reefs and mud flats and is buffered by Point Cook Coastal Park and Cheetham Wetlands on the shore. (Photo: James Fitzsimons)

The fourth *Nature Conservation Review*, while being state-wide again, focused again on marine as well as potential freshwater protected areas (VNPA 2014). There was considerable marine research done (Edmunds *et al.* 2010), with a final report identifying 20 hot spots for consideration for potential additions to the MPA network. The response from both major political parties to the proposals could only be described as muted at best. The author's appraisal of this political position is that the parties are aware that any MPA extensions will bring a swarm of letters and protests from the recreational fishing lobby groups and, despite widespread, but not concentrated, general community support for MPAs (see Chapters 20, 21) the political cost of a well-organised vocal anti-park lobby against a 'silent majority' in support means that political parties see this is an issue to avoid rather than draw attention to.

Key factors for success

There were several key factors along the way that influenced the eventual success of the 2002 declaration of a network of marine national parks and marine sanctuaries (see Wescott 2006) and may be important in any revitalised campaign for an extension to the current system of reserves. They include:

- the role of conservation NGOs, and in particular the VNPA's, never wavering 25-year commitment to the establishment of the MPA network, assisted in the 1990s and early 2000s by the now defunct Marine and Coastal Community Network
- the continuous opposition to the establishment of the system by the peak commercial and recreational fishing groups, although the initial absolute rigidity of opposition did decrease marginally in the late 1990s as some kind of system became politically inevitable. This rigid opposition had the effect of galvanising what may otherwise had been neutral observers to support the package
- the 'winning over' of what may be regarded as neutral groups (e.g. tour operators, divers) to the MPA system in the later stages of the process
- the role of some champions in garnering support across both major political parties (eventually), including an extraordinary joint letter in support of the ECC recommendations by an ex-Labor Premier and an ex-Liberal Premier in April 2002 (see Wescott 2006)
- the leadership role of some government agencies, notably the Victorian Coastal Council, in publicly fully supporting a system of MPAs in Victoria
- the sheer number and diversity of the supporters (including a considerable majority of the general public) of the system
- the apparent narrow-mindedness and inflexibility of opposing sectors, which probably over-reached in their attacks on the proposals including the use of push polling and unsubstantiated claims of potential losses (see Wescott 2006).

Challenges confronting Victoria's MPAs

The challenges facing the Victorian MPA estate today is informed by the chequered history of establishing MPAs in that state. Like many other jurisdictions, Victoria commenced its declarations of MPAs in the late 1970s in a sporadic way: propose a single reserve and lobby for its declaration, then propose another one and lobby for its declaration.

After a brief flurry of activity in the mid 1980s, there were no further declarations of any MPA from 1987 to 2002, when the outcome of a systematic study was passed in legislation establishing the current system of MPAs.

The single most clear cut challenge in Victoria is to convince the major political parties of the strong, but largely silent, support across the general community for more MPAs (as evidenced in successive state-wide surveys before production of each Victorian Coastal Strategy (e.g. Ipsos 2012)).

The opposition to MPAs by recreational fishing bodies (e.g. VRFish) in Victoria to no-take MPAs continues. The reasons for this opposition to new MPAs and for lowering protection in existing ones (e.g. Tomazin 2012) in light of the independent assessments showing widespread general community support for marine parks is neither obvious nor transparent. The question of how representative the views of this peak body are of all Victorian recreational fishers (see Chapter 21) has not been explored.

Other challenges confronting Victoria's MPAs include:

- a lack of clarity around reservation and purpose of multi-use marine parks and designations such as marine and coastal parks
- compliance and enforcement issues, such as the control of dogs in coastal stretches in marine national parks and marine sanctuaries (e.g. Weston *et al.* 2014)
- how to incorporate new knowledge on now extensive marine habitat mapping
- further implementation of the VAGO (2011) recommendations
- the lack of any policies on MPAs of the major Victorian political parties.

Future directions

Since the declaration of the MPA network in 2002, there has been little political interest in expanding the estate. This is despite the independent assessments in public surveys commissioned by the Victorian Coastal Council during preparation of the fourth Victorian Coastal Strategy (the state-wide statutory strategic planning document for Victoria's coastal zone including coastal waters), which showed that over 80% of people surveyed supported marine parks – not only the existing parks but extending the system (Ipsos 2012).

One the other hand, VRFish was on the record in February 2014 as having called for a trial in three marine national parks or sanctuaries of regulated fishing: that is, the removal of the 'no-take' high protection status of these reserves whose objective is primarily biodiversity conservation under the *National Parks Act 1975* (e.g. Arup 2014). The Minister at the time said he was not considering opening marine parks up to fishing (Arup 2014).

These interchanges highlight the future directions for MPA policy in Victoria will be contested, at least in the short term. The recommendation in the Victorian Government's Coastal Strategy (Victorian Government 2014) to develop an integrated marine planning system for Victoria's coastal waters and the development of a new Marine and Coastal Act (which started in July 2015) are two processes where the expansion of the MPA system might become a future possibility. But until the majority of the Victorian public in favour of more MPAs is motivated to become a strong constituency and advocate for an extended and connected MPA system there may be little change.

Conclusion

The Victorian MPA scene is characterised by the 'fits and starts' of MPA declarations separated by long periods of no action and complacency. Yet despite this history, Victoria was a world leader up to 2002 in its attempt to design and implement a comprehensive, adequate and representative system of high protection (no-take) MPAs across its entire coastal waters. Since then, in the absence of a more vocal constituency in favour of MPA expansion and the presence of a very vocal minority in opposition to MPAs, it is highly likely that only a state-wide marine spatial planning system zoning all Victorian coastal waters is likely to identify new opportunities – and galvanise support for a genuinely comprehensive, representative, adequate and well connected MPA system in Victoria.

Acknowledgements

The author would like to thank the VNPA for all their efforts and publications in the marine sphere, Rebecca Koss for research background information and Joan Phillips and Grimaldo Villanueva for supplying the map.

Biography

Geoff Wescott is Associate Professor of Environment at Deakin University's Melbourne Campus. He is a previous Chair of the Marine and Coastal Community Network who he also represented on the National Oceans Advisory Group. He was a Deputy Chair of Parks Victoria and has served on the steering committee of the past three Nature Conservation Reviews of the VNPA. He is current President of the Australian Coastal Society, a Director of Zoos Victoria and in mid-2015 was appointed to VEAC. He has published over 100 articles on marine and coastal matters, including a popular field guide to rocky shores in south-eastern Australia.

References

ALP (2014) *Our Environment, Our Future.* Australian Labor Party (Victorian Branch), Melbourne.

Arup T (2014) Push to reopen Victorian marine parks to recreational fishing. *The Age* 24 February 2014, <http://www.theage.com.au/victoria/push-to-reopen-victorian-marine-parks-to-recreational-fishing-20140223-33agt.html>.

Barton I, Pope A, Howe S (2012) *Marine Natural Values Study Summary.* Parks Victoria, Melbourne.

Coffey B, Fitzsimons JA, Gormly R (2011) Strategic public land use assessment and planning in Victoria, Australia: four decades of trailblazing but where to from here? *Land Use Policy* **28**, 306–313. doi:10.1016/j.landusepol.2010.06.011.

ECC (2000) *Marine, Coastal and Estuarine Investigation Final Report.* Environment Conservation Council, Melbourne.

Edmunds M, Mustoe S, Stewart K, Ong J (2010) *VNPA Nature Conservation Review: Marine Conservation Priorities and Issues for Victoria.* Australian Marine Ecology, Melbourne.

Fitzsimons JA, Wescott G (2008) Getting the measure of marine protected areas: surface area or volume as measures for reserve system auditing? *Aquatic Conservation: Marine and Freshwater Ecosystems* **18**, 518–526. doi:10.1002/aqc.867.

Fyfe M (2002) A world-first as state sanctuaries of the deep open. *The Age* 16 November 2002. <http://www.theage.com.au/articles/2002/11/15/1037080914128.html>.

Ipsos (2012) *Executive Summary Coastal and Marine Environment Community Attitudes and Behaviour.* (Wave 4) Report prepared for the Victorian Coastal Council. Ipsos Australia, Melbourne.

LCC (1982a) *South Gippsland Area 2 Proposed Recommendations.* Land Conservation Council, Melbourne.

LCC (1982b) *South Gippsland Area 2 Final Recommendations.* Land Conservation Council, Melbourne.

LCC (1993) *Marine and Coastal Special Investigation. Descriptive Report.* Land Conservation Council, Melbourne.

Parks Victoria (2003) *Victoria's System of Marine National Parks and Marine Sanctuaries: Management Strategy 2003–2010.* Parks Victoria, Melbourne.

Parks Victoria (2010) *Parks Victoria Marine Protected Areas Program Plan 2012–2014.* Parks Victoria, Melbourne.

Tomazin F (2012) Conservation fears put marine parks under the microscope. *The Age* 1 April 2012, <http://www.theage.com.au/environment/conservation-fears-put-marine-parks-under-the-microscope-20120331-1w5el.html>.

Traill B, Porter C (2001) *Nature Conservation Review 2001.* Victorian National Parks Association, Melbourne.

VAGO (2011) *Environmental Management of Marine Protected Areas.* Victorian Auditor General's Office, Melbourne.

VEAC (2014) *Marine Investigation Final Report.* Victorian Environmental Assessment Council, Melbourne.

Victorian Government (2014) *Victorian Coastal Strategy 2014.* Victorian Government, Melbourne.

VNPA (2014) *Natural Victoria: Conservation Priorities for Victoria's Natural Heritage. Nature Conservation Review 2014.* Victorian National Parks Association, Melbourne.

Wescott G (1988) A study in environmental planning and management: the coast of Victoria, Australia. PhD thesis, Deakin University, Melbourne.

Wescott G (1990) Victorian Coastal and Marine Planning and Management. Victoria College, Rusden Campus, Melbourne.

Wescott G (1995) Victoria's National Park System: can the transition from quantity of parks to quality of management be successful? *Australian Journal of Environmental Management* **2**, 210–223. doi:10.1080/14486563.1995.10648332.

Wescott G (2006) The long and winding road: the development of a comprehensive, adequate and representative system of highly protected marine protected areas in Victoria, Australia. *Ocean and Coastal Management* **49**, 905–922. doi:10.1016/j.ocecoaman.2006.08.001.

Wescott G, Fitzsimons J (2011) Stakeholder involvement and interplay in coastal zone management and marine protected area planning. In *Marine Resources Management.* (Eds W Gullett, C Schofield and J Vince) pp. 225–238. LexisNexis Butterworths, Sydney.

Weston MA, Fitzsimons JA, Wescott G, Miller K, Ekanayake KB, Schneider T (2014) Bark in the park: a review of domestic dogs in parks. *Environmental Management* **54**, 373–382. doi:10.1007/s00267-014-0311-1.

Chapter 10

Marine protected areas in Tasmania: moving beyond the policy void

Lorne Kriwoken

Introduction

The island state of Tasmania has an internationally renowned protected area system on land and the Tasmanian Wilderness World Heritage Area (TWWHA) is the jewel in the crown. The total terrestrial reserved area is nearly 3.5 million ha, representing just over half the land mass of the state. Few jurisdictions in the world have this extensive level of reserved area. In stark contrast to the terrestrial reserve system, however, is the Tasmanian marine reserve estate. The area of marine protected areas (MPAs) in Tasmania is only 135 000 ha, with sub-Antarctic Macquarie Island accounting for 55% of this total. Excluding Macquarie Island, only 1.12% of the Tasmanian State Waters is designated a no-take zone, with all marine reserve classes representing 2.67%. From these figures alone, it is obvious that the conservation of Tasmanian marine biodiversity is severely lacking and the state does not meet the minimum guidelines adopted by all Australian governments as part of the National Representative System of Marine Protected Areas (ANZECC 1998). Moreover, the present Tasmanian Government (as at 2015) has shown no interest in supporting new MPAs and has placed a moratorium on the expansion of the MPA system (Tasmanian Liberals 2014).

The aim of this chapter is to determine why progress in the declaration and management of Tasmanian MPAs has been so glacially slow. The chapter commences with an introduction of the history and development of the Tasmanian MPA network paying particular attention to overlapping jurisdictional arrangements. This is followed by a review of the Tasmanian marine reserve estate and an assessment of the bioregions that fail to meet minimum MPA representation. The current MPA policy void is introduced and the Bruny bioregion inquiry into the establishment of MPAs assessed to shed light on hurdles facing future MPA designations. The chapter concludes with an assessment of the challenges facing future MPA designations and suggests incremental changes that could assist in supporting existing MPAs.

History and development of the Tasmanian MPA network

Under the *Coastal Waters (State Powers) Act 1980* (Cth), the State of Tasmania has jurisdiction over land, sea and the seabed out to 3 nautical miles. The Commonwealth of Australia is responsible from 3 to 200 nautical miles in what is known as the Exclusive Economic Zone.

Unlike several other Australian states, Tasmania does not have dedicated MPA legislation (see also Chapter 14). Instead, MPAs are declared and managed under several pieces of legislation administered by different divisions within the Department of Primary Industries, Parks, Water and Environment (DPIPWE). The *Nature Conservation Act 2002* (NCA), administered by the Resource Management and Conservation Division of DPIPWE, provides for the conservation and protection of fauna, flora and geological diversity of the state and for the declaration of national parks and other reserved land (both marine and terrestrial). The *National Parks and Reserves Management Act 2002* (NPRMA), administered by Parks and Wildlife Service (PWS), provides for the management of national parks and other reserved land. The *Living Marine Resources Management Act 1995* (LMRMA), administered by Water and Marine Resources Division, regulates commercial and recreational fisheries, marine pollution, fish disease and pests, aquaculture and can declare a 'marine resources protected area' under section 105. The Act declares total allowable catch limits, minimum sizes for fish species, gear types and bag limits. 'Fish' are defined as any aquatic organism of any species, whether dead or alive, which in the normal course of events spends part or all of its life in the aquatic environment but does not include aquatic mammals, birds or freshwater fish. This includes seagrass, algae and aquatic vascular plants. The Minister has the power to make rules using the *Fisheries Rules 2009* under the LMRMA. This division of legislative and institutional responsibility reflects the often dual role in MPAs: primary industries and fisheries; and biodiversity (Kriwoken and Haward 1991; Kriwoken 2011).

Although there is provision to declare 'marine resources protected areas' under the LMRMA, to date no reserves have been proclaimed under this Act. All MPAs have been established under the NCA or previous legislation that was repealed by the NCA. As MPAs established under the NCA provide no protection for 'fish' as defined in the LMRMA, protection of marine life is provided via fishing restrictions declared under the *Fisheries Rules 2009*.

In the early 1980s, Tasmania declared 15 marine and estuarine protected areas (MEPAs) (Ivanovici 1984). These MEPAs included an *ad hoc* collection of conservation areas, historic sites, a small marine reserve for crayfish research, a portion of a national park and a nature reserve. Many of these MEPAs represented intertidal and estuarine habitats and nearshore habitats. In 1991, four marine reserves (Ninepin Point, Tinderbox, Maria Island and Governor Island) were declared with the specific objective of protecting marine habitats and biodiversity. In the late 1990s, Tasmania committed to the *Guidelines for Establishing the National Representative System of Marine Protected Areas* (NRSMPA) (ANZECC 1998) and the *Strategic Plan for Action for the National Representative System of Marine Protected Areas* (ANZECC 1999). In 2000, the focus turned to sub-Antarctic Macquarie Island, 1500 km SSE of Hobart. A marine extension to the Macquarie Island Nature Reserve was declared thus creating the single largest MPA in Tasmania (see Chapter 6). Two marine extensions to terrestrial protected areas were declared in 2004: Kent Group National Park in the Bass Strait and the South-west National Park. The South-west National Park extension declared the largely pristine marine estate of Port Davey-Bathurst Channel in the TWWHA. The most recent declarations were in 2009 when 14 marine conservation areas (MCAs) were finalised in the Bruny bioregion (Figs 10.1, 10.2). MCAs

Fig. 10.1. Sea caves in Waterfall-Fortescue Marine Conservation Area, south-east Tasmania. (Photo: James Fitzsimons)

are MPAs that provide for the sustainable use of natural resources. Fishing is permitted in MCAs and the rules and regulations pertaining to recreational fishing are found in the *Recreational Sea Fishing Guide 2014–15* (Tasmanian Government 2014).

Significant marine research is undertaken in Tasmania through the Institute for Marine and Antarctic Studies at the University of Tasmania and CSIRO. The University of Tasmania scientists have actively researched and monitored the effectiveness of MPAs. For instance, specific research on MPAs has involved: monitoring of biotic change in Tasmanian marine reserves (Edgar and Barrett 1997); effects of the declaration of marine reserves on reef fish, invertebrates and plants (Edgar and Barrett 1999); changes in fish assemblages over 10 years of protection in a MPA (Barrett *et al.* 2007, 2009); and resilience of protected area reef fish communities (Bates *et al.* 2014). Research has been undertaken on seabed habitat mapping and its role in MPA planning (Jordan *et al.* 2005) and mapping of inshore marine habitats for MPA planning and management (Barrett *et al.* 2001). However, some researchers have been more open in their criticism of MPAs. For instance, Buxton *et al.* (2006) have argued that considerable socio-economic impacts on Tasmanian fishers were not adequately determined in the Australian Government's proposed MPAs for the Commonwealth waters of the South-east region, surrounding Tasmanian state waters. The study argued for boundary and zoning changes to decrease the impact of the MPA network on the fishing industry. Kearney *et al.* (2012) have argued that interpretation of the precautionary principle in no-take MPAs does not meet international best practice and that the present definition reduces conservation outcomes.

Fig. 10.2. Black-faced Cormorants and kelp forests, Waterfall-Fortescue Marine Conservation Area, south-east Tasmania. (Photo: James Fitzsimons)

Wild capture fisheries are economically important for the state (Phillips *et al.* 2002). Abalone (*Haliotis* species) and Rock Lobster (*Jasus edwardsii*) fisheries are worth $97 million and $65 million, respectively. It is perhaps not surprising that the main opposition to an expansion MPAs has largely come from the commercial and recreational fishing sectors (Stump and Kriwoken 2006). Fishers are concerned about declining fish stocks in areas outside MPAs due to increased effort and fishing mortality. They argue that additional management measures such as no-take MPAs are not needed because fisheries harvesting practices are largely sustainable in Tasmania (Stump and Kriwoken 2006). Fishers also argue about the potential cost to the industry associated with reduced access and displaced fishing effort (see also Chapters 21, 22).

The main focus for Tasmanian environmental non-government organisations (ENGOs) has historically focused on terrestrial biodiversity issues including saving the tall eucalyptus forests that border the TWWHA (Buckman 2008). However, in 1989 the Tasmanian Conservation Trust was the first ENGO to establish a Marine Reserves Working Group and was active in lobbying for increased MPA coverage. More ENGOs are now actively involved in marine biodiversity issues. For example, Environment Tasmania, a consortium of 22 conservation groups, supports an active and diverse marine program (Environment Tasmania 2014). They have supported publications on marine natural values (e.g. Parsons 2011), campaigned against a super trawler in Australian waters, raised issues on the ecological and social impacts of the aquaculture industry and run marine discovery weekends.

Tasmania's MPA estate

Tasmanian state waters are ~2 356 000 ha including sub-Antarctic Macquarie Island. No-take zones represent 4.25% (100 200 ha) of Tasmanian waters, including Macquarie Island. Including unzoned and restricted-take zones brings the MPA total to 5.75% (135 400 ha) of Tasmanian waters. By excluding Macquarie Island, no-take MPAs represent only 1.12% (25 400) of Tasmanian waters. Including unzoned and restricted take zones brings the MPA total surrounding the island of Tasmania to 2.67% (60 700 ha) (DPIPWE 2014) (Table 10.1).

MPA declarations according to marine bioregions highlight areas of no representation (Fig. 10.3). Nine bioregions occur in Tasmanian state waters (Boags, Bruny, Davey, Flinders, Franklin, Freycinet, Otway, Twofold Shelf, Macquarie) yet there are no MPAs on the north or west coasts or surrounding Flinders or King Islands. As a result, four bioregions (Boags, Flinders, Franklin, Otway) have no MPA representation in Tasmanian waters.

Of the 21 MPAs, many are small no-take areas. The four MPAs designated in 1991 are all particularly small: Governor Island Marine Nature Reserve (60 ha); Ninepin Point Marine Nature Reserve (732 ha); Tinderbox MNR (144 ha); and a marine extension of Maria Island National Park (1282 ha). The Macquarie Island Nature Reserve, extended in 2000, is the largest MPA in the state at 74 738 ha, with the Kent Group National Park in the Bass Strait the second largest at 29 000 ha. The marine extension declared for the Port Davey-Bathurst

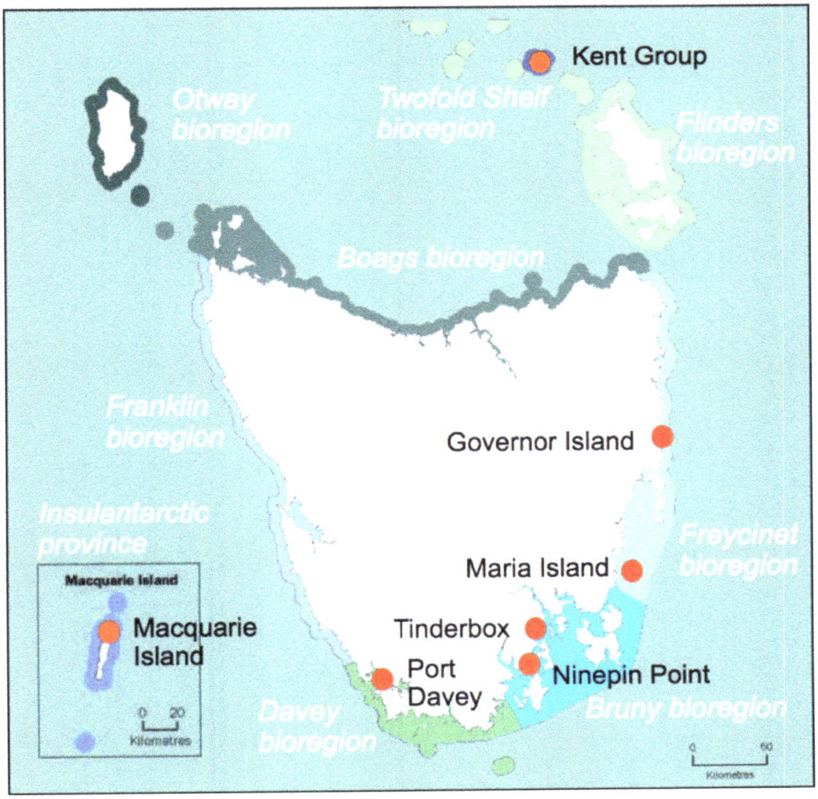

Fig. 10.3. Location of Tasmanian marine bioregions and marine protected areas (red dots). (Source: DPIPWE 2014)

Table 10.1. Tasmanian marine protected areas

(NCA = *Nature Conservation Act 2002*; LMRMA = *Living Marine Resources Management Act 1995*; IUCN = International Union for Conservation of Nature)

Marine protected area	NCA proclamation[a]	LMRMA proclamation[b]	IUCN category[c]	Area (ha)[d]	No-take area (ha)	Restricted take area (ha)	Unzoned area (ha)[e]
Governor Island Marine Nature Reserve	1991	1991	IV	60	60	0	0
Maria Island National Park	1991	1991	II/VI	1249	661	0	588
Ninepin Point Marine Nature Reserve	1991	1991	IV	731.8	731.8	0	0
Tinderbox Marine Nature Reserve	1991	1991	IV	144.1	144.1	0	0
Macquarie Island Nature Reserve	2000	2000	Ia	74 715	74 715	0	0
Kent Group National Park	2004	2005	II/IV	28 870	13 837	15 048	0
Port Davey (South-west National Park)	1976	2005	II/IV	17 753	9943	7810	0
Blackman Rivulet Marine Conservation Area	2009	No	VI	263.1	0	0	263.1
Central Channel Marine Conservation Area	2009	No	VI	3443	0	0	3443
Cloudy Bay Lagoon Marine Conservation Area	2009	No	VI	480.1	0	0	480.1
Hippolyte Rocks Marine Conservation Area	2009	No	VI	534.9	0	0	534.9
Huon Estuary Marine Conservation Area	2009	No	VI	694.2	0	0	694.2
Monk Bay Marine Conservation Area	2009	No	VI	399.2	0	0	399.2
Opossum Bay Marine Conservation Area	2009	No	VI	555.5	0	0	555.5
Port Cygnet Marine Conservation Area	2009	No	VI	103.1	0	0	103.1
River Derwent Marine Conservation Area	2009	No	VI	1636	0	0	1636
Roberts Point Marine Conservation Area	2009	No	VI	138.6	0	0	138.6
Simpsons Point Marine Conservation Area	2009	No	VI	573.9	0	0	573.9

Sloping Island Marine Conservation Area	2009	No	VI	940.3	0	0	940.3
South Arm Marine Conservation Area	2009	No	VI	772.4	0	0	772.4
Waterfall-Fortescue Marine Conservation Area	2009	No	VI	1230	0	0	1230
Total reserve area				**135 287**	**100 092**	**22 858**	**12 352**
Total reserve area (excluding Macquarie Island)				**60 572**	**25 377**	**22 858**	**12 352**

[a]Date of declaration under the *Nature Conservation Act 2002* or the previous *National Parks and Wildlife Act 1970*
[b]Date of declaration via fisheries rules under the *Living Marine Resources Management Act 1995* or before 1995 regulations under the *Fisheries Act 1959*
[c]Categories specified here may differ from previous documents, but the categories assigned reflect the way reserves are currently managed
[d]As proclaimed
[e]Unzoned areas are areas within the boundaries of MPAs that are either not proclaimed as no-take or restricted-take zones under the *Living Marine Resources Management Act 1995* or not proclaimed under the *Living Marine Resources Management Act 1995*

Fig. 10.4. Location of Tasmanian Marine Conservation Areas. (Source: DPIPWE 2014)

Channel area in the TWWHA is 17 732 ha. All 14 marine conservation areas declared in 2009 represent a total of 11 766 ha. These marine conservation areas declared in the Bruny bioregion (Fig. 10.4) and the north-east section of the Maria Island Marine Reserve are not protected under the LMRMA. As a result, these MPAs provide little security for marine life because the majority of marine plants and animals are not protected.

The only IUCN Category Ia MPA surrounds Macquarie Island and this represents the largest no-take MPA in the state. Governor Island, Ninepin Point and Tinderbox Marine Nature Reserves are IUCN Category IV (Table 10.1). Maria Island National Park, Kent Group National Park and Port Davey (South-west National Park) are classified at IUCN Categories II and IV. These MPAs have both no-take zones and fishing zones (but prohibit netting, dredging and other destructive forms of fishing).

Current policy

In response to the NRSMPA, the Minister for Primary Industries, Water and Environment established the Marine and Marine Industries Council (MMIC) in August 1999. The first role of MMIC was to develop a policy framework to guide the implementation of a system of Tasmanian MPAs. The *Tasmanian Marine Protected Area Strategy* was adopted in August 2001 (Tasmanian Government 2001), which followed a strategy background report (Tasmanian Government 2000). The primary goal of the Strategy was 'to establish and manage a comprehensive, adequate and representative system of MPAs, to contribute to the long-term ecological viability of marine and estuarine systems, to maintain ecological processes and systems, and to protect Tasmania's biological diversity' (Tasmanian Government 2001, p. 6). The Strategy outlined the goals, principles and outcomes of a Tasmanian Representative System of MPAs and proposed guidelines for identification and selection. Several strategic approaches to establishing MPA systems were canvassed. However, the Strategy was deficient on many grounds. The Strategy did not broadly outline the means by which a Tasmanian representative system would be created. There was no indication about whether Tasmania would adopt a system of smaller no-take area such as in Victoria or a system of larger multiple use MPAs with different zones (including no-take sanctuary zones and other zones allowing recreational fishing, etc.). There was no overall bioregional gap analysis undertaken to identify neither bioregional priorities nor specific targets on MPA representation for each bioregion. There were no timelines on developing a representative MPA system. No candidate MPAs were identified. Thus it was not surprising that the Strategy lacked political support for overall implementation.

However, on 6 June 2005 the Labor Minister for Environment and Planning issued a reference to the Resource Planning and Development Commission (RPDC) to conduct an inquiry and make recommendations on the establishment of MPAs in the Bruny bioregion in the south-east of Tasmania (Figs 10.3 and 10.4). The process took 2.5 years. The inquiry was conducted in accordance with Part 2 of the *Public Land (Administration and Forests) Act 1999* and followed steps outlined by the MPA Strategy (RPDC 2008). Three public reports provided information to the public and encouraged public consultation (Aiton and Kriwoken 2012). The background report identified areas of ecological significance in the bioregion and described the nine biounits (RPDC 2006). Forty-five potential areas for MPAs were identified, of which 31 were recognised as priority areas. The draft recommendations report (RPDC 2007) recommended 21 priority areas representing 38 407 ha. The final recommendations report recommended 14 MPAs covering 12 072 ha, with a mixture of no-take and restricted take areas (RPDC 2008). The RPDC noted that '[t]he resulting 14 recommended areas is as lean a system of MPAs as the Commission could determine whilst still complying with the terms of reference and the Strategy' (RPDC 2008, p. 3).

As a result of the inquiry, 14 marine conservation areas were declared, representing 11 766 ha. These MCAs were proclaimed under the NCA but were not declared under the LMRMA. All areas fall under IUCN Category VI and allow sustainable harvesting.

Finally, in 2014 any hope of further expanding a MPA system ground to a halt. The Tasmanian Liberals released a pre-election policy 'Supporting a world class Tasmanian fisheries and seafood sector' (Tasmanian Liberals 2014) and announced additional funding for biosecurity in the agricultural, aquaculture and wild fisheries sectors. Increased research funding for the salmonid industry was flagged. The document also included two short paragraphs under the heading 'Moratorium on Marine Protected Areas' outlining the newly elected state Liberal Government's policy direction on MPAs: 'Resource security is vital for investment and jobs in our wild fisheries sector. Currently, more than 135 000 ha of Tasmania's coastal waters are within MPAs, equating to nearly 8% of coastal water (*sic*)' and that '[a] majority Hodgman Liberal Government will impose a moratorium on any new MPAs in the State's waters' (Tasmanian Liberals 2014). This is the full extent of the present moratorium released by the Tasmanian Government. As a result, there are currently no goals or targets for MPA declaration and bioregional coverage in Tasmanian waters and a moratorium on establishing new MPAs is in place.

Challenges

Tasmanian society is highly polarised on issues regarding biodiversity protection and commercial resource use (Buckman 2008). The long-standing forest battles surrounding the TWWHA are a case in point. Common ground between ENGOs and the forest industry has not been resolved and these tensions continue to characterise the state. It is likely that this polarisation will now be extended to the marine environment and future designations of MPAs. The marine program outlined by Environment Tasmania (2014) points towards increased ENGO lobbying in this marine space. This seems inevitable where advocates promoting increased marine biodiversity protection and MPAs have little in common with the commercial and recreational fishing sector. The commercial and recreational fisheries sector has a very significant economic role in Tasmania and their political influence in the MPA debate cannot be underestimated. The strength of recreational fishing sector was recently evident in a successful campaign to stop a super trawler fishing in south-east Australia (Tracey *et al.* 2013). If MPAs have any chance of expansion then this increasing polarisation will need to be resolved. This remains an enormous challenge facing the state.

The MPA system is in need of legislative and administrative reform and the overall governmental responsibility for biodiversity conservation remains a challenge. The split between marine resource management and biodiversity divisions in DPIPWE needs rationalisation. One proposal is that biodiversity legislation (NCA and NPRMA) and fisheries legislation (LMRMA) be amended to allow biodiversity conservation of marine flora and fauna to be managed by the Director of National Parks and Wildlife (e.g. RPDC 2008). In order to meet the objectives of the National Representative System of Marine Protected Areas, the Tasmanian Government needs to also adequately resource DPIPWE to manage MPAs and plan for future expansion to ensure they are representative. The Tasmanian Parks and Wildlife Service does not have adequate resourcing to manage existing MPAs and additional funds would be needed to expand the MPA system. The only marine reserves officer position within DPIPWE has now been terminated.

The existing 2001 MPA Strategy is now 13 years old and faces several challenges. There is concern that the 'selection criteria' outlined in the Strategy is only a list of 'considerations' that must be assessed to determine new MPAs (RPDC 2008). More specific details are required to implement MPA selection criteria and prioritise MPAs in all Tasmanian waters. If a state-wide MPA system is to be realised, then the type of inquiry that identified the Bruny bioregion MPAs could be broadened to include all bioregions. Limiting the MPA inquiry to one bioregion had several negative consequences. For example, the Bruny bioregion inquiry had limited involvement from the commercial fishing sector. The reason was that commercial fishers could not adequately determine impacts of MPAs in the Bruny bioregion without having some understanding of the range of MPAs in the other eight bioregions. Inquiries focusing on only one bioregion will continue to limit credible input from the commercial fishing sector (see also Chapter 22).

Future directions

Given that political support for expanding the MPA system in Tasmania will not be forthcoming in the near future, increased support needs to be focused on existing MPAs. This includes providing funding to complete management plans for all MPAs. Management plans need to embrace best practice community consultation and be informed by a wide variety of marine stakeholders. The identification of the boundaries of existing MPAs is also a critical issue for all marine users. This is particularly important for enforcement undertaken by the Tasmanian Marine Police. For example, better boundary definition can be supported using poles and buoys, land marks and GPS coordinates on charts. It is important to provide workable boundaries that are unambiguous and easy to follow for users. Poor enforcement can result in illegal extraction from MPAs going unchecked and reduce the biodiversity values of MPAs.

The Tasmanian MPA Strategy includes a provision for ongoing assessment of MPA performance in meeting their stated objectives. This performance assessment can include the MPA system as a whole and specific individual MPAs. This is particularly important for the ongoing management of MPAs because boundaries can be adjusted or management regimes changed and improved. An adaptive management framework as adopted by PWS (Jones 2014) could be more widely supported and independent third-party assessments could be undertaken.

Increased public education and community awareness programs about Tasmania's marine biodiversity values are critically important. An educational strategy should be implemented to inform Tasmanians about the importance of identifying, selecting and establishing a representative system of MPAs. The Environment Tasmania report *Nowhere Else on Earth: Tasmania's Marine Natural Values* (Parsons 2011) highlights the important global conservation significance of Tasmanian waters and contributes to this need. The Tasmanian MPA Strategy outlines educational priorities including: greater awareness of marine biodiversity, habitats and ecosystems; a program to promote the role and benefits of MPAs; and establish community-based monitoring and awareness programs. The success of RedMap is an excellent example of community-based monitoring recording range extension of marine species (RedMap 2014).

Greater long-term community cooperation between ENGOs and the commercial and recreational fishing sector needs to be encouraged. Cooperation between ENGOs and the recreational fishing sector in stopping a super trawler from fishing in Australian waters showed that these groups can share a common vision for the marine environment. Determining the socio-economic impact data on potential MPAs is essential in order to engage both ENGOs and fishers in the long-term planning of a MPA system.

Conclusions

The history and development of the Tasmanian MPA network has been *ad hoc* and controversial. There has never been a single MPA champion in government and only recently have ENGOs taken an interest. The commercial and recreational fishing sector remains economically and politically powerful and has largely opposed any MPA expansion. As a result, the Tasmanian MPA estate fails to meet minimum representation standards outlined in the NRSMPA with four bioregions having no MPAs. The current moratorium on MPAs is a new low point for marine biodiversity protection and the future for new MPA designations looks bleak. However, several incremental changes could assist in promoting the role and relevance of MPAs. This could include increased community education on MPAs, management planning and performance assessment for existing MPAs and exploring ways to find common ground between advocates of MPAs and the fisheries sector.

Biography

Lorne Kriwoken is an environmental scientist in the Geography and Spatial Sciences Discipline, School of Land and Food and an Adjunct Research Associate of the Institute for Marine and Antarctic Studies, University of Tasmania. His research interests are interdisciplinary and include biodiversity protection, environmental management and ocean and coastal governance. He has degrees in environmental science from Simon Fraser University, Dalhousie University and a PhD from University of Tasmania.

References

Aiton K, Kriwoken LK (2012) *The Bruny Bioregion Marine Protected Area Inquiry: An Evaluation of Public Submissions.* Coastal Collaboration Cluster, Theme 3 Knowledge Systems, Working Paper. School of Geography and Environmental Studies, University of Tasmania, Hobart.

ANZECC (1998) *Guidelines for Establishing the National Representative System of Marine Protected Areas.* Australian and New Zealand Environment and Conservation Council Task Force on Marine Protected Areas, Environment Australia, Canberra.

ANZECC (1999) *Strategic Plan of Action for the National Representative System of Marine Protected Areas: A Guide for Action by Australian Governments.* ANZECC Task Force on Marine Protected Areas, Environment Australia, Canberra.

Barrett N, Sanderson JC, Lawler M, Halley V, Jordan A (2001) *Mapping of Inshore Marine Habitat in South-eastern Tasmania for Marine Protected Area Planning and Marine Management.* Marine Research Laboratories, Tasmanian Aquaculture and Fisheries Institute, University of Tasmania, Hobart.

Barrett N, Edgar GJ, Buxton CD, Haddon M (2007) Changes in fish assemblages following 10 years of protection in Tasmanian marine protected areas. *Journal of Experimental Marine Biology and Ecology* **345**, 141–157. doi:10.1016/j.jembe.2007.02.007.

Barrett N, Buxton CD, Edgar GJ (2009) Changes in invertebrate and macroalgal populations in Tasmanian marine reserves in the decade following protection. *Journal of Experimental Marine Biology and Ecology* **370**, 104–119. doi:10.1016/j.jembe.2008.12.005.

Bates AE, Barrett N, Stuart-Smith RD, Holbrook NJ, Thompson RA, Edgar GJ (2014) Resilience and signatures of tropicalization in protected reef fish communities. *Nature Climate Change* **4**, 62–67. doi:10.1038/nclimate2062.

Buckman G (2008) *Tasmania's Wilderness Battles: A History.* Allen & Unwin, Sydney.

Buxton C, Haddon M, Bradshaw M (2006) *Regional Impact Assessment for the Marine Protected Areas Proposed for the South-East Region.* Fisheries Research and Development Corporation and Tasmanian Aquaculture and Fisheries Institute, University of Tasmania, Hobart.

DPIPWE (2014) *Tasmanian Reserve Estate Spatial Layer.* Department of Primary Industries, Parks Water and Environment, Hobart. <http://dpipwe.tas.gov.au/conservation/development-planning-conservation-assessment/tools/tasmanian-reserve-estate-spatial-layer>.

Edgar G, Barrett N (1997) Short term monitoring of biotic change in Tasmanian marine reserves. *Journal of Experimental Marine Biology and Ecology* **213**, 261–279. doi:10.1016/S0022-0981(96)02769-4.

Edgar G, Barrett N (1999) Effects of the declaration of marine reserves on Tasmanian reef fishes, invertebrates and plants. *Journal of Experimental Marine Biology and Ecology* **242**, 107–144. doi:10.1016/S0022-0981(99)00098-2.

Environment Tasmania (2014) Environment Tasmania home page. Environment Tasmania, Hobart, <http://www.et.org.au/>.

Ivanovici A (Ed.) (1984) *Inventory of Declared Marine and Estuarine Protected Areas in Australian Waters.* Special Edition 12. Australian National Parks and Wildlife Service, Canberra.

Jones G (2014) Monitoring and reporting system. Parks and Wildlife Service, Hobart, <http://parks.tas.gov.au/index.aspx?base=17551>.

Jordan A, Lawler M, Halley V, Barrett N (2005) Seabed habitat mapping in the Kent Group of islands and its role in marine protected area planning. *Aquatic Conservation: Marine and Freshwater Ecosystems* **15**, 51–70. doi:10.1002/aqc.657.

Kearney R, Buxton CD, Goodsell P, Farebrother G (2012) Questionable interpretation of the precautionary principle in Australia's implementation of 'no-take' marine protected areas. *Marine Policy* **36**, 592–597. doi:10.1016/j.marpol.2011.10.018.

Kriwoken LK (2011) Australian marine protected areas: charting a course towards a representative system. In *Marine Resources Management.* (Eds W Gullett, C Schofield and J Vince) pp. 171–181. LexisNexis Butterworths, Sydney.

Kriwoken LK, Haward MG (1991) Marine and estuarine protected areas in Tasmania, Australia: the complexities of policy development. *Ocean and Shoreline Management* **15**, 143–163. doi:10.1016/0951-8312(91)90028-Z.

Parsons KE (2011) *Nowhere Else on Earth: Tasmania's Marine Natural Resources.* Report for Environment Tasmania. Aquenal, Hobart.

Phillips G, Kriwoken LK, Hay PR (2002) Private property and public interest in fisheries management: the Tasmanian Rock Lobster fishery. *Marine Policy* **26**, 459–469. doi:10.1016/S0308-597X(02)00026-X.

RedMap (2014) RedMap home page, RedMap, Brisbane, <http://www.redmap.org.au/>.

RPDC (2006) *Inquiry into the Establishment of Marine Protected Areas within the Bruny Bioregion: Background Report.* Resource Planning and Development Commission, Hobart.

RPDC (2007) *Inquiry into the Establishment of Marine Protected Areas within the Bruny Bioregion: Draft Recommendations Report.* Resource Planning and Development Commission, Hobart.

RPDC (2008) *Inquiry into the Establishment of Marine Protected Areas within the Bruny Bioregion: Final Recommendations Report.* Resource Planning and Development Commission, Hobart.

Stump NE, Kriwoken LK (2006) Tasmanian Marine Protected Areas: attitudes and perceptions of wild capture fishers. *Ocean and Coastal Management* **49**, 298–307. doi:10.1016/j.ocecoaman.2006.03.007.

Tasmanian Government (2000) *Tasmanian Marine Protected Areas Strategy: Background Report.* Department of Primary Industries, Water and Environment, Hobart.

Tasmanian Government (2001) *Tasmanian Marine Protected Area Strategy.* Department of Primary Industries, Water and Environment, Hobart.

Tasmanian Government (2014) *Recreational Sea Fishing Guide 2014-15.* Wild Fisheries Management Branch, Department of Primary Industries, Parks, Water and Environment, Hobart.

Tasmanian Liberals (2014) Supporting a world class Tasmanian fisheries and seafood sector. Tasmanian Liberals, Hobart, <https://www.tas.liberal.org.au/sites/default/files/policy/Supporting%20a%20world%20class%20Tasmanian%20fisheries%20and%20seafood%20sector.pdf>.

Tracey S, Buxton C, Gardner C, Green B, Hartmann K, Haward M, *et al.* (2013) Super trawler scuppered in Australian fisheries management reform. *Fisheries* **38**, 345–350. doi:10.1080/03632415.2013.813486.

Chapter 11

Marine protected areas in New South Wales

Pepe Clarke

Introduction

From Tweed Heads to Eden, and out to Lord Howe Island, the New South Wales (NSW) marine estate covers around 1 million ha, including a remarkable diversity of species and ecosystems:

> *'From the cool waters in the south to the warm waters of the north, the coral reefs, estuaries, seagrass meadows, sandy beaches, rocky headlands, kelp beds, sponge gardens, subtidal and deep-water rocky reefs, sandy plains and seamounts in NSW support a rich mix of tropical and temperate marine life'.* (Babcock *et al.* 2014)

Marine protected areas (MPAs) are well established as a tool for the conservation of marine species and ecosystems in NSW. Early MPAs were declared under existing fisheries management and national parks legislation. The *Marine Parks Act 1997* provided a modern legislative framework for the establishment and management of MPAs in NSW until November 2014, when it was replaced by the *Marine Estate Management Act 2014.*

The NSW system of MPAs includes six multiple use marine parks, 12 aquatic reserves and 62 national parks and reserves with relatively small marine components. In total, these MPAs cover ~367 500 ha, representing nearly 37% of state waters.

The expansion of the NSW MPA estate over the past 25 years is a substantial conservation achievement, but further progress is required to meet national criteria for a comprehensive, adequate and representative network of MPAs. Currently, less than 7% of state waters are protected in sanctuary zones, and two marine bioregions are effectively unrepresented in the NSW marine reserve system.

This chapter reviews the development of the NSW MPA network, provides an overview of the current status of the network, identifies key challenges and lessons learnt and examines current directions in NSW marine conservation policy.

Development of the NSW marine protected area network

For thousands of years, Indigenous Traditional Owners have harvested marine resources from the estuaries, foreshores and coastal waters between Tweed Heads and Eden. In the early years of European settlement, coastal fisheries were an essential source of sustenance for a young colony struggling to survive. Over time, a nascent commercial fishery began to emerge, with a particular focus on netting in coastal rivers and estuaries.

Fisheries management laws

By 1865, concerns about localised depletion of fish stocks prompted the enactment of the first fisheries management legislation in the colony.[1] In 1881, the NSW Parliament enacted further legislation to conserve marine and freshwater fish stocks more effectively, including provision for spatial closures, establishing a precedent for future area-based management measures.[2] Fishermen played a key role in lobbying parliament for the introduction of these early fisheries management laws, to prevent exhaustion of coastal and estuarine fish stocks.

Today, the *Fisheries Management Act 1994* provides the legislative framework for fisheries management in NSW, including spatial management tools. In 2002, the Minister for Primary Industries used this legislation to declare 30 'recreational fishing havens' on the NSW coast, including Lake Macquarie, Port Botany and the Bermagui River.[3] Commercial fishing is prohibited in these areas to enhance recreational fishing opportunities. There is anecdotal evidence to suggest that the number and size of fish increases over time within recreational fishing havens, but limited scientific research has been conducted to date to assess their long-term conservation benefits (NSW Legislative Council 2010).

Fisheries laws have also been used to declare 12 aquatic reserves, covering ~2000 ha. These aquatic reserves include popular visitor destinations such as Long Reef (declared in 1980), Shiprock (1982) and Cabbage Tree Bay (2002). In November 2014, legislative provisions relating to aquatic reserves were moved from the *Fisheries Management Act 1994* to the *Marine Estate Management Act 2014*. Under these new arrangements, the Minister for Primary Industries and the Minister for the Environment share the power to declare aquatic reserves and adopt management rules and management plans. The primary purpose of aquatic reserves is 'to conserve biological diversity' (s.33). In practice, the level of protection provided by aquatic reserves is highly variable, ranging from targeted protection of invertebrate species to strict no-take reserves (NSW Department of Primary Industries 2015).

National parks and nature reserves

The first permanently protected marine area in NSW was the Port Hacking estuary, which became part of the Royal National Park when the park was declared in 1879.

Since that time, more than 850 national parks and reserves have been declared in NSW, reserved under the *National Parks and Wildlife Act 1974* and managed by the National Parks and Wildlife Service (NPWS). Nearly half of the state's coastline is protected within terrestrial reserves, and there are 62 national parks and reserves with marine and estuarine components. These areas include ocean coastlines, estuaries, coastal lakes, wetlands, intertidal areas, ocean beaches, rocky shores and islands (NSW Government 2012). Booderee National Park, located in federal territory at Jervis Bay and surrounded by NSW, is managed by Parks Australia and contains a marine component of 875 ha (Director of National Parks 2011).

Since its establishment in 1957, the National Parks Association has been the leading public advocate for the establishment and effective management of national parks in NSW. Successful campaigns by the Association led to the establishment of the National Parks and Wildlife Service and the enactment of the *National Parks and Wildlife Act 1974*.

Historically, expansion of the national parks estate has received bipartisan political support, but over the 4 years up to 2014 there has been a substantial reduction in the rate of national park establishment (see Department of Premier and Cabinet 2014), together with proposals to open protected areas to previously prohibited extractive activities. Encouragingly, the state government recently abandoned a proposal to permit commercial netting in protected estuaries and national parks (Hodgkinson 2014).

Marine parks

Establishment of a modern network of marine parks in NSW commenced in 1991, when the Liberal-National Government established the Solitary Islands Marine Reserve (Fig. 11.1). The reserve contained a diverse range of habitats, including estuaries, beaches, reefs, rocky shores and open ocean. The reserve was established by regulation under existing fisheries legislation, and was divided into sanctuary, refuge and recreation zones.[4]

The following year, federal and state governments committed to the establishment of a national system of MPAs (COAG 1992). In the lead up to the 1995 NSW state election, the Australian Labor Party (ALP) committed to introduce marine parks legislation and commence establishment of a representative marine parks network for the state.

This commitment was supported by key state conservation organisations, including the Nature Conservation Council, which had adopted policy in 1990 calling for the establishment of marine parks and a marine conservation authority (NCC 1990). The National Parks Association was active in lobbying to establish a statutory framework for marine parks in NSW (A. Cox *pers. comm.* 2014).

In 1997, marine parks legislation was introduced by the Labor Government and passed by the NSW Parliament with bipartisan support. The Liberal-National opposition moved

Fig. 11.1. Anemone Bay, Solitary Islands Marine Park. (Photo: Hamish Malcolm)

Table 11.1. Marine parks in New South Wales at June 2015

(NSW Government 2015)

Marine park name	Established	Total area (hectares)
Solitary Islands Marine Park	1998	72 200
Jervis Bay Marine Park	1998	21 100
Lord Howe Island Marine Park	1999	48 000
Cape Byron Marine Park	2002	22 000
Port Stephens–Great Lakes Marine Park	2005	98 000
Batemans Marine Park	2006	85 000
Total		**346 300**

multiple amendments to the *Marine Parks Bill 1997*, but did not oppose the legislation, on the basis that it implemented an ALP election promise and was consistent with the Liberal-National coalition's commitment to marine conservation (NSW Parliament 1997).

The following year, the legislation was used to reclassify the Solitary Islands Marine Reserve as a marine park and establish a new marine park at Jervis Bay (NSW Government Gazette 1998). Informed by bioregional assessments and community consultation, a further four marine parks were declared – Lord Howe (1999), Cape Byron (2002), Port Stephens-Great Lakes (2005) and Batemans (2006) Marine Parks – increasing the total area of marine parks in NSW waters to 346 300 ha (Table 11.1; Fig. 11.2).

The establishment of these marine parks was championed by Ministers for the Environment Pam Allan (1995–1999) and Bob Debus (1999–2007) and supported by Premier Bob Carr (1995–2005). No new marine parks have been declared in NSW since 2006.

Current status of the NSW marine protected area network

In November 2014, the *Marine Parks Act 1997* was repealed and replaced with the *Marine Estate Management Act 2014*. The new legislation provides for management of the NSW marine estate, including 'declaration and management of a comprehensive system of marine parks and aquatic reserves'. The *Marine Estate Management Act 2014* is administered by the Minister for Primary Industries and the Minister for the Environment.

The NSW system of MPAs currently includes six multiple use marine parks, 12 aquatic reserves and 62 national parks and nature reserves with marine components, covering nearly 37% of state waters (Table 11.2, Fig. 11.2). The remainder of this chapter deals

Table 11.2. Marine protected area types in New South Wales at June 2015

(NSW Government 2015; NSW Department of Primary Industries 2015)

Protected area type	Legislation	Total area (approximate) (ha)	Proportion of state waters (%)
Marine parks	*Marine Estate Management Act 2014*	346 300	34.6
National parks and nature reserves	*National Parks and Wildlife Act 1974*	20 000	2
Aquatic reserves	*Marine Estate Management Act 2014*	2000	0.2
Total		**368 000**	**36.8**

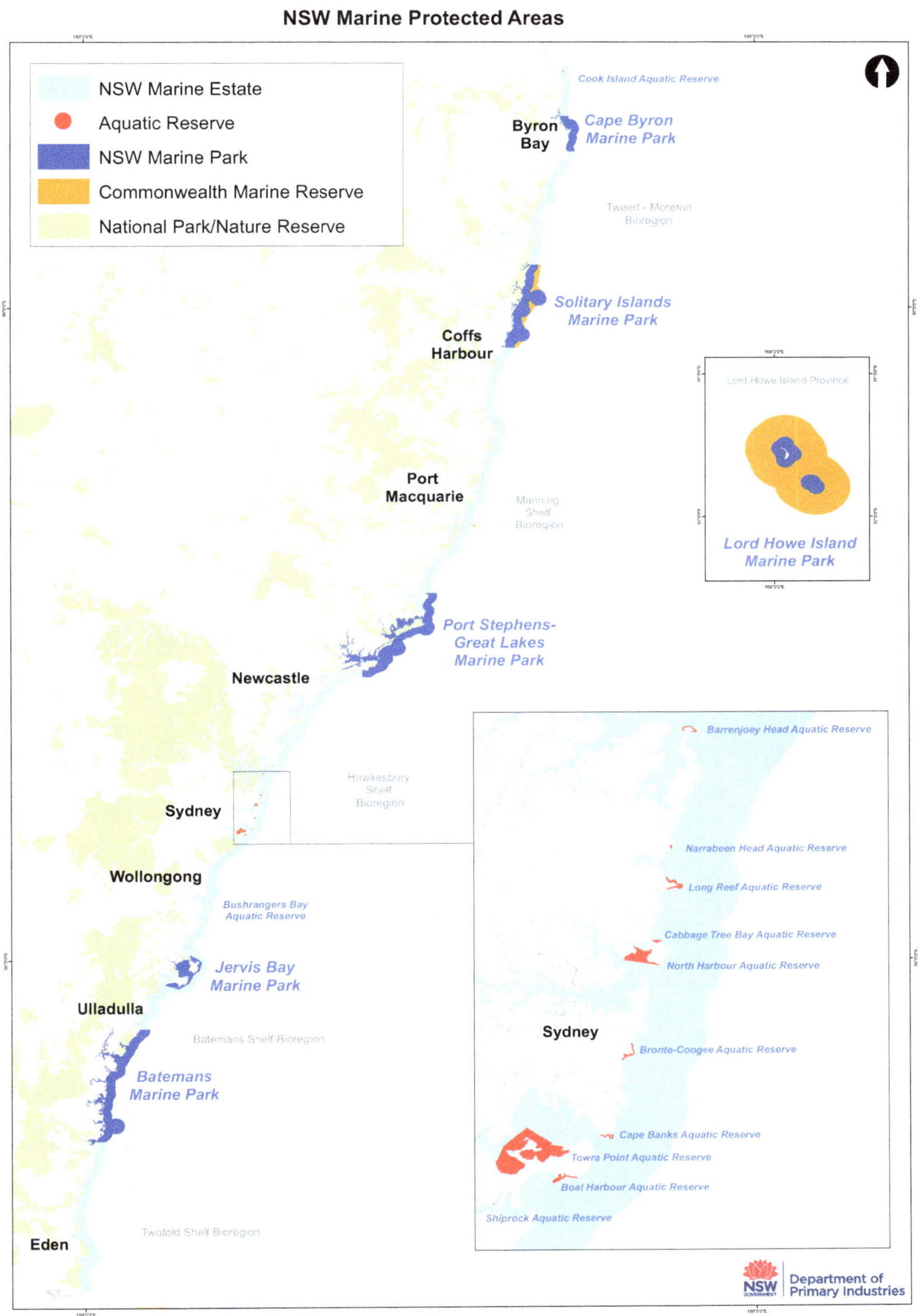

Fig. 11.2. Marine protected areas in New South Wales. (Source: NSW Marine Parks Authority, www.mpa.nsw.gov.au/pdf/NSW-Marine-protected-areas.pdf)

Table 11.3. Marine park zones in New South Wales

(NSW Government 2015; Department of the Environment 2014)

Zone	IUCN protected area category[a]	Total area	Proportion of marine park network (%)	Proportion of state waters (%)
Sanctuary	II	64 922	19	6.3
Habitat protection	IV	167 194	48	16.2
General use	VI	112 226	33	10.9
Special purpose	-	743	0.22	0.07
Total		**345 085**	**100**	**34**

[a]Source: CAPAD 2014, www.environment.gov.au/land/nrs/science/capad/2014. Note that this characterisation of HPZ (IUCN IV) and GUZ (IUCN VI) is contested – see Winn (2007).

primarily with marine parks established under the *Marine Parks Act 1997*, which are now managed under the *Marine Estate Management Act 2014*.

Management regulations and zoning plans

Marine parks in NSW are zoned to conserve biodiversity, maintain ecological processes and provide for a range of uses such as recreational and commercial fishing, diving, boating, snorkelling and tourism.

The *Marine Estate Management Act 2014* provides for the preparation of marine park management regulations and sets out the process for the preparation of zoning plans. The *Marine Estate Management (Management Rules) Regulation* provides for the declaration of four types of zone: sanctuary zones, habitat protection zones, general use zones and special purpose zones (see Fig. 11.3 for an example of zoning at Cape Byron Marine Park).

The purpose and characteristics of each zone are:

- **Sanctuary zones** provide the highest level of protection by prohibiting harm to animals, plants or habitats, including fishing. Recreational, educational and scientific activities that do not harm plants, animals and habitats are permitted.
- **Habitat protection zones** conserve marine biodiversity by protecting habitats and reducing high-impact activities. Recreational fishing and some forms of commercial fishing are permitted.
- **General use zones** permit a wide range of activities, including commercial and recreational fishing provided they are ecologically sustainable.
- **Special purpose zones** are used when there are special management needs, including protection of cultural values, or for marine facilities.

To date, less than 20% of the marine park network has been zoned as sanctuaries, representing ~6.3% of state waters (Fig. 11.4). Nearly half of the marine park network is zoned for habitat protection with a third of the network zoned for general use (Table 11.3). Two marine bioregions – the Hawkesbury Shelf and Twofold Shelf – are entirely unrepresented in the marine park network (NSW Government 2012) (Figs 11.2, 11.5).

Progress towards a comprehensive, adequate and representative network

The period from 1997 to 2006 marked significant progress towards the establishment of a comprehensive, adequate and representative marine park network in NSW. However, progress has since stalled, leaving the marine park network incomplete. The low level of

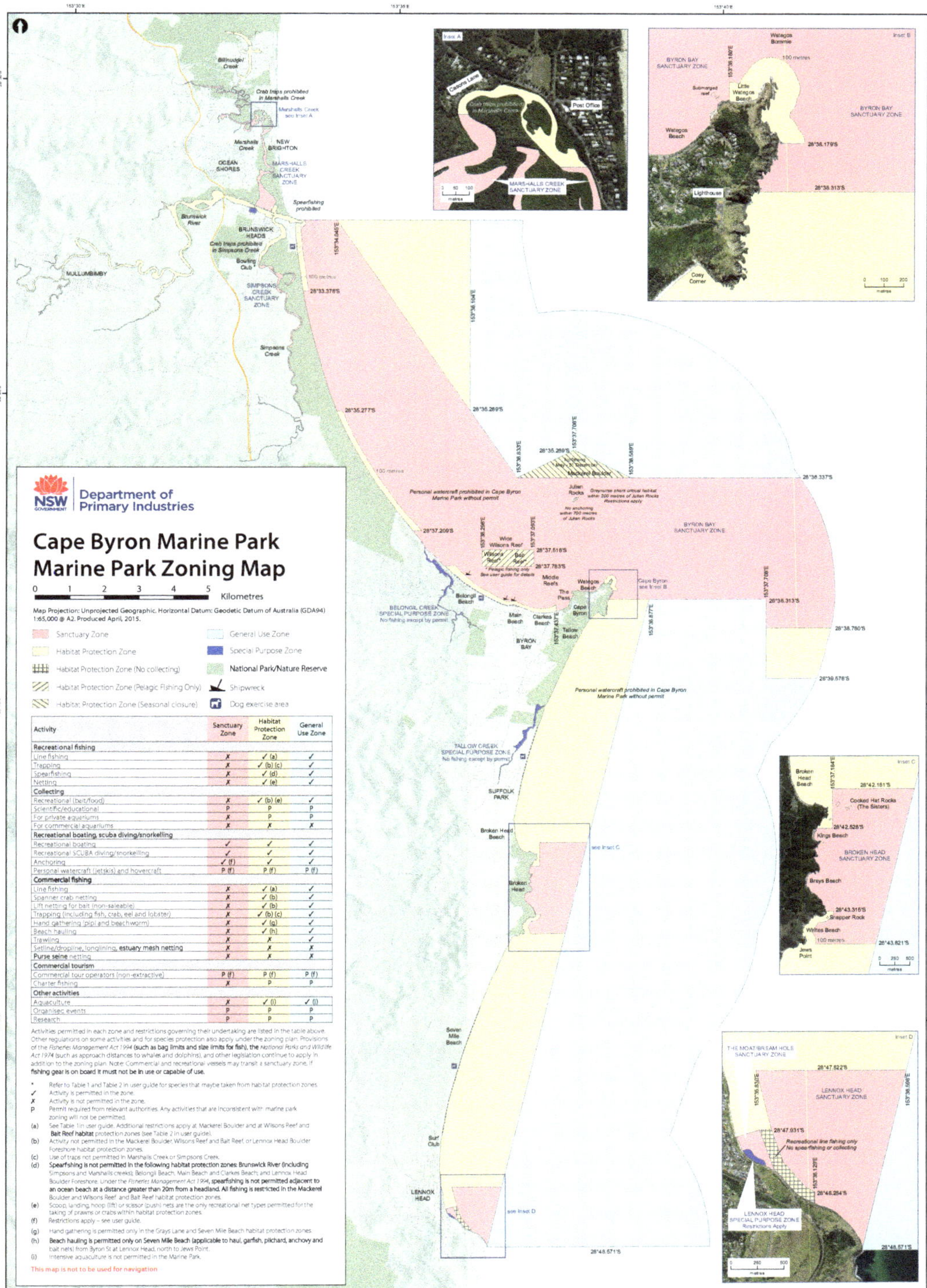

Fig. 11.3. Zoning scheme for the Cape Byron Marine Park.

Fig. 11.4. Marine park zones as a proportion of NSW waters. (Source: NSW Government 2015)

sanctuary protection (6.3%) and the absence of marine parks in the Hawkesbury Shelf and Twofold Shelf bioregions are particularly notable gaps in the network (Fig. 11.5).

Recent developments in marine park zoning

The *Marine Parks Act 1997* established a process for the preparation and periodic review of zoning plans, including requirements for public consultation. The zoning plans for Solitary Islands and Jervis Bay Marine Parks were reviewed most recently, with public consultation occurring from 2008 to 2010 and the final zoning plans gazetted in March 2011. Overall, these new plans resulted in a modest increase in the level of protection, expanding the total area contained within sanctuary zones and enhancing protection for key species.[5]

Two months later, following the 2011 state election, the newly elected Liberal-National NSW Government disallowed the new management regulations, returning to the previous zoning arrangements. Conservation organisations reacted strongly to this unprecedented reversal of marine conservation measures, and dismissed claims by the new government that there had been inadequate consultation (Australian Broadcasting Corporation 2011).

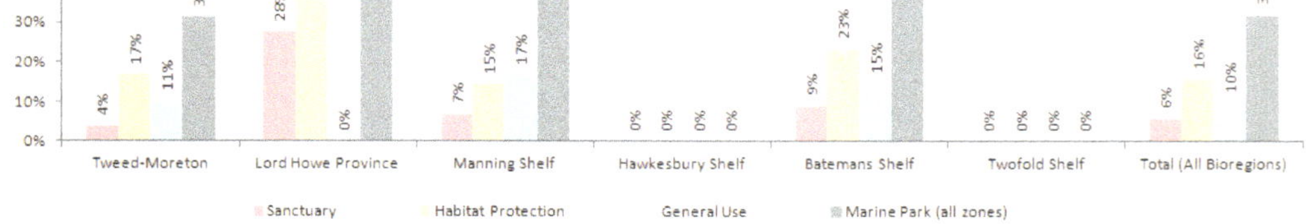

Fig. 11.5. Proportion of bioregions in NSW waters covered by marine park zones. (Source: NSW Government 2012)

Overall, there has been little progress towards a comprehensive, adequate and representative marine park network in NSW since 2006.

Policy and political context

Commitment to establish a marine park network

The commitment to the establishment of a comprehensive, adequate and representative network of MPAs contained in the 1992 Intergovernmental Agreement on the Environment was reaffirmed in a succession of national and state policy documents (COAG 1992; Commonwealth of Australia 1992, 1996, 1998a, 1998b, 1998c, 1999; ANZECC 1998; NSW Government 1997, 1999; NPWS 1999), and reflected in the statutory aims of the *Marine Parks Act 1997*.

In 2001, the NSW Government released a policy document entitled *Developing a Representative System of Marine Protected Areas in New South Wales*, which confirmed the state's commitment to the establishment of 'a comprehensive, adequate and representative system of MPAs that includes a full range of marine biodiversity at ecosystem, habitat and species levels' (NSW Marine Parks Authority 2001).

The policy set out principles for the establishment of the MPAs system, together with guidance on identification, declaration and management of MPAs, public consultation, research and evaluation. However, marine parks have since become a highly contested issue in NSW, resulting in a progressive departure from the principles and priorities set out in the 2001 policy on MPAs.

Political opposition leads to policy reversals

Fishing organisations, including the NSW Recreational Fishing Alliance and the Professional Fishermen's Association, have strongly opposed new marine parks and sanctuary zones (PFA 2014). Environmental organisations, including the National Parks Association and Nature Conservation Council, continue to advocate for progressive expansion of the marine park network and increased representation of marine ecosystems in sanctuary zones (NCC and NPA 2014).

This strong division has contributed to an increasing politicisation of marine conservation in NSW. In contrast to the 1997 parliamentary debates on the *Marine Parks Act 1997*, in which both sides of politics highlighted their commitment to marine conservation, marine parks are today a highly partisan issue.

Opposition to marine sanctuaries is a key policy platform of the Shooters and Fishers Party, a minor party with two representatives in the NSW Legislative Council (Shooters and Fishers Party 2013). Under electoral pressure from the Shooters and Fishers Party, the National Party have been increasingly active in promoting the interests of recreational and commercial fishing organisations. In the lead up to the 2011 state election, the National Party campaigned actively against new marine parks and sanctuaries (NSW Parliament 2010). Following the election, the newly elected Liberal-National Government moved quickly to implement election commitments made by National Party candidates, including:

- reversing the new zoning plans for Solitary Islands and Jervis Bay Marine Parks
- rolling back new protections for critically endangered Grey Nurse Shark populations at Fish Rock and Green Island
- transferring responsibility for management of marine parks to the Department of Primary Industries.

Fig. 11.6. Wagonga Inlet, Batemans Marine Park. (Photo: Geoff Kelly)

During the 2011–2015 term of government, the Shooters and Fishers Party held the balance of power in the Legislative Council, and used this influence effectively to limit expansion of marine parks. In September 2011, the Shooters and Fishers Party successfully introduced a legislated moratorium on the declaration of new marine parks, creation of new sanctuary zones or alteration of existing sanctuary zones.[6]

Following commencement of the moratorium, the Minister for Primary Industries ordered an 'independent scientific audit' of the marine parks network. Following the release of the audit report (Beeton *et al.* 2012), the following policies were announced (Hodgkinson and Parker 2013):

- establishment of a Marine Estate Management Authority
- development of a threat and risk assessment framework
- a new approach to management of the marine estate, including marine parks
- a review of zoning plans for Solitary Islands and Batemans Marine Parks (Fig. 11.6)
- an amnesty on recreational fishing from beaches and headlands in sanctuary zones.

The decision to open coastal sanctuary zones to recreational fishing was welcomed by fishing advocates, but attracted strong criticism from environmental organisations and marine scientists. In January 2014, more than 200 marine scientists wrote to Premier O'Farrell, expressing concern that 'the integrity of the NSW marine parks network … will be severely undermined by moves to allow recreational fishing in sanctuary zones' (Babcock *et al.* 2014).

Recent developments

In April 2014, following the resignation of Premier O'Farrell, the newly appointed Minister for the Environment, Rob Stokes, signalled a potential shift in government policy on marine conservation, expressing support for improved marine conservation measures in NSW (Patty 2014).

In November 2014, the *Marine Estate Management Act 2014* was passed with bipartisan support. The new legislation repeals the *Marine Parks Act 1997*, provides a statutory basis for Marine Estate Management Authority, and provides for declaration and management of marine parks and aquatic reserves. The marine park provisions of the new legislation largely replicate the relevant provisions of the *Marine Parks Act 1997*.

Interestingly, the new legislation removed the 5-year moratorium on new marine parks and changes to sanctuary zones. In December 2014, following commencement of the new legislation, the Ministers for Primary Industries and the Environment announced that:

- 10 coastal sanctuaries would be permanently rezoned to permit recreational line fishing from the shore
- the fishing amnesty would be immediately removed for the remaining 20 coastal sanctuaries, restoring full protection to those sites
- a comprehensive bioregional assessment would be completed for the Hawkesbury Shelf, to inform future marine conservation measures (Hodgkinson and Stokes 2014a).

The Hawkesbury Shelf Marine Estate Assessment will be coordinated by the Marine Estate Management Authority, and cover state waters from Newcastle to Wollongong, including the Greater Sydney region. The Minister for the Environment has indicated that the assessment will provide ecological, social and economic data to inform decision making about the creation of an 'integrated marine protected area' for the region by 2016 (Hodgkinson and Stokes 2014b).

In March 2015, the Liberal-National Government was re-elected with a reduced majority in the Legislative Assembly (54 seats out of 93, down from 69 seats). In the Legislative Council, the Liberal-National Coalition secured a net increase of one seat (20 seats out of 42, up from 19 seats). The Shooters and Fishers Party and Christian Democrats maintained their level of representation in the Legislative Council, with two seats each. This result is likely to reduce the influence of the Shooters and Fishers in the 2015–2019 term, because the government will be able to pass legislation with the support of the Christian Democrats.

Key challenges and lessons learnt

In the decade following the passage of the *Marine Parks Act 1997*, NSW made strong progress on marine conservation: six new marine parks were established, covering more than a third of state waters. Bioregional assessments were completed and zoning plans were prepared, providing protection for a diverse range of marine habitats (Fig. 11.7).

Key factors contributing to the establishment of the state's marine park network include:

- a commitment by NSW and Australian Governments to the establishment of a comprehensive, adequate and representative marine park network
- public advocacy and community education by environmental organisations
- introduction of a modern legislative framework, including provisions related to:
  - declaration of marine parks
  - regulation of activities within marine parks
  - preparation and review of management plans
  - compliance and enforcement powers
- establishment of a dedicated Marine Parks Authority and advisory bodies
- high-level political support from the Premier and Ministers for the Environment.

Fig. 11.7. Rock platform, Lord Howe Island Marine Park. (Photo: Geoff Kelly)

However, progress has since stalled, leaving an incomplete and inadequate marine park network, with less than 7% of state waters protected in sanctuary zones. Key factors that appear to have limited progress since 2006 include: public advocacy and lobbying by fishing organisations; the increased influence of the Shooters and Fishers Party, including election of representatives to NSW Parliament; and, active opposition to marine parks and sanctuaries by the National Party.

Future directions

The recent experience with marine parks in NSW underscores the importance of securing and sustaining high-level political support for marine conservation. Despite the presence of a robust legislative framework, management capacity and scientific expertise, progress towards long-term conservation goals has been compromised by an increasingly politicised and partisan approach to marine parks and sanctuaries.

Long-term progress towards an effective network of marine parks in NSW will require concerted action by a range of stakeholders. Environmental organisations, marine scientists and the tourism industry can contribute by:

- building and demonstrating community support for marine parks and sanctuaries
- educating elected representatives about the environmental, social and economic benefits of marine parks
- encouraging community members to communicate their support for marine parks directly to their elected representatives (Box 11.1).

Despite promising public commentary by the Minister for Environment in 2014, government policy in relation to marine parks remains uncertain and ill-defined. The overall level of protection for the marine environment has been reduced over the past 4 years, and the Liberal-National Coalition remains internally divided on future conservation measures.

Box 11.1. Public support for marine parks and sanctuaries

The perception that marine parks are unpopular is inconsistent with public opinion polls, which consistently demonstrate strong support for marine parks. Most recently, polling commissioned by the Dive Industry Association found that 93% of respondents supported marine sanctuaries, including 91% of recreational fishers surveyed (Galaxy Research 2014).

In 2009, state-wide polling commissioned by the NSW Government found that 85% supported MPAs (NSW Government 2009). Government polling from the Solitary Islands Marine Park found 82% of resident recreational fishers supported their local marine park (NSW Marine Parks Authority 2009). In Jervis Bay, 82% of recreational fishers say they support sanctuary zones in the marine park (McGregor Tan Research 2008) (see also Chapters 20, 21).

Political leaders have an opportunity to respond to strong public support for marine conservation by:

- outlining a clear vision for marine conservation in NSW, including a renewed commitment to the establishment of a comprehensive, adequate and representative marine parks network
- engaging effectively with the scientific community and promoting the involvement of local communities (including recreational fishers) in marine park monitoring
- remedying key gaps in the MPA network by establishing a Sydney Marine Park and expanding marine sanctuaries
- enhancing protection for threatened marine species and supporting effective ecosystem-based fisheries management
- increasing funding for research, monitoring and management in the marine estate.

Conclusion

Establishment of a representative network of MPAs is a long-term process, involving a complex interaction of scientific, legal, policy and political considerations. In NSW, a well-established legislative and policy framework provided a strong foundation for the identification, establishment and management of marine parks between 1997 and 2006. During this period, high-level political leadership played an important role in sustaining progress towards a representative network of MPAs.

However, several political developments – including the loss of leading conservation champions within the Labor Party, the increased influence of the Shooters and Fishers Party, a hardening policy position within the National Party and, ultimately, a change in government – have since led to a series of backward steps on marine conservation. It remains to be seen whether Premier Baird's personal support for MPAs will translate into meaningful conservation outcomes.

Despite the highly partisan political debate surround marine parks, community support for marine parks and sanctuaries remains strong. Mobilising this strong base of public support, and educating elected representatives about the benefits of MPAs, will be essential to rebuilding bipartisan support for marine conservation in NSW.

Acknowledgements

Many thanks to Andrew Cox, former Executive Director of the National Parks Association of NSW (2000–2009) for providing historical background on the development of the marine parks network and the *Marine Parks Act 1997*.

Biography

Pepe Clarke is currently Head of Policy for BirdLife International. From 2010 to 2014, he was Chief Executive Officer of the Nature Conservation Council, the peak environmental organisation for NSW. He is a member of the IUCN World Commission on Protected Areas and the IUCN World Commission on Environmental Law. He has worked as a consultant to WWF, Conservation International and the Wildlife Conservation Society and was regional legal advisor to the IUCN Regional Office for Oceania and Programs Director for the Environmental Defenders Office.

Endnotes

1 *An Act to protect the Fisheries of New South Wales*, 1 June 1865.

2 *An Act to provide for the development and regulation of the Fisheries of the Colony*, 6 April 1881.

3 *Fisheries Management Act 1994*, s.20. *Fisheries Management (General) Regulation 2010*, Schedule 3 Fish and waters protected from commercial fishing.

4 *Fisheries and Oyster Farms (Solitary Islands Marine Reserve) Regulation 1991.*

5 *Marine Parks (Zoning Plans) Amendment (Solitary Islands and Jervis Bay Marine Parks) Regulation 2011.*

6 *Marine Parks Amendment (Moratorium) Act 2011*, Schedule 1.

References

ANZECC (1998) *Guidelines for Establishing the National Representative System of Marine Protected Areas.* Australian and New Zealand Environment and Conservation Council, Task Force on Marine Protected Areas. Environment Australia, Canberra.

Australian Broadcasting Corporation (2011) *New Zoning Scrapped for NSW Marine Parks.* 25 May 2011. ABC, Sydney, <www.abc.net.au/news/2011-05-25/new-zoning-scrapped-for-nsw-marine-parks/2729920>.

Babcock R, Bates M, Bathgate R, Beck H, Becker A, Bennett M *et al.* (2014) *Science statement on Marine Park Zoning in New South Wales,* 13 January 2014, <www.sydney.edu.au/science/biology/news_and_events/documents/ScienceStatement-on-marine-park-zoningNSW.pdf>.

Beeton RJS, Buxton CD, Cutbush GC, Fairweather PG, Johnston EL, Ryan R (2012) *Report of the Independent Scientific Audit of Marine Parks in New South Wales.* NSW Department of Primary Industries and Office of Environment and Heritage, Sydney, <www.marineparksaudit.nsw.gov.au/imagesDB/wysiwyg/MPksAuditReport_2012_web_witherratum.pdf>.

Commonwealth of Australia (1992) *National Ecologically Sustainable Development Strategy.* Commonwealth of Australia, Canberra.

Commonwealth of Australia (1996) *National Strategy for the Conservation of Australia's Biological Diversity.* Commonwealth of Australia, Canberra.

Commonwealth of Australia (1998a) *Australia's Oceans Policy*. Commonwealth of Australia, Canberra.
Commonwealth of Australia (1998b) *Guidelines for Establishing the National Representative System of Marine Protected Areas*. Commonwealth of Australia, Canberra.
Commonwealth of Australia (1998c) *Interim Marine and Coastal Bioregionalisation for Australia*. Commonwealth of Australia, Canberra.
Commonwealth of Australia (1999) *Strategic Plan of Action for the National Representative System of Marine Protected Areas*. Commonwealth of Australia, Canberra.
Council of Australian Governments (COAG) (1992) *Intergovernmental Agreement on the Environment*. Council of Australian Governments, Canberra.
Department of Premier and Cabinet (2014) *NSW 2021 Performance Report 2014–15*. Budget Related Paper No. 1. NSW Government, Sydney.
Department of the Environment (2014) *Collaborative Australian Protected Areas Database 2014*. Marine Protected Area Data (NSW Summary), Department of the Environment, Sydney, <www.environment.gov.au/land/nrs/science/capad/2014>.
Director of National Parks (2011) *Booderee National Park Draft Management Plan 2011–2021*. Director of National Parks, Canberra.
Galaxy Research (2014) *Community Attitude Survey prepared for the Dive Industry Association of Australia*. January 2014. Galaxy Research, Sydney.
Hodgkinson K (2014) 'Concerns addressed on commercial mesh netting'. Media release: Minister for Primary Industries, 14 August 2014, <www.dpi.nsw.gov.au/__data/assets/pdf_file/0005/526280/ media_release_140814_concerns_mesh_netting.pdf>.
Hodgkinson K, Parker R (2013) 'A new approach to managing the NSW marine estate'. Media release, 12 March 2013, <http://www.dpi.nsw.gov.au/__data/assets/pdf_file/0018/460620/20130312-MARINE-PARKS-MEDIA-RELEASE-final.pdf>.
Hodgkinson K, Stokes R (2014a) 'New era for marine estate management: landmark Bill passes NSW Parliament'. Media release, 5 November 2014, <http://www.dpi.nsw.gov.au/__data/assets/pdf_file/0005/533903/media_release_141105_marine_estate_management_bill_passes_parliament.pdf>.
Hodgkinson K, Stokes R (2014b) 'Study to assess Sydney marine region'. Media release, 15 December 2014, <http://www.dpi.nsw.gov.au/__data/assets/pdf_file/0003/537501/media_release_141215_study_assess_sydney_marine_region.pdf>.
McGregor Tan Research (2008) *Jervis Bay Marine Park Community Survey Final Report*. McGregor Tan Research, Sydney, <www.mpa.nsw.gov.au/pdf/jbmp-community-survey-2008.pdf>.
National Parks and Wildlife Service (1999) *New South Wales Biodiversity Strategy*. National Parks and Wildlife Service, Sydney.
Nature Conservation Council (NCC) and National Parks Association (NPA) (2014) *A Vision for Nature Conservation in New South Wales*. 13 November 2014. Nature Conservation Council of NSW and National Parks Association of NSW, Sydney, <www.nature.org.au/media/2005/a_vision_for_nature_conservation_in_new_south_wales_november_2014.pdf>.
Nature Conservation Council of NSW (NCC) (1990) *Marine Policy*. Annual Conference Minutes. Nature Conservation Council of NSW, Sydney.
NSW Department of Primary Industries (2015) *Marine Protected Areas*. NSW Department of Primary Industries, Sydney, <www.dpi.nsw.gov.agazettu/fisheries/habitat/protecting-habitats/mpa>.
NSW Government (1997) *New South Wales Coastal Policy*. NSW Government, Sydney.

NSW Government (1999) *New South Wales Biodiversity Strategy*. NSW Government, Sydney.

NSW Government (2009) *Who Cares About the Environment?* Department of Environment and Climate Change, Sydney.

NSW Government (2012) *New South Wales State of the Environment 2012*. NSW Environment Protection Authority, Sydney.

NSW Government (2015) *NSW Marine Parks*. NSW Government, Sydney, <www.mpa.nsw.gov.au>.

NSW Government Gazette (1998) *Marine Parks Act – Proclamation*, 2 January 1998. NSW Government, Sydney, <http://www5.austlii.edu.au/au/legis/nsw/num_reg/mpa1997pdatbmp19985515.pdf>.

NSW Legislative Council (2010) *Recreational Fishing in New South Wales*. Select Committee on Recreational Fishing, Sydney.

NSW Marine Parks Authority (2001) *Developing a Representative System of Marine Protected Areas in New South Wales – An Overview*. NSW Marine Parks Authority, Sydney.

NSW Marine Parks Authority (2009) *Solitary Islands Marine Park: Zoning Plan Review Report*. Marine Parks Authority, Coffs Harbour. <www.mpa.nsw.gov.au/pdf/SolitaryIslands-ReviewReport.pdf>.

NSW Parliament (1997) *Hansard*, 18 June 1997. NSW Parliament, Sydney.

NSW Parliament (2010) 'Marine Parks' (Hon. Duncan Gay). *Hansard*, 10 March 2010. NSW Parliament, Sydney.

Patty A (2014) 'Rob Stokes vows to help protect marine areas alongside Mike Baird'. *Sydney Morning Herald*, 29 April 2014, <www.smh.com.au/nsw/rob-stokes-vows-to-help-protect-marine-areas-alongside-mike-baird-20140430-zr18j.html>.

Professional Fishermen's Association (PFA) (2014) 'NSW Commercial Fishing Community condemns Labor's unnecessary marine park proposal'. Media release, 30 September 2014, <www.nswpfa.com.au/pfa-condemns-proposed-marine-park>.

Shooters and Fishers Party (2013) *Position Statement and Action Plan on Fair Recreational Access*. Shooters and Fishers Party, Sydney, <www.shootersandfishers.org.au/files/1/206476483/recreational-access-19-oct.-2013-.pdf>.

Winn P (2007) *The Torn Blue Fringe: Marine Conservation in NSW*. National Parks Association of NSW, Sydney.

Chapter 12

Marine protected areas in Queensland: past and present

Peter Ogilvie

Introduction

Queensland has a strong and lengthy record in the declaration and management of marine protected areas, both in state waters and waters that were subsequently removed from state jurisdiction and administered by the Australian Government. To appreciate the present-day arrangements for marine protected areas and the complexities of offshore jurisdiction, some understanding of the Great Barrier Reef Marine Park (GBRMP) and its establishment is essential (see also Chapter 5).

History and development of the marine protected area network in Queensland

The first marine protected area

The first action to protect an offshore area in Queensland, and possibly the first such action in Australia, took place under the provisions of *The Fish and Oyster Acts, 1914 to 1935*, which contained a capacity to restrict the take of marine organisms in specified areas.

On 16 October 1937, the *Queensland Government Gazette* (No. 100, p. 1079) published the 14 October Order in Council which 'doth hereby prohibit the taking of fish, coral or other marine product in the waters of the area described in the Schedule hereto; provided that this prohibition shall not apply to the taking of fish by means of a hand-line or a rod-line with not more than two hooks attached'. The Schedule specified the area as 'The foreshore and reef surrounding Green Island, near Cairns, and within a radius of one mile from low-water mark around that island'. At that time, Green Island was the most heavily used island on the Great Barrier Reef (GBR). The island itself had been declared a national park under the provisions of *The State Forests and National Parks Act of 1906* on 25 March 1937 (Fig. 12.1).

Green Island and its reef had been a centre for collection and preparation of bêche-de-mer (sea cucumbers) as well as a source for collection of other marine organisms for private and commercial purposes. A passenger ferry service had operated since 1924 and a tourist resort was constructed in the 1940s.

Fig. 12.1. Green Island and jetty in 1932, just before the adjacent coral reef and waters being declared the first marine protected area in Queensland (and one of Australia's first) in 1937. (Photo: Oxley Library, Brisbane)

Fish habitat areas

The *Fisheries Act 1957* made provision for action to be taken to protect certain marine resources, particularly mangroves. However, it was not until *The Fisheries Habitat Reserves Regulation of 1968* was proclaimed on 23 January 1969 that a capacity to declare this new category of marine protected area was provided. This followed a Queensland Cabinet Decision (No. 12515 of 14 January 1969) which approved 'That action be taken to create Fisheries Habitat Reserves in Moreton Bay in the localities delineated on the plan produced by the Hon. the Treasurer to Cabinet and that appropriate regulations and Order in Council be promulgated'. The first seven approved reserves were declared in Moreton Bay on 23 January 1969, which established Fisheries Habitat Reserves at the following Moreton Bay locations: Myora, Pine River and Hays Inlet, Deception Bay (two areas), Jumpinpin, Pumicestone Passage and Moreton Banks. As the name implied, the areas are designed to protect certain habitats, primarily mangroves, that are important for nurturing and maintaining fish populations. These reserves still permitted certain forms of fishing.

The name was subsequently changed to fish habitat area (FHA), and the current legislation, the *Fisheries Act 1994*, provides for the declaration of a FHA under a regulation (s. 615, *Fisheries Regulation 2008*). The Act also clearly establishes a capacity to control any development inside the FHA and to control the removal, destruction or damage of any marine plant inside the area.

The existing network of FHAs consists of 70 declared areas covering more than 1 million ha and is spread across the state (Department of Agriculture, Fisheries and Forestry 2012) (Fig. 12.2). Declared FHAs are presently administered by the Department of National Parks, Sport and Racing (which also administers Queensland's marine parks). However, development applications in these areas are assessed by Fisheries Queensland via an administrative agreement between the National Parks and Fisheries Departments.

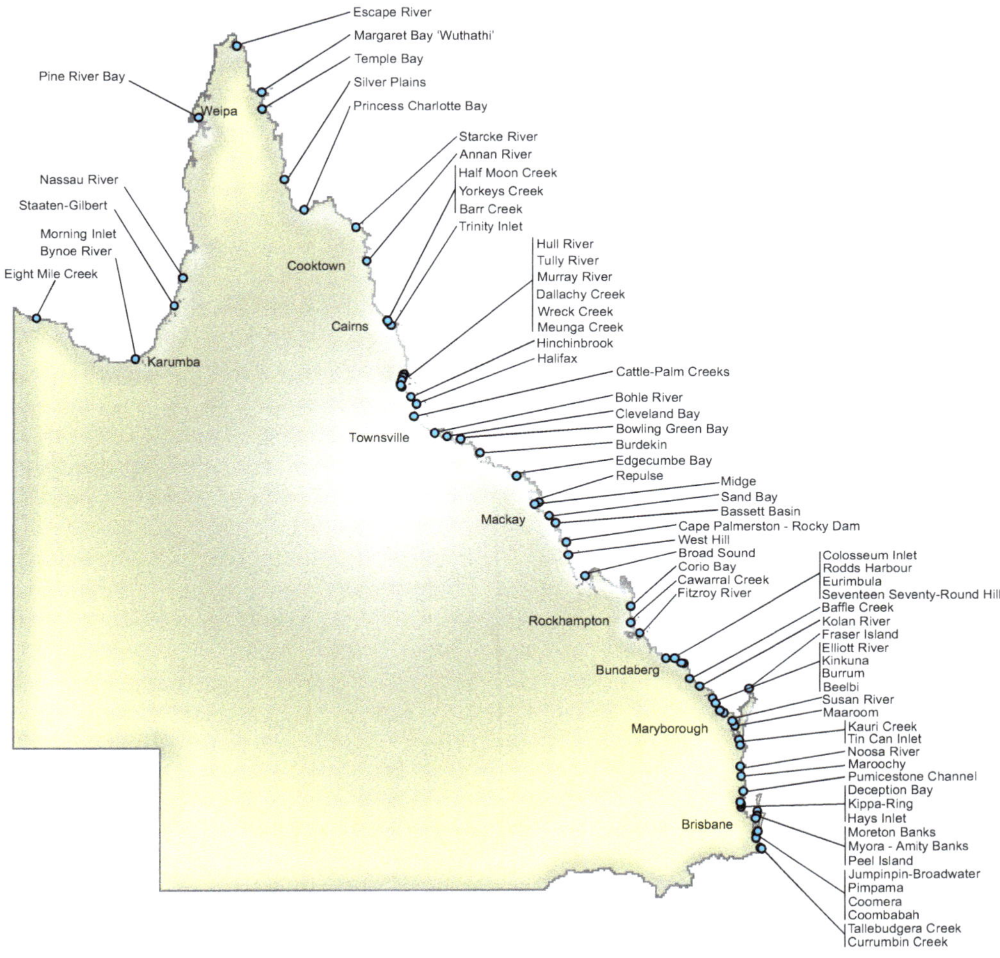

Fig. 12.2. Location of declared Fish Habitat Areas in Queensland. (Source: Department of National Parks, Sport and Racing 2012. Detailed maps of each declared fish habitat area are available at http://www.npsr.qld.gov.au/managing/habitat-areas/area-plans.html)

The first marine national parks

In a submission, dated 30 May 1968, to Premier Jo Bjelke-Petersen, the National Parks Association of Queensland (NPAQ) proposed three substantial offshore areas in the GBR for consideration by the government as marine national parks. These were:

1 A northern park 'to be established to include approximately the whole of the reefs from 14°30′ South – reference Admiralty Chart No. 2344. This will extend approximately 150 miles northward, along the line of Barrier Reefs from Trinity Opening'.
2 A central park 'to be established to extend south easterly along the line of the outer Barrier approximately 100 miles from a point level with Flinders Passage – i.e. all the Reefs between 18°45′ South Latitude and 19°14′ South, east of 148° longitude – reference Admiralty Chart No. 348'.

3 A southern park 'to be established to extend along the inner edge of the Reef from the most southerly extremity of Swain Reefs to Herald Reef Prong, and along the outer edge of the Reef 120 miles, the extremities being joined by a straight line – reference Admiralty Chart No. 1024'.

The NPAQ also sent the same proposal to the Commonwealth Government to cover those waters in the proposal that were outside the 3-mile territorial sea administered by Queensland. The proposal was based on extending the terrestrial national park concept into the marine environment. In other words, these would be areas of strong protection for biodiversity.

At that time, Queensland's national parks were administered by the Department of Forestry. The department was receptive to the proposal. The officer responsible for national parks (the department's Secretary, Bill Wilkes) had recently attended the inaugural World Parks Congress in Seattle (USA) and had participated in the resolution on marine reserves. However, the proposal clearly raised several offshore jurisdictional issues. Only part of the proposal lay within what was then regarded as Queensland territorial waters – 3 nautical miles from low-water mark on both the mainland and any island that was part of Queensland.

The matter was presented to Cabinet, which decided (26 August 1968) to establish a committee of five departmental heads to make recommendations relating to (i) the marine national park proposal, (ii) offshore jurisdiction, and (iii) the report on the GBR prepared by Dr Harry S Ladd.

The report by the Cabinet-appointed committee relating to the three matters was delivered on 31 December 1969 and was subsequently accepted by Cabinet. It made several recommendations, among which were the following:

- Provision should be made for the gazettal of marine national parks that could only be revoked by resolution of Parliament. They should be administered by the agency responsible for national parks.
- No direct action can be taken on the NPAQ proposals because only a small proportion was subject to Queensland jurisdiction.
- All forms of mining should be prohibited in such parks.
- Because of the *Petroleum (Submerged Lands) Act 1967* and the state/Commonwealth jurisdictional issues, a complicated process will be required for establishing such parks.

As a consequence, the *Forestry Act 1959* was amended in April 1971 to provide for national parks below low-water mark. The actual declaration involved two distinct steps. First, the Governor-in-Council had to declare an area or areas extending seawards from the mean high-water mark at spring tides along the maritime boundary of any part of Queensland to be a 'Marine Park Area'. This was, in essence, a legal statement asserting Queensland's offshore jurisdiction. Having done that, a national park could be declared over all or part of the Marine Park Area. Any park so declared was required to be assigned a name that included the word 'marine'. This ensured that the relevant existing regulations for terrestrial (and fresh water) parks would automatically apply.

The first two marine national parks in Queensland waters – Heron-Wistari Reefs Marine National Park (9700 ha) and Green Island Reef Marine National Park (3000 ha) – were declared adjacent to two prominent Barrier Reef coral cays, Heron Island and Green Island, on 14 February 1974. Despite the preparation of several proposals for additional parks, these were the only ones declared at this time. The government chose to transfer the legislative provisions and administrative responsibility to the Queensland Fisheries Service in 1975 to coincide with the establishment of the Queensland National Parks and Wildlife Service (QNPWS) that year. No new marine national parks were declared by the Queensland Fisheries Service under that transferred legislation in the following 6 years during which that Service had responsibility for administering the legislation.

A new approach to marine conservation

Around this time (early 1970s), a new approach to marine conservation was being contemplated by the Commonwealth Government for the GBR region. Public interest in the well-being of the GBR was prompted by a number events, but two of them in the late 1960s and early 1970s stood out as rallying points for public response. The first was the 1967 application by the Cairns District Canegrowers Executive to mine Ellison Reef, off Innisfail, for limestone. The second was the granting, jointly by Queensland and Commonwealth governments, of petroleum prospecting licences over virtually all of the GBR in 1969. One of the founders of the Wildlife Preservation Society of Queensland, poet Judith Wright, was a leader in arguing the case against mining. She subsequently recorded the course of events in her book *The Coral Battleground* (Wright 1977).

Given the strong public sentiment, a change of Australian Government in 1972, the offshore jurisdictional arrangements in the 1973 Commonwealth legislation, and an active tourist industry, the circumstances were favourable for some protective action by the Australian Government. The matter of offshore jurisdiction was brought to a head with the passing of the *Seas and Submerged Lands Act 1973* (Cth), which established that 'sovereignty in respect of the territorial sea ... is vested in the Crown in right of the Commonwealth'. This was challenged by all Australian states; nevertheless, its validity was upheld by the High Court in 1975. That left the way clear for the Commonwealth Government to assert control over all of the GBR outwards from low-water mark or from baselines that delineated Queensland's internal waters (see also Chapter 5). Heron-Wistari Reefs Marine National Park and Green Island Reef Marine National Park were effectively superseded by the High Court ruling.

The result was the Australian Government passing the *Great Barrier Reef Marine Park Act 1975* (GBRMP Act), which provided a new multi-purpose approach to marine conservation over all waters, seabed and natural resources in an area established in that legislation and entitled the Great Barrier Reef Region (see Fig. 5.1). By that means, a hierarchy of zones could be established, extending from a zone providing total protection (equivalent to the marine national parks already established in Queensland, but which were superseded by the federal action) to a category that effectively accepted the status quo. Only one activity, mining, was excluded from any declared marine park. It was a bold and innovative move and one that had considerable influence on, and consequences for, the subsequent establishment and management of marine protected areas in Queensland. The development of the GBRMP is recounted in another chapter (Chapter 5), but its close relationship with marine conservation in Queensland needs to be understood. It is not possible to comprehend the recent development of marine protected areas in Queensland without that appreciation.

The legislation caused considerable tension between the Queensland Bjelke-Petersen and Federal Whitlam governments, particularly in relation to what part Queensland would play in administering and managing the proposed marine park. This was partly resolved by the dismissal of the Whitlam Government in November 1975 and the subsequent election of the Fraser Government. Although the new government was not prepared to repeal the

GBRMP Act, it was prepared to negotiate joint management arrangements. The 1979 Offshore Constitutional Settlement between all states and the Northern Territory and Commonwealth governments returned the 3 nautical mile territorial sea to all states for management purposes, with one exception: the GBR region remained under direct federal control, thereby confirming the validity of the *Great Barrier Reef Marine Park Act 1975* (see Haward 1989 for more details on the Offshore Constitutional Settlement).

The framework for joint management of the GBR was decided in the Queensland town of Emerald on 14 June 1979 when a 3-page agreement, known as the Emerald Agreement, was signed by Prime Minister Malcolm Fraser and Premier Joh Bjelke-Petersen. The following are relevant extracts from the document (see Appendix E in Commonwealth of Australia 2006):

> *'It is recommended that, subject to the Authority, Queensland be assigned the day-to-day management role and that the necessary preparatory steps be taken for arrangements to be put in place on a basis to be agreed by Ministerial Council.*
>
> *The day-to-day management to be undertaken by officers of the Queensland National Parks and Wildlife Service, who, in discharging these responsibilities, will be subject to the Great Barrier Reef Marine Park Authority; relevant State legislation to be brought in line with the Great Barrier Reef Marine Park Act'.*

The agreement also confirmed that the proposed Offshore Constitutional Settlement would not be returning the territorial sea within the GBR region to Queensland, and no changes would be made to the *Great Barrier Reef Marine Park Act 1975*. These matters were formally established the following year in the Commonwealth *Coastal Waters (State Title) Act 1980*.

The QNPWS was clearly identified as the state agency responsible for marine park matters. As a consequence, the Marine National Parks Section in the Queensland Fisheries Service, which had been extracted from QNPWS in 1975, was returned in December 1981. New Queensland legislation, the *Marine Parks Act 1982*, was enacted, thereby bringing state declarations and zoning procedures in line with the Commonwealth Government provisions. This ensured that complementary zoning and management arrangements could be undertaken in state waters between high- and low-water marks and inside baselines across bays and estuaries. Waters inside those baselines (which are established under international rules) are classified as internal waters and are under state jurisdiction.

The complementary arrangements were also designed to sidestep issues arising in areas where jurisdiction was in dispute. For example, the exposed outer edge of Heron Island reef at low tide is, in some places, 10 km from high-water mark. The Queensland Government asserted that the whole reef flat inside that drying outer edge lay between high- and low-water marks and was therefore subject to state jurisdiction. The Commonwealth Government argued that the low-water mark was merely a few metres down the sand on the island to where lagoonal water was encountered at low tide. Identical zoning arrangements would ensure that all bases were covered in the event of a legal challenge.

The Emerald Agreement was replaced by the Great Barrier Reef Intergovernmental Agreement 2009,[1] which was signed by then Australian Prime Minister Kevin Rudd and Queensland Premier Anna Bligh. The preamble clearly establishes the function and purpose of the agreement. It states:

> *'The Commonwealth and Queensland Governments have long demonstrated their commitment to working together collaboratively for the long-term protection and*

conservation of the Great Barrier Reef ecosystem, to ensure it is preserved as an important feature of the nation's and the world's heritage. The Emerald Agreement formalised this cooperative approach in 1979 and a series of collaborative arrangements has evolved over time. The Commonwealth and Queensland governments have agreed to update these arrangements, recognising the challenges of protecting the Great Barrier Reef into the future arising from pressures from the land and marine environments and the impacts of climate change. This agreement establishes these new arrangements, clearly articulating objectives, respective functions and accountabilities'.

The Agreement contains schedules relating to the GBR World Heritage Area (which has a slightly different boundary from the GBR Marine Park), Great Barrier Reef Ministerial Forum, joint field management program, climate change, fishing and reef water quality.

Leaving aside inevitable political differences, the joint arrangements are working well. The areas of jurisdictional dispute based on the location of low-water mark have been adequately catered for by complementary zoning, such that any breach of zone conditions would be met with identical penalties by both state and Commonwealth legislative provisions. It provides a reasonable working model for any circumstance where more than one jurisdiction claims legal responsibility in offshore areas.

State marine parks within the Great Barrier Reef Region

Sections of the GBRMP were declared progressively by the Commonwealth, commencing with the Capricornia Section, which was proclaimed and published in the *Commonwealth Government Gazette* on 21 October 1979. In order to achieve complementary zoning in Queensland waters adjacent to those sections, the Queensland Government took action to declare marine parks over state waters adjacent to the sections declared under the GBRMP Act. The following state marine parks were declared under the provisions of the *Marine Parks Act 1982*:

- Townsville/Whitsunday Marine Park – gazetted 3 October 1987; zoning plan came into force on same date
- Mackay/Capricorn Marine Park – gazetted 27 August 1988; zoning plan came into force on same date
- Cairns Marine Park – gazetted 18 February 1989; zoning plan came into force on same date
- Trinity Inlet/Marlin Coast Marine Park – gazetted 2001.

These parks were declared to a depth of 1000 m below the seabed or tidal land, and to a height of 915 m above the surface, those being the same depth and height provisions applying to the GBRMP under the Commonwealth legislation.

A major zoning review of all sections in the GBRMP (see also Chapter 5) led to the abolition of those sections and their section-specific zoning plans, and the development of a single zoning plan for the whole GBRMP. This took effect on 1 July 2004, and was done in concert with a representative areas program, which sought to ensure strong protection for representative samples of all biogeographic components of the reef. It resulted in some 33% of the GBRMP being assigned to no-take 'green zones' (Marine National Park Zone), a substantial increase from the 10% that previously lay within that particular zoning category. A colour coding system for individual zones had been developed over time and reinforced in maps, documents and discussions. The zones proclaimed in 2004 were: Preservation (pink), Marine National Park (green), Scientific Research (orange), Buffer (olive green), Conservation Park (yellow), Habitat Protection (dark blue) and General Use (light blue) (see Fig. 5.4).

The current marine protected area estate in Queensland

Queensland marine park inside the Great Barrier Reef Region

In order to maintain complementary zoning and management across Commonwealth and state waters in the GBR region, Queensland undertook a rezoning exercise to establish a single state marine park that incorporated the existing state marine parks referred to above, plus all other state waters that had not yet been given marine park status in the region. The same zoning categories and colour coding as for the GBRMP were employed, as well as the same management criteria for those zones.

The Great Barrier Reef Coast Marine Park (GBRCMP) was declared and the zoning plan commenced on 5 November 2004. From that date, all Queensland state waters adjacent to the mainland and islands between the northern and southern boundaries of the GBR region (defined in the GBRMP Act) had Queensland Marine Park status and a zoning regime that complemented the adjacent GBRMP zoning. While the total declared area of the GBRCMP is 63 263 km^2, it is important not to consider the marine park in isolation from the GBRMP. The use of complementary zoning ensures a seamless approach to management with joint Australian and Queensland Government mapping reflecting the true extent of marine park in the GBR region (see also Fig. 5.6).

The eight zones used in the GBRCMP are as follows: (i) Preservation Zone, (ii) Scientific Research Zone, (iii) Marine National Park Zone, (iv) Buffer Zone, (v) Conservation Park Zone, (vi) Estuarine Conservation Zone, (vii) Habitat Protection Zone, and (viii) General Use Zone. The respective areas and percentages for each zone are provided in Table 12.1 and their location in Fig. 12.3. The Estuarine Conservation Zone is not matched in the Commonwealth zoning regime because all estuarine areas are internal waters of Queensland.

Queensland marine parks outside the Great Barrier Reef Region

Outside the GBR region, responsibility for the 3-nautical mile territorial sea had been returned to Queensland under the *Coastal Waters (State Title) Act 1980* (Cth). That authority was first exercised under state marine park legislation with the gazettal of the Hervey Bay Marine Park on 14 September 1989. Hervey Bay is an important stop-off point for migrating Humpback Whales, and whale-watching was becoming an important tourist attraction in the area (Fig. 12.4). The associated zoning plan, which came into force on 14 September 1989, provided the capacity to manage the whale-watching activities, including limiting (by permit) the number of boats and prescribing minimum viewing distances, along with other requirements to limit the level of disturbance to the whales (DEHP 2013). This helped stimulate other whale-watching protocols in other parts of Australia.

A nearby area, Woongarra Marine Park, was gazetted on 14 December 1991. Turtle conservation was the primary objective in this instance, as the park bordered the important Flatback Turtle rookery at Mon Repos Beach. It also enclosed some small coral outcrops.

These two marine parks were incorporated into a much larger Great Sandy Marine Park in 2006 (Fig. 12.5). The park encompasses all Queensland waters and the 3-mile territorial sea from Baffle Creek in the north (at which point it joins the GBRCMP and GBRMP) to Double Island Point in the south. In so doing, it surrounds Fraser Island and includes Hervey Bay, Great Sandy Strait (a Ramsar site) and Tin Can Bay Inlet. It covers some 5933 km^2 and contains a wide range of marine habitats including mangroves, seagrass meadows, rocky shores, sandy beaches, coral outcrops and estuaries. Sections of the

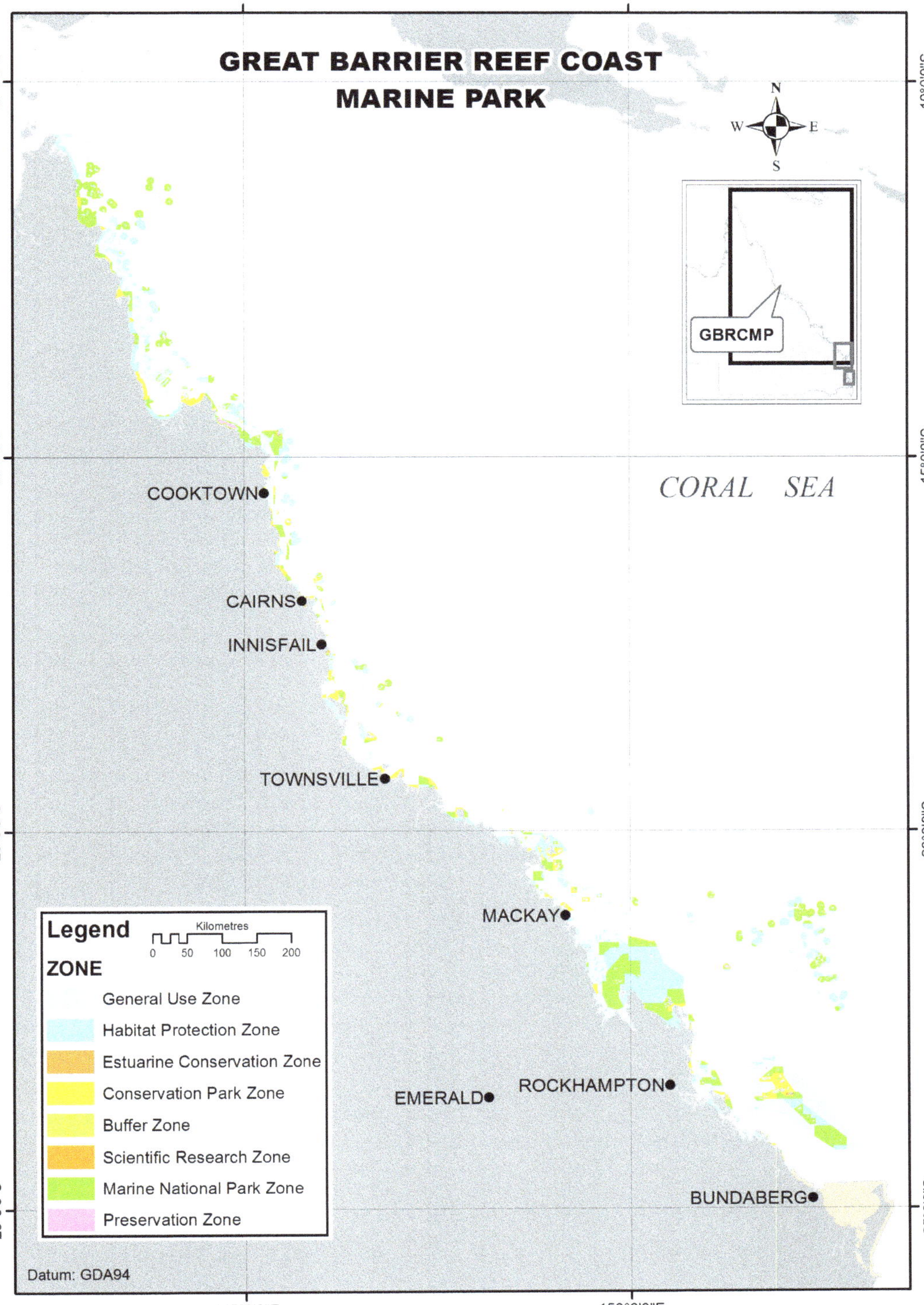

Fig. 12.3. Great Barrier Reef Coast Marine Park. (Source: Department of National Parks, Sport and Racing)

Fig. 12.4. Commercial whale-watching in Great Sandy Marine Park. Humpback Whales being viewed in Hervey Bay in July 2015 during their annual migration. (Photo: Steve Parish, NatureConnect)

park are particularly important for whales, marine turtles, Dugong, Grey Nurse Sharks and migratory shorebirds.

The Marine Parks (Great Sandy) Zoning Plan 2006 commenced on 31 August 2006. It made provision for five zones: (i) Marine National Park Zone, (ii) Buffer Zone, (iii) Conservation Park Zone, (iv) Habitat Protection Zone, and (v) General Use Zone. The respective areas and proportions for each of the zones are provided in Table 12.1.

The other present-day marine park declared under Queensland legislation is located right next to Queensland's major centres of population – the cities of Brisbane, Redcliffe, Logan and the Gold Coast. Its location adjacent to a substantial proportion of Queensland's population ensures that the park's management is subject to considerable public attention and scrutiny. The Moreton Bay Marine Park (Fig. 12.6) was declared in 1993, though its first zoning plan did not commence until December 1997. That plan was revised after 10 years, as required by legislation, and was replaced by the *Marine Parks (Moreton Bay) Zoning Plan 2008* after a protracted and difficult consultation period. Not surprisingly, many groups strongly resisted any constraints on resource extraction, particularly fishing (recreational and commercial). It took strong arguments and persuasive advice from a scientific advisory group to ensure that a reasonable component of the park was given green zone (marine national park zone) status. Pressure to reduce or remove green zones continues.

The zoning plan commenced on 1 March 2009. The park's total area of 3463 km^2 was allocated to four zones: (i) Marine National Park Zone; (ii) Conservation Park Zone; (iii) Habitat Protection Zone; and (iv) General Use Zone. An entry and use guide for the four zones is provided in Table 12.2. The respective areas and percentages of those zones in the park are provided in Table 12.1.

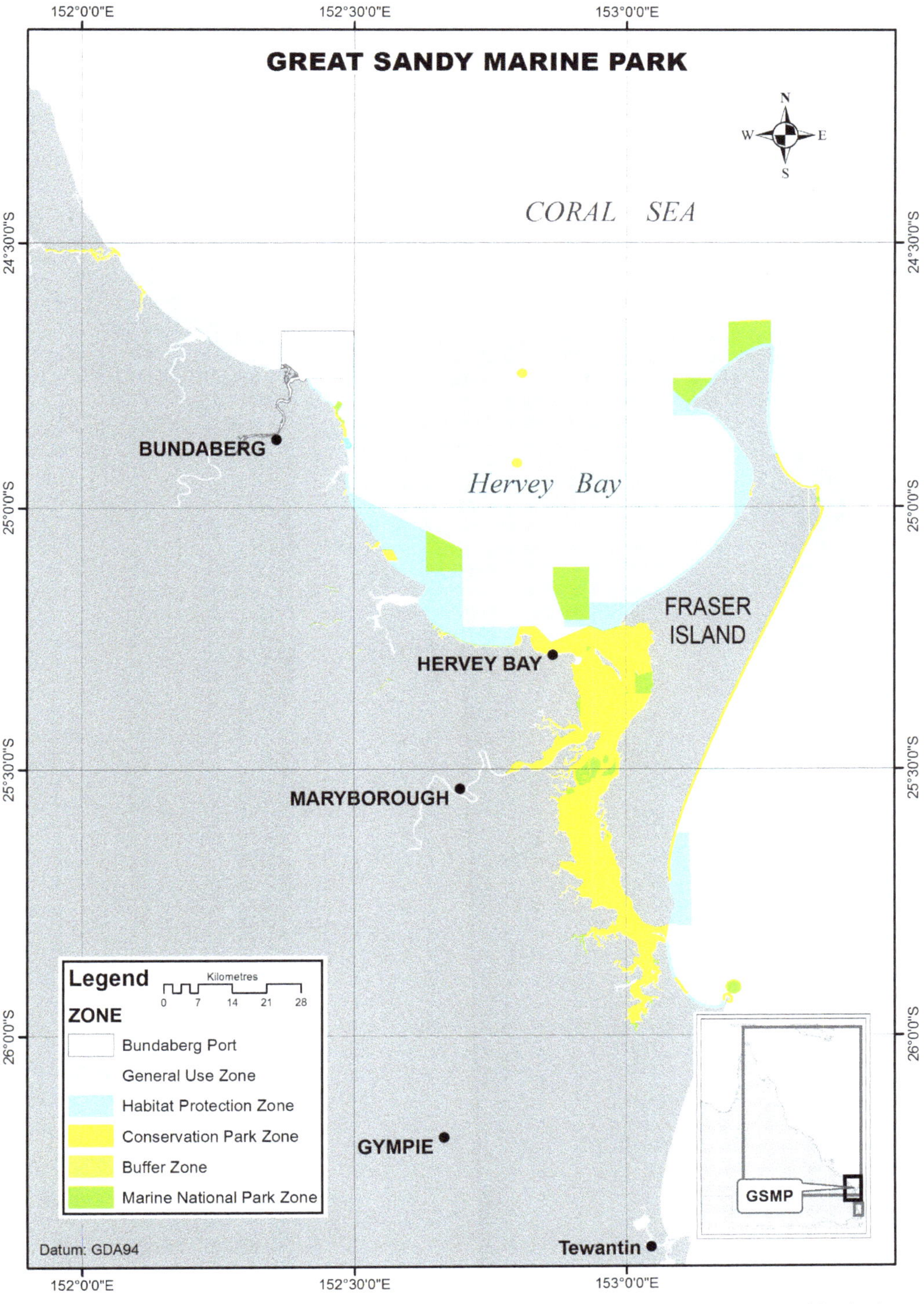

Fig. 12.5. Great Sandy Marine Park. (Source: Department of National Parks, Sport and Racing)

Table 12.1. Zones and their areas and percentages of total area in each of the three Queensland marine parks as at 2015

(Source: Department of National Parks, Sport and Racing)

Qld marine park	Preservation zone (km^2)	Scientific research zone (km^2)	Marine national park zone (km^2)	Buffer zone (km^2)	Conservation park zone (km^2)	Estuarine conservation zone (km^2)	Habitat protection zone (km^2)	General use zone (km^2)	Totals (km^2)
Great Barrier Reef Coast	298.0 *(0.5%)*	118.3 *(0.2%)*	15 838.0 *(25%)*	40.0 *(0.06%)*	4088.2 *(6.4%)*	119.4 *(0.2%)*	17 371.2 *(27.5%)*	25 389.2 *(40.2%)*	63 262
Great Sandy	–	–	241.6 *(4%)*	2.6 *(0.04%)*	787.5 *(13.3%)*	–	461.5 *(7.8%)*	4440.2 *(74.8%)*	5933
Moreton Bay	–	–	543.0 *(16%)*	–	260.0 *(8%)*	–	1040.0 *(30%)*	1620.0 *(46%)*	3463
Totals	**298**	**118**	**16 623**	**43**	**5136**	**119**	**18 873**	**31 449**	**72 659**

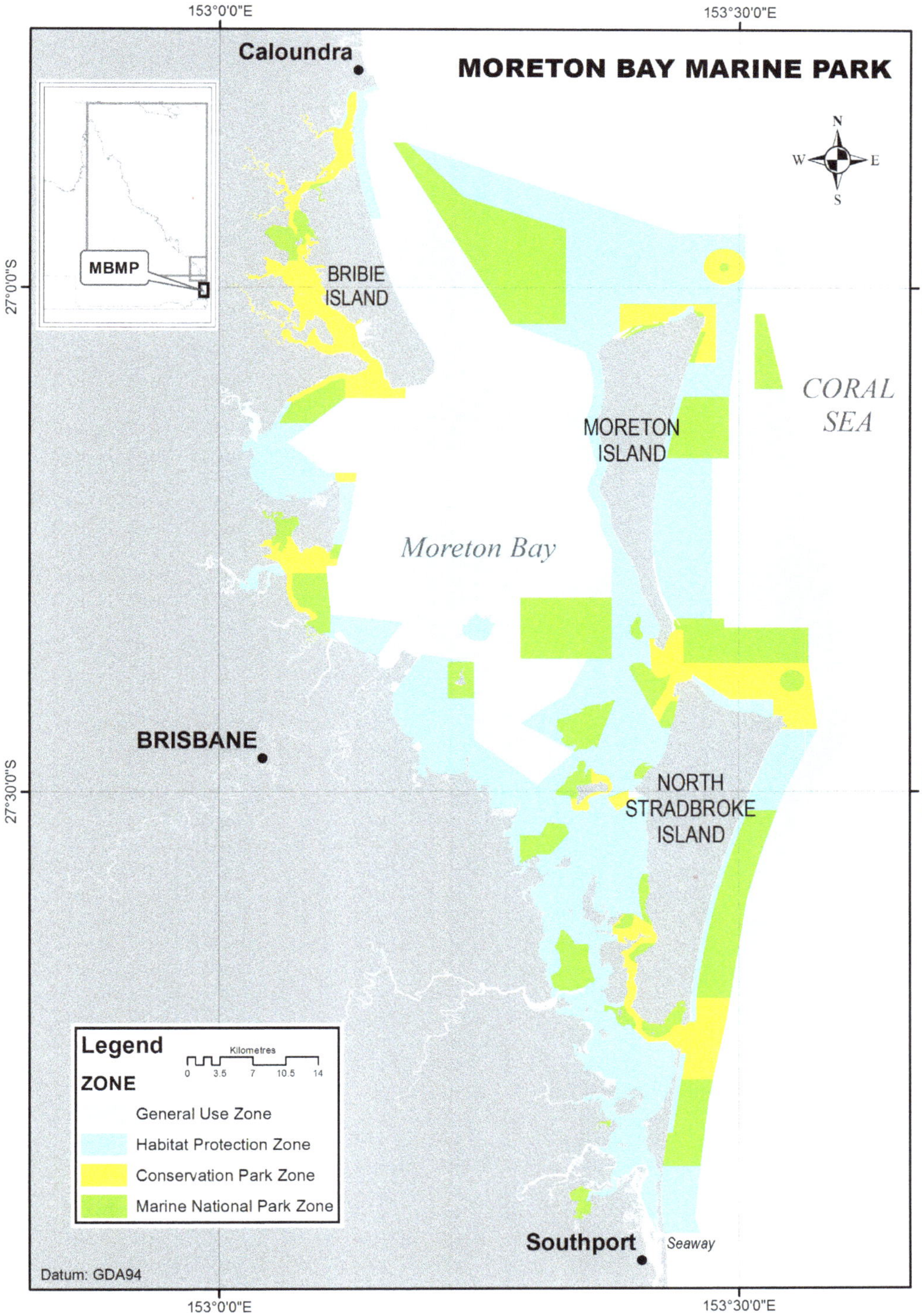

Fig. 12.6. Moreton Bay Marine Park. (Source: Department of National Parks, Sport and Racing 2015)

Table 12.2. Entry and use guide to Moreton Bay Marine Park

(From *Moreton Bay Marine Park User Guide 2012* (DERM 2012). This is a summary only. Refer to the *Marine Parks (Moreton Bay) Zoning Plan 2008* for full details.)

Entry and use guide	General use zone	Habitat protection zone	Conservation park zone	Marine national park zone
Boating	Yes	Yes	Yes	Yes
Line fishing	Yes	Yes	Yes[a]	No[b]
Trolling	Yes	Yes	Yes[a]	No[b]
Crabbing	Yes	Yes	Yes[c]	No[b]
Bait gathering	Yes	Yes	Yes	No[b]
Bait netting	Yes	Yes	Yes	No[b]
Netting (other than bait netting)	Yes	Yes	No[b]	No[b]
Limited spearfishing	Yes	Yes	Yes	No[b]
Limited collecting	Yes[d]	Yes[d]	Yes[d]	No[b]
Trawling	Yes	No[b]	No[b]	No[b]
Fisheries (fisheries legislation still applies)				
Harvest fishery (not including coral fishery)	Permit[e]	Permit[e]	Permit[f]	No
Aquaculture (not involving addition of feed)	Permit[g]	Permit[g]	Permit[g]	No
Traditional use				
Traditional use of marine resources	Permit[h]	Permit[h]	Permit[h]	Permit[h]
Research and education				
Research (limited impact research, non-extractive and extractive)	Permit[i]	Permit[i]	Permit[i]	Permit[i]
Limited educational program	Yes	Yes	Yes	Permit
Tourism and charter				
Tourism program	Permit	Permit	Permit	Permit

Vessels				
Navigate a managed vessel (e.g. hovercraft)	Permit	Permit	Permit	Permit
Aircraft				
Flying an aircraft below an altitude of 500 feet	Permit	Permit	Permit	Permit
Landing or taking off in the marine park	Permit	Permit	Permit	Permit
Other activities				
Works	Permit[j]	Permit[j]	Permit	Permit
Operate a facility	Permit	Permit	Permit	Permit
Mineral, petroleum or geothermal exploration	Permit[k]	No	No	No

[a]Up to two lines and a maximum of two hooks per angler
[b]All equipment must be stowed or secured
[c]Maximum of four pots per person
[d]No coral. Limits apply to species and equipment used.
[e]A permit is required unless the harvest fishery is accredited
[f]Limited to aquarium fish fishery and beach worm fishery. A permit is required unless the fishery is accredited.
[g]A permit is required unless it is an accredited activity
[h]A permit is not required for activities undertaken under a traditional use of marine resources agreement (TUMRA)
[i]A permit is required unless the institution conducting the research is accredited. Research in marine national park zones is restricted to priority research that cannot reasonably be conducted elsewhere.
[j]Works that are not consistent with the objects of the zone may be allowed in designated works areas. Private structures may only be allowed in a general use zone.
[k]The person must hold a lease, licence, permit or other authority under the appropriate legislation

The park encompasses a wide range of habitat types, including mud flats, rocky shores, coral reefs, mangroves, seagrass meadows, sandy beaches and estuaries. It surrounds four major islands, each of which contains a significant area of national park – Bribie, Moreton, North Stradbroke and South Stradbroke Islands. As well as being a feeding and breeding location for Dugong and marine turtles, part of the park has been listed as a Ramsar site because of its international significance as a feeding and roosting area for migratory shorebirds.

Six artificial reefs have been established in the marine park – Harry Atkinson, West Peel, East Coochie, Wild Banks, North Moreton and South Stradbroke – employing a variety of items, including car tyres, concrete pipes, shopping trolleys, concrete 'reef balls', fabricated steel 'fish caves' and even a derelict tuna fishing vessel. The reefs have been developed specifically for recreational fishing, though spearfishing is only allowed on three of them. Scuba diving and surface-supplied-air diving are not permitted, and commercial fishing is prohibited. For further information visit: http://www.nprsr.qld.gov.au/parks/moreton-bay/zoning/pdf/artificial-reefs-locality-map.pdf

There are also designated 'go slow areas' to protect turtles and Dugong. These particularly relate to seagrass areas that are feeding locations for Green Turtles and Dugong. All vessels must travel off the plane or in displacement mode to minimise boat strike. Vessels more than 8 m are restricted to less than 10 knots.

The *Marine Parks Act 2004* seeks to provide complementary management provisions with the GBRMP. This is reflected in the use and entry provisions for the various zones, whereby consistent zone objectives and entry and use provisions occur across all three state marine parks. Nevertheless, some minor variations do occur, reflecting local circumstances.

Although, no specific marine protected areas under Queensland or Commonwealth legislation have been declared in the Torres Strait region due to the Treaty between Australia and the Independent State of Papua New Guinea (see Parliamentary Paper No. 101/1979, The Parliament of the Commonwealth of Australia), there are a number of other MPAs or marine conservation zone categories in this region (see Chapter 18).

There are also three Indigenous Protected Areas (IPA) declared in 2013 in Queensland that incorporate substantial areas of sea country. The Thuwathu/Bujimulla IPA around the Wellesley Islands region, in the Gulf of Carpentaria protects 1.7 million ha of islands and sea. It is supported by an extensive sea country plan (Carpentaria Land Council Aboriginal Corporation 2006). North of Cairns, the Eastern Kuku Yalanji IPA protects 70 135 ha, taking in parts of two of Australia's World Heritage Areas – the GBR and the Wet Tropics – and overlaps with part of the GBRCMP and GBRMP. Finally, the Girringun Region IPA, between Cairns and Townsville takes in some 1.2 million ha and overlaps with ~9 fish habitat areas (26 000 ha), and 729 300 ha of the GBRCMP and GBRMP (Girringun Aboriginal Corporation *et al.* 2013). Both Eastern Kuku Yalanji and Girringun Region IPAs are some of the first IPAs in the country to overlap with legislated protected areas on Crown land (terrestrial and marine) and establish joint management arrangements. It is likely more such IPAs will be declared in the future in Queensland.

National parks along the Queensland coast are normally declared to high-water mark. However, a very small number (for example Cape Upstart National Park and part of Girringun National Park) have been declared to low-water mark, primarily to protect important mangrove areas. Consequently, those national park areas between the tide marks can be regarded as marine protected areas, though, in every instance, those areas have also been given marine park status.

The ultimate marine protected area – World Heritage

Perhaps the strongest protection possible for any marine area is a combination of marine park status under Commonwealth and/or state laws overlaid with the strength of international law under the World Heritage Convention. That circumstance applies to the GBR.

The GBR was inscribed on the World Heritage List at the 5th Session of the World Heritage Committee meeting in Sydney from 26 to 30 October 1981. The nomination used the description of the GBR Region in the GBRMP Act to define the boundaries of the property. In so doing, it complicated even further the boundary issues that are referred to earlier in terms of state and Commonwealth jurisdiction. Unlike the GBRMP, which was unable to be declared over Queensland islands plus the internal waters of Queensland and certain ports and harbours, the GBR World Heritage Area incorporated all waters, seabed, islands (of which there are ~1000), ports and harbours from low-water mark along the Queensland coast between the northern-most extremity of Cape York Peninsula and a point where the parallel of latitude 24° 30' South intersects the Queensland coast. In responding to, and dealing with, a range of threats to the GBR, it is the World Heritage status that has achieved the most leverage in terms of responding to those threats.

World Heritage status, coupled with the World Heritage provisions of the Commonwealth *Environment Protection and Biodiversity Conservation Act 1999* (EPBC Act), prompted the development of the Reef Water Quality Protection Plan, known as Reef Plan, which was initially established in 2003 and updated in 2009 and 2013. The latest version was endorsed by the Queensland and Australian governments at the Great Barrier Reef Ministerial Forum in July 2013 (Australian Government and Queensland Government 2013). The plan presents a range of actions designed to tackle runoff effects of pollution from broad-scale land use adjacent to the GBR.

These effects, coupled with proposed port development and dumping of dredge spoil within the World Heritage property, have triggered an enhanced level of scrutiny by NGOs and the international bodies responsible for monitoring World Heritage properties – UNESCO (World Heritage Centre) and IUCN. The last two organisations undertook a joint Reactive Monitoring Mission in 2012 and have raised several concerns with the Australian Government. In response to the possibility of the GBR being listed as 'In Danger' by the World Heritage Committee, the Reef 2050 Long-term Sustainability Plan was developed by the Australian Government in association with the Queensland Government (Australian Government 2015).

Another World Heritage property just to the south of the GBR, the Fraser Island World Heritage Area, incorporates the whole island and extends 500 m offshore from high-water mark. The offshore area within the property also lies within the Great Sandy Marine Park.

Current policy

Although it had been a stated policy of the Bligh Government to declare marine parks over the remaining east coast area, along with state waters in the Gulf of Carpentaria (e.g. Queensland Government 2010), this did not eventuate. Neither is there any apparent desire to act on those segments of the Queensland coastline.

Several operational policies relate to such matters as high speed vessel events in marine parks, diving and snorkelling, assessment of works in the Estuarine Conservation Zone of the GBR Coast Marine Park, and the issue of permits and other authorities. However, the zoning plans, and any associated management plans, are the primary policy documents. Because the zones vary so widely in their objectives, which extend from total protection to controlled extraction, many policies are very zone specific.

Queensland plays a major role in the day-to-day operational management of the GBRMP. The policy template for that is the 2009 Intergovernmental Agreement referred to earlier. On-park management policies are developed in concert with the Great Barrier Reef Marine Park Authority.

Challenges

There are many challenges associated with managing vast expanses of the marine environment. Perhaps the most difficult is maintaining the momentum for marine conservation and generating public support for a system of MPAs, including fish habitat areas. That includes generating public awareness of the benefits of such protected areas with particular emphasis on disseminating scientific evidence that supports the effectiveness of such areas and the need for some highly protected no-take zones.

Mechanisms to monitor the effectiveness of the zones and their associated management procedures are essential. This includes the development of habitat mapping, within the framework of marine bioregions, which can identify areas of high ecological significance. As a result of ongoing research, monitoring and resource mapping, there needs to be a strong element of flexibility, which can lead to modifications of zone boundaries based on new data. This is particularly pertinent where the establishment of no-take areas has been restricted by public pressure when a park is initially declared. For example, only 4% of Great Sandy Marine Park is designated as Marine National Park Zone (Table 12.1).

Having political will and commitment is also a key challenge. The GBRMP and the Moreton Bay Marine Park demonstrate the effectiveness of political support through effective planning and improved conservation outcomes.

In the process of developing management models and procedures, it is important that appropriate recognition is given to Traditional Owner values and uses of the area. And, bearing in mind the offshore jurisdictional issues, there needs to be continuing discussions between state and Commonwealth governments and agencies to maximise conservation outcomes.

Perhaps one of the major challenges for any system of protected areas in the marine environment is convincing the public that the constraints on resource extraction are actually achieving something. This is particularly pertinent in relation to the green zones (marine national park zone) where commercial and recreational fishing are not permitted. A broadly held attitude that the sea is an area of common ownership, coupled with obvious difficulties in assessing and appreciating resource damage, have tempered people's attitudes to marine conservation. These matters also have an effect on people's approach to complying with the entry and use constraints established for the protected area zones. In general terms, people tend not to have the same level of respect for marine organisms that they do for terrestrial species. Marine mammals, particularly whales and dolphins in this part of the world, are exceptions to that generality (Fig. 12.7). Recreational fishing is the last form of hunting readily available to the public (Fig. 12.8). Any constraints on recreational fishing are therefore prone to strong resistance.

Policing compliance with the entry and use provisions of the various zones is not a simple task, particularly when it is spread across more than 72 000 km^2 of state marine park alone, without factoring in the vastly greater expanse of the GBRMP for which state officers have a compliance responsibility. And it is an expensive task: one that has the capacity to consume disproportionate amounts of time and money in relation to other management requirements.

Fig. 12.7. Juvenile Humpback Whale in Platypus Bay, Great Sandy Marine Park, July 2015. This part of the marine park is a favoured stop-off location during the annual migration to and from Antarctic waters. Fraser Island is in the background. (Photo: Steve Parish, NatureConnect)

The challenges therefore lie in:

- encouraging and supporting the scientific research and monitoring that can justify the protective measures
- disseminating and explaining the results of such studies to the public
- educating the public about the importance of marine conservation and the necessity for a compliance regime to enforce certain constraints
- helping people appreciate the life histories of selected marine organisms (both animals and plants)
- countering community attitudes that the sea supports an endless supply of organisms that can be extracted at will
- explaining that the sea is an excellent transporter of onshore pollutants and therefore marine conservation has a lot to do with the management of adjacent land-based activities
- anticipating the potential effects of climate change on marine organisms and habitats.

Fig. 12.8. Moreton Bay Marine Park Rangers talking to park users. (Photo: Adam Creed ©Queensland Government)

Conclusion

Queensland has a proud record of initiating a procession of MPAs from 1937 to the present under several different statutes. The state has also been, and continues to be, closely involved in day-to-day management of the GBRMP under the provisions of Commonwealth legislation. The multi-purpose management procedures and arrangements for the GBRMP have been used in the administration of the three present marine parks established under the Queensland *Marine Parks Act 2004* – Moreton Bay (1993), Great Barrier Reef Coast (2004) and Great Sandy (2006). The combined area of the three parks is 72 659 km^2. In combination, they provide marine park status for more than 95% of the eastern Queensland coastline, including islands, from the tip of Cape York to the NSW border.

World Heritage status over marine areas and islands of the Great Barrier Reef has added another potent dimension to marine conservation along a substantial component of the Queensland coastline.

While much has been done, there are still marine areas awaiting assessment (Gulf of Carpentaria and parts of Torres Strait). Also, zoning plans for existing marine parks must be reviewed regularly to ensure they take into account changing circumstances offshore and onshore, including predicted impacts resulting from climate change.

Acknowledgements

Particular thanks are due to James Murphy of the Queensland Parks and Wildlife Service (Department of National Parks, Sport and Racing) for valuable assistance with tables, figures and advice relating to MPAs administered by his Department.

Biography

Peter Ogilvie retired from the Queensland Parks and Wildlife Service in December 2011 after 46 years involvement in the management and administration of national parks, marine parks, World Heritage Areas and wildlife in Queensland. During that time he held several state-wide positions, including Director of Marine Parks and Director of Park Management. He has undertaken research on several seabird species on the Great Barrier Reef. Peter has had close involvement with the development of nature conservation legislation in Queensland and was a member of the Australian negotiating team for the Convention on Biological Diversity. He was awarded the Australian Public Service Medal in 2002 for 'outstanding public service in the field of conservation in Queensland'. He is currently President of the Wildlife Preservation Society of Queensland.

Endnote

1 http://www.environment.gov.au/marine/gbr/protecting-the-reef/intergovernmental-agreement

References

Australian Government (2015) *Reef 2050 Long-Term Sustainability Plan.* Commonwealth of Australia, Canberra.

Australian Government and Queensland Government (2013) *Reef Water Quality Protection Plan 2013.* Department of Environment and Heritage Protection, Brisbane.

Carpentaria Land Council Aboriginal Corporation (2006) *Thuwathu/Bujimulla Sea Country Plan: Aboriginal Management of the Wellesley Islands Region of the Gulf of Carpentaria.* Carpentaria Land Council Aboriginal Corporation, Cairns.

Commonwealth of Australia (2006) *Review of the Great Barrier Reef Marine Park Act 1975.* Commonwealth of Australia, Canberra.

DEHP (2013) *Rules for Watching Marine Mammals.* Department of Environment and Heritage Protection, Brisbane, <http://www.ehp.qld.gov.au/wildlife/watching/marine-mammals/rules.html>.

Department of Agriculture, Fisheries and Forestry (2012) *Declared Fish Habitat Area Network Assessment Report, 2012.* Department of Agriculture, Fisheries and Forestry, Brisbane.

DERM (2010) *Moreton Bay Marine Park User Guide.* Department of Environment and Resource Management, Brisbane, <http://www.nprsr.qld.gov.au/parks/moreton-bay/zoning/pdf/marine-park-user-guide.pdf>.

Girringun Aboriginal Corporation, Bandjin, Djiru, Girramay, Gugu Badhun, Gulnay, Nywaigi, Warrgamay and Warungnu Traditional Owners and Regional Advisory & Innovation Network (RAIN) Pty Ltd (2013) *Girringun Region Indigenous Protected Areas Management Plan.* Girringun Aboriginal Corporation, Cardwell.

Haward M (1989) The Australian offshore constitutional settlement. *Marine Policy* **13**, 334–348. doi:10.1016/0308-597X(89)90018-3.

Queensland Government (2010) *Building Nature's Resilience: A Draft Biodiversity Strategy for Queensland.* Department of Environment and Resource Management, Brisbane.

Wright J (1977) *The Coral Battleground.* Thomas Nelson, Melbourne.

Chapter 13

Marine protected areas and marine conservation in the Northern Territory

Karen Edyvane and Stuart Blanch

Introduction

At the national level, the Northern Territory (NT) ranks last among Australian jurisdictions on progress towards the establishment (and effective management) of a comprehensive, adequate and representative network of marine protected areas (MPAs) (Lavers 2010; NMPAWG 2010) – in 2014, contributing < 0.1% to the total national MPA estate in Australia (Department of the Environment 2014a).

Although the NT's coastal and marine ecosystems are largely ecologically intact and recognised as globally significant (Halpern *et al.* 2008), the Territory's MPA system is far from comprehensive, adequate and representative. Seven of the Territory's 13 marine bioregions receive no protection from NT MPAs, whilst the five declared MPAs in the Territory protect just 6.4% of NT waters, with only an estimated 1.6% in highly protected areas. Only two significant Northern Territory Government-declared Marine Parks exist: Cobourg and Limmen Bight. As such, the NT has to date failed to make substantial progress towards contributing to national and global goals of establishing and effectively managing a network of MPAs by 2020 (CBD 2010).

Although the Northern Territory Government has not embargoed the creation of new Marine Parks, it has no MPA Strategy, no plans or targets for new MPAs, no public servants responsible for MPA planning, and no plan of management for Limmen Bight Marine Park.

However, the NT arguably leads Australia on Indigenous 'sea country' management, including collaborative and cooperative approaches to marine conservation and fisheries management. In contrast to many other jurisdictions in Australia, the NT's sparse population, and relatively undeveloped coastline, has resulted in its coastal and marine ecosystems and associated biodiversity being considered 'least disturbed' and globally significant (Halpern *et al.* 2008). The NT itself contains some of the largest, near-pristine catchments, tidal estuaries (~137) and intact coastal wetlands in Australia, including the largest area of mangroves (4120 km^2), and among the highest diversity (i.e. 49 species, including the endemic, *Avicennia integra*). NT waters also contain: five of the world's seven sea turtle species (including the largest rookeries of Olive Ridley Turtles *Lepidochelys olivacea* in Australia and South East Asia-Western Pacific); nationally and internationally significant nesting colonies of shorebirds and seabirds; and globally significant populations of threatened species – such as sharks, rays, sawfish, sea snakes, pipefishes and seahorses (Ward *et al.* 2007).

Unlike many other jurisdictions in Australia, Aboriginal people are the major custodians of the Territory's coastal and marine ecosystems and biodiversity. Indigenous Territorians represent more than 30% of the Territory's total population and own around 85% of the coastline, including intertidal lands and waters to the low-water mark, under inalienable Aboriginal freehold tenure granted under the *Aboriginal Land Rights Act (Northern Territory) 1976* (Ganter 1996). Systems of traditional law in the sea remain strong over most of the coastline and have major influence over contemporary Aboriginal use, occupation and aspirations in the marine and coastal zone. Aboriginal people still make significant use of subsistence resources in many marine and coastal areas and have binding cultural relationships with particular places, flora and fauna and natural phenomena in the sea (Billyard 1993; Ganter 1996; Smyth and O'Leary 2007; Smyth 2008; see also Chapter 18).

The extent of Indigenous coastal tenure and interests makes the NT unique in Australia, providing challenges for establishing MPAs, but also opportunities and obligations, for Aboriginal people with rights and interests in coastal areas, to be involved in MPAs, and coastal and marine management (Billyard 1993; Ganter 1996; Smyth and O'Leary 2007; Ward *et al.* 2007; Smyth 2008; Chapter 18). For MPAs in the NT, Indigenous saltwater people have been successfully managing vast areas of coastal and marine estate or 'sea country' (including wildlife and habitats) according to their traditional laws and customs for an estimated 47 000 years.

With legal certainty, from the *Blue Mud Bay* decision of the High Court confirming coastal tenure (and the necessity for co-management), future MPAs in the NT will need to be developed and managed through formal cooperative management arrangements with Aboriginal people. Further, MPAs cannot only strengthen Indigenous Protected Areas (IPAs) (which are non-legal, voluntary declarations), but also (like IPAs) have the potential to provide a range of regional economic development and training opportunities for remote Aboriginal communities, Indigenous Sea Ranger groups and Indigenous enterprises, and much needed, improved economic outcomes for Aboriginal people.

History and development of the MPA network in the Northern Territory

Formal protection of the Territory's coastal areas under 'white man's laws' commenced in 1983 with the declaration of the Cobourg Marine Park (under the *Territory Parks and Wildlife Conservation Act 1983*). This Act is the primary legislative mechanism for developing and managing protected areas in the Territory and enables the creation of protected areas including parks and reserves on land or sea, for a range of purposes, including the protection of representative areas (Billyard 1993). In 1995, joint management arrangements (and the Cobourg Peninsula Sanctuary and Marine Park Board) were established under the *Cobourg Peninsula Aboriginal Land, Sanctuary and Marine Park Act 1996* to give management responsibility for the Marine Park to the Board. The Marine Park is part of the larger Garig Gunak Barlu National Park. Garig Gunak Barlu National Park includes both the Marine Park and the Cobourg Sanctuary. The Sanctuary extends to low-water mark and includes the intertidal zone and waters covering the peninsula while the Marine Park extends seaward from the low-water mark.

In 1992 and 1993, two small Aquatic Life Reserves (Doctors Gully and East Point) under the NT *Fisheries Act 1988* were declared to protect aquatic life in Darwin Harbour.

The NT *Fisheries Act 1988* provides for the conservation, enhancement, protection, utilisation and management of fish and aquatic life in the NT and also regulates fishing activities within a MPA through the declaration of Fishery Management Areas.

Although the Northern Territory Government received Australian Government funding to develop a NT MPA Strategy to guide the planning and establishment of a MPA network, the finalised *Northern Territory Marine Strategy* (NRETA 2008) was never publicly released. Successive governments have also initiated, but failed to finalise and implement, two major proposed MPAs, including the Beagle Gulf Marine Park (from 1991 to 1996), and the Bynoe Harbour Marine Park (during 2006–2007). In July 2012, the NT declared its second Marine Park, the Limmen Bight Marine Park, almost three decades after the declaration of Cobourg Marine Park.

The major drivers for the NT's attempts to establish a comprehensive, adequate, representative (or CAR) system of MPAs, have primarily been the Territory's commitment to existing national policy agreements and specifically commitments to establish a National Representative System of Marine Protected Areas (NRSMPA). Specifically, these include agreements under the Intergovernmental Agreement on the Environment (1992), and subsequent national commitments to protect marine biodiversity and the long-term viability of ecological processes and ecosystems such as the *National Strategy for Ecologically Sustainable Development* (1992) and the *National Strategy for the Conservation of Australia's Biological Diversity* (1996) and, specifically to the NRSMPA, through the *Strategic Plan of Action for Establishing the National Representative System of Marine Protected Areas* (ANZECC TFMPA 1998) and the *Guidelines for Establishing the National Representative System of Marine Protected Areas* (ANZECC TFMPA 1999). As such, the Northern Territory Government is a formal signatory to the above arrangements and as such, is committed – in writing – to the establishment of the NRSMPA.

Despite commitments by both major political parties when in government – the Country Liberal Party to the NRSMPA in 1999 (ANZECC TFMPA 1999), and election commitments by the Labor Party in 2005 (Scrymgour 2007) – to develop a 'world class' network of MPAs, the Northern Territory Government has failed to significantly increase or make more representative its MPA estate.

In contrast to the MPA network, there has been rapid progress by Indigenous custodians in establishing Indigenous Protected Areas (IPAs), over large areas of the NT, including 'sea country'. This has been assisted by increased Australian Government support (and funding) for IPAs and 'sea country' planning, particularly under the 'Caring for our Country' program, and also, ongoing support, training, and 'on the ground' capacity building by the Territory Parks and Wildlife Service, which has pioneered joint management of protected areas (Moyses and Panton 2008). Under the *Parks and Reserves (Framework for the Future) Act 2005*, joint management provisions and Indigenous Land Use Agreements (ILUAs), are now in place for 27 parks and reserves across the Territory (Moyses and Panton 2008), many of which also include coastal ecosystems (down to the low-water mark). As such, Traditional Owners are now established as primary joint management partners across the entire protected area estate in the NT.

Legal uncertainty surrounding the tenure and governance of coastal resources has been a major factor in the failure to progress formal MPAs in the NT. To this end, legal decisions by the High Court of Australia, since 2008, have had profound implications for the directions and management of the Territory's coastal, estuarine and marine ecosystems (Edyvane 2010). Indigenous people own 85% of the Territory's coastline (down to the low-water mark), and another 13% is under claim. Under the *Blue Mud Bay* legal decision of 2008, the High Court of Australia confirmed Indigenous ownership and control of the seabed and waters in the intertidal zone of the NT and also, confirmed that the *NT Fisheries Act* is valid in all Territory waters. Thus the *Blue Mud Bay* decision has confirmed the necessity for co-management of the Territory's coastal, estuarine and marine natural resources (Edyvane 2010).

Indigenous 'sea country' planning and management has been considerably progressed since 2006, through the development of formal sea country plans and also, more recently, Indigenous Protected Areas that include sea country. In the Territory, several formal sea country plans and/or sea country IPAs have been finalised or are in development for coastal communities, including Dhimurru, li-Anthawirriyarra, Djelk, Laynhapuy, Marthakal, Anindilyakwa and Bugul (Edyvane 2010; and see Chapter 18).

Over the past three decades, Australian Government funding for the NRSMPA has been the primary source of funding to the NT to progress MPA research, planning and management. This includes the proposed Beagle Gulf Marine Park (Smit *et al.* 2000), NT MPA Strategy and projects identifying areas of significance for potential MPAs (Edyvane and Dethmers 2010), sites of significance for biodiversity conservation (Harrison *et al.* 2009; Ward and Harrison 2009), and also, community-based, systematic marine conservation planning adjacent the Dhimurru Indigenous Protected Area (using Marxan) (Dhimurru Land Management Aboriginal Corporation 2011).

Northern Territory Marine Protected Areas Strategy

On 9 March 2007, the Minister for Natural Resources, announced the 'development of a Strategy to establish a system of Marine Parks in the NT', creating a 'bigger, better world class parks system', giving effect to a 2005 pre-election commitment (Scrymgour 2007). The Minister also announced the establishment of a key stakeholder committee to draft a Strategy and define the 'roadmap' for implementing the system of MPAs, 'in particular, identifying the scientific and consultation processes required to establish such a system' (Scrymgour 2007).

The Strategy was developed during 2007 and 2008, by then NT Department of Natural Resources, Environment and The Arts (NRETA), with input from an independently chaired, 22-person, multi-stakeholder, NT MPA Advisory Committee. The Strategy committed the NT to developing a comprehensive, adequate and representative network of MPAs consistent with the NRSMPA, and outlined technical and planning processes to establish a MPA network (NRETA 2008). Independent technical advice was sought on Indigenous engagement and participation in MPA planning (Smyth and O'Leary 2007) and applying science-based MPA planning in the NT (Ward *et al.* 2007).

Significantly, the NT MPA Strategy (NRETA 2008) adopted a best practice, science-based approach to MPA planning – including the development of transparent, decision-rules and 18 operational planning principles, criteria and guidelines (8 biophysical; and 10 social, cultural and economic). In identifying, selecting and establishing MPAs for the NT Representative System of MPAs (or NTRSMPA), the Strategy identified the 12 meso-scale bioregions of the NT as the basis and template for MPA planning, and recommended scientific review of individual MPA bioregional plans by an independent MPA Scientific Advisory Group. The Strategy also identified a process for community nominations and pathways for cooperative management with Indigenous custodians as well as a process to consider, recognise and include sea country IPAs in the NTRMSPA. In this way, the NT MPA Strategy provided a strategic framework for MPA planning, but with local flexibility.

The NT MPA Strategy, unanimously endorsed by the NT MPA Advisory Committee, also supported an aspirational timeline for establishment of the NTRSMPA by 2012, including, significantly, bioregional targets in establishing the NTRSMPA (i.e. for each bioregion, a minimum representative area of 10% per bioregion, to be reserved as MPAs). Although no targets for strict protection ('no-take') were specified for the NTRSMPA, the Strategy acknowledged that 'highly protected areas are a fundamental component of an effective MPA network and will be defined through application of the operational principles and criteria of the NTRSMPA' (NRETA 2008).

The NT MPA Strategy (and media and communications strategy) was finalised in early 2008 by the Advisory Committee and NRETA. However, agency and political commitment to the NT MPA Strategy waned, following increasing opposition from fishers and also, significantly, Ministerial and senior agency staff changes. Despite calls from conservationists throughout 2008 for the public release of the NT MPA Strategy (ABC 2008, 2009a), the finalised Strategy never went to Cabinet (for endorsement and public release).

In March 2009, with increasingly trenchant opposition from the recreational fishing lobby and pro-fishing Ministers, the government's opposition to 'no-take areas' in NT waters was outlined, when the then NT Fisheries Minister, Kon Vatskalis, stated at the Annual General Meeting of the Amateur Fishermen's Association of the Northern Territory:

> *'I would like to say yes the Government is prepared to close areas but the Government is not prepared to have the same issues like in other states. If areas are to be closed, amateur fishermen will go fishing and I will go with them. So no marine protected areas closed off to amateur fishermen. The only intention is to close areas to the commercial sector and, for obvious reasons, for the protection of species'.* (Anon. 2009)

The public release of the NT MPA Strategy was also significantly adversely affected by the concurrent, Australian Government's MPA planning process for the North Marine Region in the offshore waters of the NT. With the release of the North Marine Bioregional Profile by the Australian Government in November 2008 (DEWHA 2008) (and stakeholder consultations throughout the NT, on MPA planning in the North Marine Bioregion), there was widespread concern regarding the potential proliferation of MPAs in the Territory – in both inshore and offshore waters – and, the scale and size of proposed Commonwealth marine parks.

Relations between the Northern Territory Government and the Australian Government deteriorated politically (and publicly) on the issue of the Commonwealth marine parks, and exacerbated by the lack of cooperation on MPA planning between the two jurisdictions (ABC 2009b,c). Despite the development of a formal Memorandum of Understanding (MOU) between the Northern Territory and Australian governments in 2006 to foster cooperation on MPA and integrated regional marine planning in the NT and adjoining Commonwealth waters, the primary focus of the MOU was information sharing. Significantly, the MOU provided no provision for input on MPA boundaries or levels of protection or the timing of stakeholder consultation (despite requests by the NT). As a result, the MOU afforded very limited opportunity for effective inter-jurisdictional cooperation and complementary trans-boundary (i.e. inshore–offshore) MPA planning. This is starkly evident in the non-alignment of Limmen Commonwealth Marine Reserve and the Limmen Bight Marine Park (NT Government) (see Fig. 13.1).

The marine protected area estate in the Northern Territory

A total of 5 MPAs have been declared in the Territory, protecting 459 670 ha, or ~6.4%, of the Territory's jurisdictional waters. As such, the MPA network is far from complete, and lags considerably behind the efforts of other Australian states/territory jurisdictions (NMPAWG 2010).

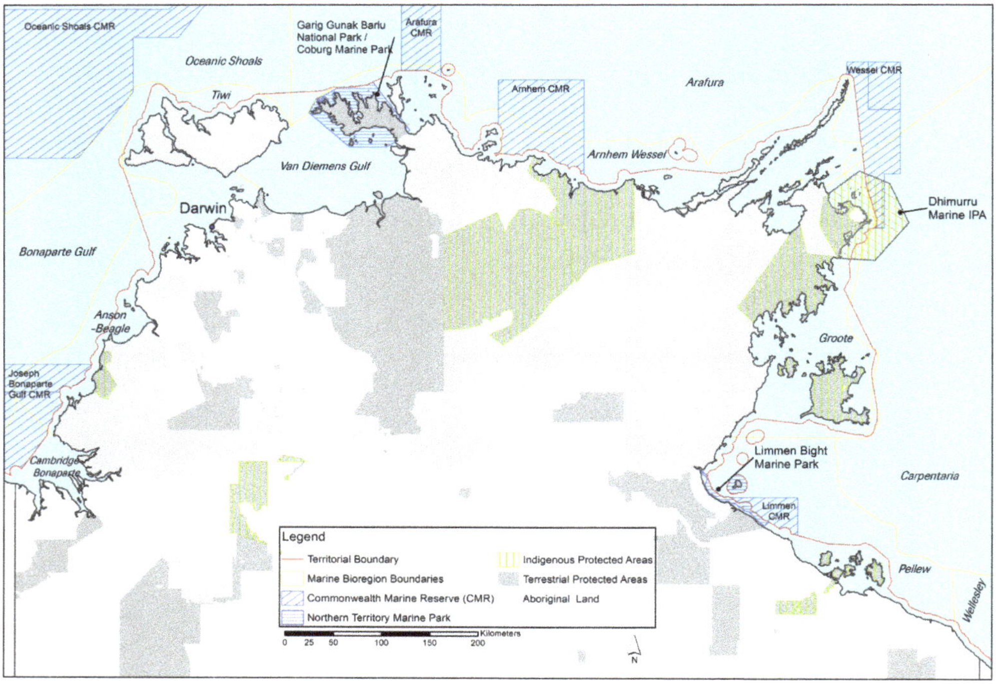

Fig. 13.1. Marine protected areas and Indigenous Protected Areas in the Northern Territory, 2015.

MPAs declared in the Territory under parks and fisheries legislation, and Indigenous Protected Areas established by Indigenous Traditional Owners, are shown in Table 13.1 and Fig. 13.1. Declared MPAs in the Territory's jurisdictional waters currently include:

- two Marine Parks declared under the *Territory Parks and Wildlife Conservation Act 1993* (Cobourg and Limmen Bight Marine Parks)
- one Indigenous Protected Area over sea country, off the north-east Arnhem Land (Dhimurru IPA) (see Figs 18.2, 18.3 and 18.4), which has no legislative basis, but is strongly supported by Traditional Owners and stakeholders
- two small Aquatic Life Reserves around Darwin, declared under the *NT Fisheries Act 1988* (East Point and Doctors Gully).

An estimated one-half of the total area of Cobourg Marine Park (223 374 ha) occurs within highly protected zones (Conservation, Scientific Reference, Port Essington and Coral Bay Buffer Zones), protecting ~112 000 ha, and representing 1.6% of NT waters. Of the NT's two established Marine Parks, Cobourg Marine Park is joint managed with Traditional Owners (Cobourg Peninsula Sanctuary and Marine Park Board and Parks and Wildlife Service 2011), and Limmen Bight Marine Park is expected to be so when a plan of management is negotiated.

The Limmen Bight Marine Park (88 000 ha) was declared in June 2012, 2 months before a NT election. It followed a 2-year campaign by environment organisations, Indigenous Traditional Owners (Marra and Wandarrang people) and commercial and recreational fishermen to prevent a proposed deep water port and iron ore export plant proposed for construction on Kururrulinya (or Maria Island) in Limmen Bight, south-west Gulf of

Table 13.1. Current status of marine protected areas in the waters of the Northern Territory

(Source: areas and zoning for Marine Parks from CAPAD 2014[a]. Aquatic Life Reserves are currently not included in CAPAD)

Marine protected area type	Area (ha)	Zoning	IUCN protected area category	Declared	Management plan
Marine Parks *Territory Parks and Wildlife Conservation Act 1976*					
Cobourg Marine Park	223 374	Zoned	II	July 1983, re-declared Nov 2002	2011
Limmen Bight Marine Park	77 016	Unzoned	VI	July 2012	None
Sub-total	300 390				
Aquatic Life Reserves *NT Fisheries Act 1988*					
East Point	265	Unzoned	IV	1992	None
Doctors Gully	14	Unzoned	IV	1993	None
Sub-total	279				
Total NT Government parks and reserves	**300 669**				
Sea Country Indigenous Protected Areas (IPAs)					
Dhimurru IPA	~159 000	Unzoned	V	November 2000, expanded April 2013	Dhimurru Aboriginal Corporation (2015)
Total MPAs	459 669				
Total NT territorial waters	**7 183 900**				
Total parks and reserves (proportion of NT waters)	**4.2%**				
Total Sea Country IPAs (proportion of NT waters)	**2.2%**				
Total MPA (proportion of NT waters)	**6.4%**				

[a]Collaborative Australian Protected Area Database (CAPAD) (Department of Environment 2014a)

Carpentaria. However, the then NT Labor Government provided significant concessions to anglers by stating the marine park would be 'fishing friendly' and unlikely to include no-take zones (Australian Labor Party of the Northern Territory 2012). Although the new NT Country Liberal Party Government has not rescinded Limmen Bight Marine Park since winning office, no efforts have been made to develop a management plan, negotiate different zonings with Traditional Owners and stakeholders, nor has additional budget been provided for its management.

While Aquatic Life Reserves in the NT are currently not listed as MPAs under the Collaborative Australian Protected Area Database (CAPAD) (Department of Environment 2014a), the authors believe that two of the three declared reserves should be recognised, due to their size and specific management objectives. Similarly, a range of other spatial, marine management mechanisms also significantly contribute to marine conservation efforts, such as 'sacred sites' (declared under the *NT Sacred Sites Act 1989*), 'sea closures' (declared under the *NT Aboriginal Land Act 1992*), and also, 'fisheries closures' (declared under the *NT Fisheries Act 1988*) (see Table 13.2). Significantly, the *NT Sacred Sites Act 1989* provides Indigenous Traditional Owners and managers, through the support of Aboriginal Land Councils, the opportunity to protect both cultural and biological values of seascapes, sacred sites and dreaming tracks associated with marine features (such as reefs and islands), and by extension, associated marine life. Like sea country IPAs, these important Indigenous marine management approaches need to be supported and recognised (where appropriate and desired by Traditional Owners) as contributing to the national MPA estate.

National parks, coastal reserves and other terrestrial conservation reserves also afford some level of protection over marine, estuarine and coastal habitats in the NT. These include Kakadu National Park, a Commonwealth protected area reserved under the *Environment Protection and Biodiversity Conservation Act 1999* (Cth), and three coastal reserves (Channel Island, Shoal Bay, Casuarina), two coastal national parks (Mary River and Charles Darwin, Fig. 13.2) and one conservation area (Tree Point), declared under the *Territory Parks and Wildlife Conservation Act* (NT) (Fig. 13.1).

The Dhimurru IPA was first declared in November 2000, over 101 000 ha on the north-east coast of Arnhem Land and included 9000 ha of sea country. In 2013, this sea country component of the IPA was expanded to over 400 000 ha in Commonwealth and Northern Territory waters. In November 2013, the Thuwatha/Bujimulla IPA, a large terrestrial and marine IPA (~1.7 million ha) was declared around the Wellesley Islands and Gulf coast in the Queensland portion of the shared Wellesley bioregion.

However, while there has been much progress (particularly in the NT), sea country IPAs continue not to be formally recognised as contributing to the NRSMPA due to the lack of legislative basis – a higher threshold than the IUCN protected area definition and a higher threshold than for terrestrial areas to be added to the National Reserve System (Rose 2012). As an increasingly major component of the MPA estate in the Territory (and Northern Australia), we strongly support calls for formal recognition of sea country IPAs under the NRSMPA (e.g. Smyth 2008).

Significantly, the MPA network in the Territory has low levels of comprehensiveness, adequacy and representativeness. Levels of protection range for bioregions occurring in Territory waters range from < 0.01 to 7.5% (only 2 bioregions $> 5\%$). Similarly, NT MPA size range from 0.91 to 1268 km^2 (only 3 MPAs are > 100 km^2). Of the 12 meso-scale marine bioregions recognised in NT waters (Commonwealth of Australia 2006), only three receive significant protection from NT MPAs (Cobourg, Van Diemen Gulf and Pellew) (Table 13.3). The three largest MPAs by area – Cobourg Marine Park, Dhimurru Indigenous Pro-

Fig. 13.2. Diverse mangrove communities and intertidal systems comprise almost two-thirds of the Charles Darwin National Park, on the outskirts of Darwin city (pictured in background). (Photo: James Fitzsimons)

tected Area and Limmen Bight Marine Park – together, represent 99.9% of the NT's MPA estate, or 6.4% of the Territory's jurisdictional waters.

Of the 13 recognised bioregions occurring in Territory waters, many include large areas of Commonwealth waters, particularly offshore bioregions (Oceanic Shoals, Arafura, Carpentaria and Bonaparte Gulf) (Fig. 13.1). To this end, the Commonwealth Marine Reserves have recently been established in offshore waters across all bioregions that occur in Territory waters, except Van Diemens Gulf and Groote bioregions (which largely comprise NT waters). However, the coastal, inshore waters of many bioregions largely remain unprotected, with the Territory failing to establish MPAs in seven bioregions: Arnhem Wessel, Bonaparte Gulf, Oceanic Shoals and Tiwi and three bioregions it also shares with Queensland (Carpentaria and Wellesley) and Western Australia (Cambridge-Bonaparte) (see Table 13.3). As a result, the Territory's current MPA system excludes many coastal ecosystems and coastlines of national and global significance (particularly for threatened species and biodiversity), including large, sheltered, gulf ecosystems (Gulf of Carpentaria and Sir Joseph Bonaparte Gulf), major offshore islands (i.e. Wessell Islands, Pellew Islands and Groote Eylandt group) and the extensive rocky and sandy shores of the Arnhem Land coast. Many of these remote, intact ecosystems contain globally significant populations of threatened marine megafauna (Nursey-Bray and Pedder 2004; Ward *et al.* 2007; DEWHA 2008). Although the Commonwealth Marine Reserve network in the region is far from adequate (or comprehensive), it has at least recognised these significant ecological values, particularly protecting offshore sites for threatened and migratory species (including

Table 13.2. Current types of marine protected areas and other marine managed areas in the Northern Territory

MPAs and marine managed areas	Legislation	Primary management goal	Management objectives
National parks, reserves	*Territory Parks and Wildlife Conservation Act 1993*	Establishment and management of parks and reserves and sustainable utilisation of wildlife.	The *Territory Parks and Wildlife Conservation Act 1993* provides for the establishment of national parks and other parks and reserves and the study, protection, conservation and sustainable utilisation of wildlife.
Fisheries closures (e.g. protection of species, areas etc.)	*NT Fisheries Act 1988*	Conserve, enhance, protect, utilise and manage aquatic resources and habitats.	The *NT Fisheries Act 1988* provides for the ecological sustainable management of the aquatic resources of the Territory, whether managing a single fish species or an ecosystem, to ensure the protection of fish and aquatic life resources, and fish habitats. Fisheries closures may be enacted by way of fishery regulations or fishery management plans.
Aquatic Life Reserves	*NT Fisheries Act 1988*	Conserve, enhance, protect, utilise and manage aquatic resources.	In addition to closures, the *NT Fisheries Act 1988* provides for the ecological sustainable management of the aquatic resources of the Territory, whether managing a single fish species or an ecosystem, to ensure the protection of fish and aquatic life resources, and fish habitats. Aquatic Life Reserves may be declared as management areas under the Act, and managed by way of a management plan.
Indigenous Protected Areas	None; Agreement between Traditional Owners and Australian Government	Indigenous Protected Area are governed by the continuing responsibilities of Aboriginal and Torres Strait Islander peoples to care for and protect lands and waters for present and future generations.	Indigenous Protected Areas include areas of land and/or waters over which Aboriginal and Torres Strait Islanders are custodians, and which are managed for cultural biodiversity and conservation, permitting customary sustainable resource use and sharing of benefit. As IPAs are voluntary declarations by Indigenous people, with no legal basis, they cannot legally impact on the activities of any non-Indigenous group unless by the consent of those interests, or by imposing restrictions on access to land or through other legal mechanisms. Existing laws, regulations and responsibilities continue to apply in any sea country Indigenous Protected Area – including existing bag limits and fisheries management arrangements. Decisions about the management of land-based Indigenous Protected Areas are made by Indigenous people over land that they own or have management arrangements over.

Sacred sites	*NT Sacred Sites Act 1989*	Protection of cultural values, customary practices and interests.	The *NT Sacred Sites Act 1989* allows for the recording and registering of sacred sites in the sea. All sacred sites are afforded protection by the Act whether registered or not, but registration provides *prima facie* evidence in the event that a prosecution for violating a site is brought. The Aboriginal Areas Protection Authority is a statutory authority responsible for administration of the Act.
Sea closure	*NT Aboriginal Land Act 1992*	Protection of cultural values, customary practices and interests.	The *NT Aboriginal Land Act 1992* (ALA) regulates access onto Aboriginal land consistent with the provisions of the *Aboriginal Land Rights Act 1976* (ALRA). The ALA has provisions to enable Traditional Owners of coastal areas to close seas up to 2 km from the boundary of the Aboriginal land held under the ALRA (i.e. the low-water mark). There are currently two areas around Millingimbi where successful closure applications have been made.
Heritage place/ object	*NT Heritage Act 2015*	Protection of historic and cultural heritage.	The *Heritage Act 2015* provides for the conservation of the Territory's cultural and natural heritage, through the identification, assessment, recording, conservation and protection of places and objects of prehistoric, protohistoric, historic, social, aesthetic or scientific value, including geological structures, fossils, archaeological sites, coastlines and plant and animal communities or ecosystems of the Territory.

Table 13.3. Reservation levels of mesoscale bioregions in Northern Territory waters

(Source: CAPAD 2014[a])

IMCRA meso-scale region	Inshore or offshore	Jurisdictions	Bioregion area (km^2)[a]	Commonwealth Marine Reserves	NT MPA area (km^2)[b]	Proportion of bioregion declared as NT MPAs (%)
Anson Beagle	Inshore	NT/Cth	20 499	✓	2.79	< 0.01
Arafura	Offshore	Cth/NT	154 999	✓	0.91	< 0.01
Arnhem Wessel	Inshore	NT/Cth	27 439	✓	0	0
Bonaparte Gulf	Offshore	NT/Cth	41 656	✓	0	0
Cambridge-Bonaparte	Inshore	NT/WA/Cth	14 962	✓	0	0
Carpentaria	Offshore	Cth/NT	228 965	✓	0	0
Cobourg	Inshore	NT/Cth	10 294	✓	999	9.70
Groote	Inshore	NT/Cth	19 440	—	90	0.50
Oceanic Shoals	Offshore	Cth/NT	246 404	✓	0	0
Pellew	Inshore	NT/Cth	25 506	✓	770	3.02
Tiwi	Inshore	NT/Cth	5105	✓	0	0
Van Diemens Gulf	Inshore	NT	17 010	NA	1 268	7.45
Wellesley	Inshore	NT/Qld/Cth	27 274	✓	0	0

[a]Includes state/territory and Commonwealth waters

[b]Collaborative Australian Protected Area Database (CAPAD) (Department of Environment 2014a). Does not include Dhimurru IPA sea country which occurs mostly in the Groote bioregion and a small part of the Arafura bioregion. Thus area and percentage reserved for these bioregions are an underestimate.

resting and foraging sites of turtles and shorebirds) (Table 13.4). However, arguably the more important, inshore breeding and nesting areas for these key species have not been protected by complementary, inshore MPA planning by the Territory.

Sea Country Indigenous Protected Areas

In many respects, the Territory is leading Australia in both recognising the rights and aspirations of Indigenous Australians in sea country management including through the establishment of Indigenous Protected Areas over marine ecosystems (see Chapter 18), and also in the range of training and partnership programs to assist Indigenous management of the customary coastal estate (Edyvane 2010). There are currently ~17 active Indigenous land and sea management groups, distributed around the NT coastline, often referred to as Aboriginal rangers, with direct interest in sea country management. These groups have varying levels of capacity according to their funding base, history of development and other matters relating to the local context of where they operate.

Although the Dhimurru IPA is the only IPA formally recognised as covering sea country in the NT, another five IPAs have been declared over land and include significant areas of adjacent saltwater ancestral estates for management by Indigenous Ranger groups (Yanyuwa, Anindilyakwa, Laynhapuy Stage 1, Djelk, and Marri-Jabin-Thamurrurr Stage 1) (Department of the Environment, 2014b), and at least another five IPAs, with marine estates, are in consultation or are being discussed (South East Arnhem Land, Laynhapuy Stage 2, Marthakal, Thamurrurr Stage 2, and Yugul Mangi). Significantly, several of these groups are in the process of consulting and/or developing formal key stakeholder agreements (through sea country planning), for the management of the marine components of their IPAs.

With the legal certainty of the *Blue Mud Bay* decision, all future MPA declarations in the NT need to work closely with Australian Government's Indigenous Protected Areas Program to ensure a complementary approach to exploring extensions of IPAs over sea country. Although sea country IPAs are currently being developed and implemented in the coastal waters of the NT, as voluntary declarations by Indigenous people, with no legislation, they have no legal basis, and on their own do not legally impact on the activities of any non-Indigenous group (Smyth and O'Leary 2007; Smyth 2008). To this end, legislated MPAs provide a legal mechanism to strengthen the management effectiveness of IPAs.

Significantly, saltwater Indigenous people manage many, vast, ecologically and culturally important areas of shallow tropical marine seas across the Territory, through a range of mechanisms and governance arrangements, including traditional customary closures. For instance, the Crocodile Island Rangers manage 600 000 ha of sea country within Castlereagh Bay off the north Arnhem Land coast, 'a site recognised as being of international conservation significance due to the large aggregations of migratory shorebirds, large seabird colonies and important marine turtle nesting beaches' (Crocodile Island Rangers 2013). Though not a marine IPA or Marine Park, the Traditional Owners have declared their own marine turtle sanctuary in this area.

Current policy

As outlined earlier, the Northern Territory Government is signatory to several national commitments that support the development of MPA networks, particularly the establishment of the NRSMPA (ANZECC TFMPA 1999). In addition to these national commitments to the NRSMPA, a commitment to a MPA network in the NT is also found in several framing policies, such as the Integrated Natural Resource Management Plan for the

Table 13.4. Current status of marine protected areas in Commonwealth waters off the Northern Territory

(Source: areas and zoning for Marine Reserves from CAPAD 2014[a])

Commonwealth Marine Reserve	MPA size (km²)	IUCN protected area category	Conservation values
Limmen Reserve – multiple use	1399	VI	• Important resting area for threatened Flatback Turtles between egg laying (inter-nesting area) • Example of an ecosystem of the Northern Shelf Province (including the Pellew meso-scale bioregions) • Key ecological feature: Gulf of Carpentaria coastal zone (high productivity, biodiversity and endemism, aggregations of marine life)
Wessel Reserve – Marine National Park – multiple use	1632 4276	II VI	• Important resting area for the threatened Flatback, Green, Hawksbill and Olive Ridley Turtles between egg laying (inter-nesting area) • Important foraging habitat for breeding aggregations of the migratory Common Noddy, migratory Roseate Tern, listed marine Crested Tern • Examples of ecosystems of the Northern Shelf Province (including the Arafura, Arnhem-Wessel and Carpentaria meso-scale bioregions) • Key ecological feature: Gulf of Carpentaria basin (biodiversity, aggregations of marine life)
Arnhem Reserve – special purpose	7125	VI	• Important resting area for the threatened Flatback Turtle between egg laying (inter-nesting area) • Important foraging habitat for breeding aggregations of the migratory Bridled Tern, migratory Roseate Tern, listed marine Crested Tern • Examples of ecosystems of the Northern Shelf Province (including the Arafura, Arnhem-Wessel and Cobourg meso-scale bioregions)
Arafura Reserve – multiple use	22 924	VI	• Important resting area for the threatened Flatback, Green, Hawksbill and Olive Ridley Turtles between egg laying (inter-nesting area • Important foraging habitat for breeding aggregations of the migratory Roseate Tern • Examples of the ecosystems of two provincial bioregions: the Northern Shelf Province (which includes the Arafura and Cobourg meso-scale bioregions) and the Timor Transition Province • Key ecological feature: Tributary canyons of the Arafura Depression (unique sea-floor feature)

Joseph Bonaparte Gulf Reserve – multiple use – special purpose	6345 2251	VI VI	• Important foraging area for threatened Green and Olive Ridley Turtles and the Australian snubfin dolphin • Examples of the shallow water ecosystems and communities of the North-west Shelf Transition Province, the second largest of all the provincial bioregions on the shelf, which includes the extensive banks that make up the Van Diemen Rise and the Sahul Shelf, broad shelf terraces and the shallow basin in the Joseph Bonaparte Gulf (including the Cambridge-Bonaparte, Anson Beagle and Bonaparte Gulf meso-scale bioregions) • Key ecological feature: carbonate banks of the Joseph Bonaparte Gulf (enhanced productivity, high biodiversity, unique seafloor feature)
Oceanic Shoals Reserve – multiple use	71 744	VI	• Important resting area for the threatened Flatback and Olive Ridley Turtles between egg laying (inter-nesting area) • Important foraging area for the threatened Loggerhead and Olive Ridley Turtles • Examples of the ecosystems of two provincial bioregions: the North-west Shelf Transition Province (which includes the Bonaparte, Oceanic Shoals, and Tiwi meso-scale bioregions) and the Timor Transition Province • Key ecological features (4): carbonate bank and terrace system of the Van Diemen Rise (unique sea-floor feature); carbonate banks of the Joseph Bonaparte Gulf (enhanced productivity, high biodiversity, unique sea-floor feature); pinnacles of the Bonaparte Basin (enhanced productivity, unique sea-floor feature); shelf break and slope of the Arafura Shelf (unique sea-floor feature)

[a]Collaborative Australian Protected Area Database (CAPAD) (Department of Environment 2014a)

Northern Territory (LCNT 2005) and the draft Parks and Conservation Masterplan (Northern Territory Government 2005). However, the latter document was never released as a final approved plan. Significantly, the Integrated Natural Resource Management Plan for the NT (2005) committed the Territory to the following matters for targets (MATs) and management actions (MAs):

MAT 6-7 – *By 2008, integrate management arrangements across jurisdictions and sectors so that coastal and marine management issues are addressed effectively*

MA6-40 – *Develop a stakeholder-inclusive strategy for identifying, selecting, managing and monitoring a comprehensive, adequate and representative system of Marine Protected Areas (MPAs)*

MA6-41 – *Implement MPA selection process and establish at least two MPAs, based on the strategy developed in MA-60.*

Although an NT MPA Strategy was developed under MA6-40, it was never publicly released. Similarly, establishment of the Limmen Bight Marine Park (declared in June 2012) did not follow the planning process outlined in the NT MPA Strategy.

At the time of writing, the Northern Territory Government has no formal MPA strategy or policy, no apparent plan to meet its national commitments, nor staff dedicated to planning, developing and monitoring MPAs.

Prior to the 2012 NT election, the Country Liberal Party (2012) stated:

'A Country Liberals Government will adopt a 'common sense' approach to marine parks and will not establish any marine parks where they are not needed. Any proposals for a marine protected area must be based on demonstrated need and provide clear benefits ... we will not make any decisions on marine parks without consulting the community first. We will not meddle with the environment – at a huge cost to Territory taxpayers – where interference is not necessary'.

Although not an encouraging policy, it is not a moratorium on new MPAs. However, without any dedicated government staff or resources committed to undertake MPA planning and establishment, any progress on MPAs in the NT is unlikely to be led by the current NT Country Liberal Party Government.

Challenges

With just 6.4% of NT waters protected as MPAs, the jurisdiction's network of declared MPAs is far from complete, and lags considerably behind other Australian states/territory jurisdictions (NMPAWG 2010). In 2013, Australia has reached the point where most of its marine environment has been assessed in a marine sanctuary selection and designation process at some point in the last 10 years. The exceptions to this are the NT, Tasmania, Cocos and Christmas Islands, the Torres Strait, much of Western Australia and the Hawkesbury Shelf bioregion in NSW (see Chapter 17).

From the first declaration of a MPA in 1983 (Cobourg Marine Park) by the Country Liberal Party Government, progress in establishing additional MPAs by successive government agencies has been delayed and constrained by political and bureaucratic instability, lack of resources and scientific capacity, and, significantly, until recently, unclear coastal-marine governance in respect to Indigenous coastal and marine tenure and rights.

Strong, sectoral fishing interests (particularly recreational fishing) have also continued to hamper efforts to establish MPAs in the NT. With one of the highest levels of boat ownership in Australia, fishing is a major recreational and economic activity and also, a major tourism draw card (see Chapter 21). The Territory has a pervasive, anti-regulation, 'frontier' culture, commonly associated with developing nations and regions, coupled with a widely held view that there is plenty of fish left in the sea. As such, a strong anti-Marine Parks lobby exists among many recreational and commercial fishing sectors, fishing tour operators, fishing tackle businesses and also, significantly, local media outlets. Hence any proposed restrictions on access to popular fishing spots through creation of Marine Parks, particularly with no-take zones, has been met by strenuous (and public) resistance (e.g. Fishing World 2011).

The 'fishing vote' is considered by politicians (of both major parties) to be strong in key seats in Darwin's northern suburbs, and much stronger than the pro-Marine Parks' vote in marginal seats. Further, when the need for protection from 'over-fishing' is acknowledged, the NT Government has opted for temporary fisheries closures declared under the *Fisheries Act 1988*, rather than declare long-term, secure MPAs. Importantly, within these fisheries closures, only commercial fishers – not anglers – have been excluded through fisheries management regulations (as evidenced by five recent closures to protect Red Snapper and Black Jewfish on reefs near Darwin: Department of Primary Industries and Fisheries 2013).

Significantly, the media has played a major role as a voice for anti-Marine Park sentiment in the NT. The NT has one daily newspaper, the *NT News*, and over the past decade the paper has actively supported recreational fishers and pro-fishing politicians, providing inaccurate, misinformed and highly partisan, 'anti-Marine Park' commentary and views (see Gosford 2011) – epitomised by the paper's vigorous public support and organisation of the 'I fish and I vote' campaign and culminating in a pre-election, public rally in 2011.

Lack of community education, public awareness and perceived dangers of the NT's marine environment have also hampered marine environmental advocacy in the NT. While the Territory supports intact and diverse coastal ecosystems, including coral reefs, abundant marine wildlife and warm waters, its waters are commonly turbid and home to dangerous animals, particularly Saltwater Crocodiles, Box Jellyfish and Irukanji Jellyfish. There is also limited public (and political) awareness of the globally significant marine biodiversity values of the Territory's seas. As a consequence, the dive tourism sector – a major supporter and economic driver of MPAs – is poorly developed. Within 200 km of Darwin, many more tourism businesses earn an income from fishing than they do from snorkelling, diving and marine tourism. As such, politicians believe more jobs are threatened by Marine Parks, particularly no-take zones, than are created by them.

For the most part, the NT's seas have largely avoided 'over-fishing' due to historically low fishing pressure, relatively conservative fisheries management and, remoteness and lack of access to the coast, particularly through Aboriginal lands. Unlike other jurisdictions, it is difficult to identify any large areas of the marine estate in the Territory's waters that does not support high marine biodiversity levels, ecologically intact habitats or significant Indigenous cultural values. The large extent of intact coastal ecosystems in the Territory has inoculated decision makers against establishing MPAs, rather than emboldened them to proactively declare a CAR network of MPAs, to ensure healthy fish stocks and ecosystems remain so.

Technical capability remains a major challenge for the expansion of the NT's MPA network. The Territory has the smallest (and most transient) marine science community in

Australia. During the mid to late 2000s, expert Marine Park researchers and planners had to be recruited from southern Australia, or abroad, to lead and undertake much of the work and that capacity has since largely been lost.

Looking forward, any expansion of the NT's MPA system is likely to only succeed with the support and participation of Indigenous saltwater communities, both in planning and ongoing management. In the authors' experience, Indigenous saltwater people frequently express a strong commitment to protect their ancestral sea country estates, as well as greater involvement in economic development through fisheries, aquaculture and ecotourism (e.g. see Murdoch 2011; Protect Arnhem Land 2015). In providing a framework for conservation and ecosystem-based management (through spatial zoning), MPAs provide an ideal model for sustainable, rural Indigenous economic development. In addition to providing a tool for ecosystem-based management, MPAs created through legislation (unlike community-based Indigenous Protected Areas and locally managed marine areas), also provide long-term, legislative security and protection (through regulations) from potential threats (such as fishing, seabed mining and exploration, and production of oil and gas). For these reasons, MPAs are attractive for investment by governments, philanthropists or nature- or culture-based tourism businesses. As such, MPA declarations have the significant potential to provide legislative security, regulatory powers and promote investment, to strengthen, reinforce and support sea country IPAs.

However, in the authors' experience, many senior Indigenous Traditional Owners in the NT remain wary of government-managed MPAs. To this end, Traditional Owners can veto proposals over their inalienable freehold communal title lands, including Marine Parks. Even joint management of Marine Parks in the Territory may be viewed as too compromising, based on the 'mixed views' of the Indigenous experience in jointly managed National Parks (e.g. Christophersen 2013). In short, Indigenous elders do not want to surrender control of their sea country, and largely resile from legal declarations over their ancestral estates that may prevent them from using their marine estates as they wish, or which enable large numbers of non-Indigenous peoples to access their special places.

With these complexities, a pressing challenge is to find new planning models for identifying and protecting sea country that meet the aspirations and needs of Indigenous communities, meld western science and Indigenous ecological knowledge, and also, meet the requirements of a CAR network of MPAs. With the prevalence of Indigenous marine estates in the Territory, community-based MPA planning approaches from tropical developing countries in the region, such as locally managed marine areas (Govan *et al.* 2008), offer potentially useful planning models for the Territory.

Future directions

We propose the following recommendations to enable the NT to fully establish a comprehensive, adequate and representative network of MPAs, consistent with the requirements of the NRSMPA Action Plan (ANZECC TFMPA 1998):

1 Support the planning and formal declaration of Sea Country Indigenous Protected Areas, and marine extensions of coastal IPAs. This should include ranger training and supporting Sea Ranger groups and Aboriginal land and sea management organisations to consult Traditional Owners and key stakeholders, and prepare documents, such as sea country plans, and assist in building capacity for managing future marine IPAs.

2 At the national level, formal consideration and recognition of sea country IPAs as MPAs under the NRSMPA, conditional on satisfying the requirements of 'effective management' and appropriate security.
3 To provide sea country IPAs with legislative security and regulatory powers, the Australian and Northern Territory governments should (with the support of Traditional Owners) develop and implement formal mechanisms to coordinate complementary sea country IPA and MPA planning and management in the NT.
4 This coordination of IPA and MPA planning is best achieved through the current 'sea country planning' and consultations processes under the IPA program and should include consideration of the regulatory powers of relevant MPA legislation (i.e. parks, fisheries, petroleum and mining), where appropriate, to afford specific protection to habitats and wildlife consistent with the goals of sea country IPAs.
5 Develop and finalise a NT MPA Strategy. The most sensible approach would be to review, finalise and publicly release the existing draft Northern Territory Marine Protected Areas Strategy (NRETA 2008). This should include: (i) re-establishing a consultation processes to engage stakeholders in finalising the strategy and identifying priorities for the establishment of a network of MPAs in the NT; and (ii) a review of MPA planning and consultation processes, particularly in light of the *Blue Mud Bay* High Court Decision.
6 Expand the current MPA network in the NT through: (i) prioritising and supporting the declaration of MPAs in priority marine bioregions in which no significant MPAs currently exist: Anson Beagle, Van Diemens Gulf, Arnhem Wessel, Tiwi and Cambridge-Bonaparte bioregions; and (ii) re-commence a consultation process for declaration of Bynoe Harbour Marine Park, in the Anson Beagle bioregion.
7 Support existing MPA and marine-IPA management, through the development of: (i) a plan of management for Limmen Bight Marine Park consistent with best practice MPA planning, in consultation with Indigenous Traditional Owners and stakeholders; and (ii) systematic monitoring and research programs for MPAs and marine-IPAs to guide management, including Cobourg Marine Park.

Conclusions

Notwithstanding the major political and bureaucratic challenges, there are significant opportunities for establishing and effectively managing new MPAs, consistent with CAR principles for establishing a reserve system, across the NT's marine bioregions.

Importantly, Indigenous people manage vast areas of ecologically and culturally important, shallow tropical marine seas across the NT, through a range of mechanisms (i.e. Indigenous Protected Areas over marine areas, sea country planning and management, native title, sacred sites and sea closures) and governance arrangements (e.g. joint management, Indigenous management and traditional customary management). This has been accompanied by a rapid growth, particularly over the past decade, in Indigenous Ranger programs, Indigenous Protected Areas and Indigenous 'sea country' planning and management. Arguably, these initiatives have achieved more towards establishing and effectively managing a CAR network of MPAs, than have the two government-declared Marine Parks (Cobourg, Limmen Bight) declared over the past three decades.

Although the Northern Territory Government has national commitments to establish a representative network of MPAs as part of the NRSMPA, 85% of the Territory's coastal estate is under Indigenous ownership and customary management. With legal certainty from the *Blue Mud Bay* decision (and the necessity for co-management), future MPAs in the NT will need to be developed managed through formal cooperative management arrangements with Aboriginal people. Importantly, legislated MPAs not only strengthen sea country Indigenous Protected Areas (which are non-legal, voluntary declarations), but also provide a range of potential regional economic development and training opportunities for remote Aboriginal communities, Indigenous Sea Ranger groups and Indigenous enterprises, and much needed, improved economic outcomes for Aboriginal people. To this end, Indigenous custodians should continue to be supported to develop the NT's MPAs system, in consultation with key government agencies and stakeholders.

Acknowledgements

We would like to thank the Australian Government's Indigenous Protected Areas Program for providing current advice on sea country within declared Indigenous Protected Areas within the Northern Territory. Ben Carr kindly produced Fig. 13.1.

Biography

Karen Edyvane is a marine ecologist with expertise and experience in MPA design and system planning, and marine planning. Her career highlights include: the establishment of South Australia's first marine park, the Great Australian Bight Marine Park; the first large-scale systematic survey, mapping, ecological classification and conservation assessment of South Australia's underwater, inshore marine habitats and ecosystems; and identification of candidate Marine Parks and Reserves. In South Australia, Tasmania, and the Northern Territory, as a government MPA planner, she developed the first technical and policy blueprints for the establishment of Marine Park systems. Karen has also held a wide range of technical, policy and statutory positions in marine biodiversity conservation planning and management in Australia, at the state/territory and national level. She is currently an Adjunct Fellow with Charles Darwin University and The Australian National University.

Stuart Blanch is an environmental advocate and policy advisor with 20 years' experience in the community environment, government, natural resource management and research sectors. His interests include river and wetland conservation, climate change and renewable energy, landscape management, marine conservation and the future of Northern Australia. He led the community campaign and partnered with Indigenous communities that led to the declaration of Limmen Bight Marine Park, the Territory's second Marine Park. He lived for almost 9 years in Darwin where he worked for WWF-Australia and led the Environment Centre NT. He is an ecologist by training, with postgraduate qualifications in river ecology and environmental law.

References

ABC (2008) 'Govt pushed to release marine parks plan'. 18 February 2008. Australian Broadcasting Corporation, Darwin, <http://www.abc.net.au/news/2008-02-18/govt-pushed-to-release-marine-parks-plan/1046056>.

ABC (2009a) 'Govt backtracking on marine protection promise: AMCS'. 17 May 2009. Australian Broadcasting Corporation, Darwin, <http://www.abc.net.au/news/2009-05-17/govt-backtracking-on-marine-protection-promise-amcs/1685210>.

ABC (2009b) 'Minister slams "illogical, stupid" marine parks' 22 October 2009. Australian Broadcasting Corporation, Darwin, <http://www.abc.net.au/news/2009-10-22/minister-slams-illogical-stupid-marine-parks/1113184>.

ABC (2009c) 'Take a cold shower, Garrett tells NT minister' 23 October 2009. Australian Broadcasting Corporation, Darwin, <http://www.abc.net.au/news/2009-10-23/take-a-cold-shower-garrett-tells-nt-minister/1114480>.

Anon (2009) NT rejects no-fishing zones. *Fishing World.* 18 March 2009, <http://www.fishingworld.com.au/news/nt-rejects-no-fishing-zones>.

ANZECC TFMPA (Australian and New Zealand Environment and Conservation Council Task Force on Marine Protected Areas) (1998) *Strategic Plan of Action for the National Representative System of Marine Protected Areas: A Guide for Action by Australian Governments.* Environment Australia, Canberra.

ANZECC TFMPA (Australian and New Zealand Environment and Conservation Council Task Force on Marine Protected Areas) (1999) *Guidelines for Establishing the National Representative System of Marine Protected Areas.* Australian and New Zealand Environment and Conservation Council, Task Force on Marine Protected Areas. Environment Australia, Canberra.

Australian Labor Party of the Northern Territory (2012) *Greening the Territory.* Election 2012 policy document, ALPNT, Darwin, <http://territorylabor.com.au/portals/territorylabor/policy/12-Greening-the-Territory.pdf>.

Billyard R (1993) Aboriginal involvement in Northern Territory marine parks. In *Turning the Tide: Papers Presented at Conference on Indigenous Peoples and Sea Rights.* pp. 198–207. Faculty of Law, Northern Territory University, Darwin.

CBD (2010) *Strategic Plan for Biodiversity 2011–2020 and the Aichi Biodiversity Targets.* Convention on Biological Diversity, Montreal, Canada, <https://www.cbd.int/decision/cop/?id=12268>.

Christophersen P (2013) 'Contested Territory: tussle to control Kakadu, Uluru as tourism jewels fade'. Comments by Bunitj elder Peter Christophersen, *The Australian*, 14 December 2013, <http://www.mirarr.net/news_items/the-australian-contested-territory-tussle-to-control-kakadu-uluru-as-tourism-jewels-fade>.

Cobourg Peninsula Sanctuary and Marine Park Board and Parks and Wildlife Service of the Northern Territory (2011) *Cobourg Marine Park Plan of Management.* Department of Natural Resources, Environment, The Arts and Sport, Darwin.

Commonwealth of Australia (2006) *A Guide to the Integrated Marine and Coastal Regionalisation of Australia.* Version 4.0. Department of the Environment and Heritage, Canberra.

Country Liberal Party (2012) 'Country Liberals will close Chambers and Finke Bay'. Media release, Darwin, 10 August 2012, <http://fishingterritory.com/fishing-politics-f10259/clp-fishing-policy-released-t30051.html>.

Crocodile Island Rangers (2013) *About Us.* Crocodile Island Rangers, Milingimbi, Northern Territory, <http://crocodileislandsrangers.org/about-us/ >.

Department of Primary Industries and Fisheries (2013) *Proposed Fishing Closures to Rebuild Reef Fish Stocks.* Northern Territory Government, Darwin, <http://www.nt.gov.au/d/Fisheries/rfc/documents/Closure%20maps%20Fact%20Sheet.pdf>.

Department of the Environment (2014a) *Collaborative Australian Protected Areas Database (CAPAD) (2014).* Australian Government, Canberra, <http://www.environment.gov.au/land/nrs/science/capad/2014 >.

Department of the Environment (2014b) *Locations of Indigenous Protected Areas in Australia, October 2014.* Australian Government, Canberra, <http://www.environment.gov.au/indigenous/ipa/map.html>.

DEWHA (2008) *The North Marine Bioregional Plan Bioregional Profile. A Description of the Ecosystems, Conservation Values and Uses of the North Marine Region.* Department of the Environment, Water, Heritage and the Arts, Canberra.

Dhimurru Aboriginal Corporation (2015) *Dhimurru Indigenous Protected Area Management Plan 2015–2022.* Dhimurru Aboriginal Corporation, Nhulunbuy, Northern Territory.

Dhimurru Land Management Aboriginal Corporation (2011) *Systematic Conservation Planning of Dhimurru Sea Country.* Dhimurru IPA Sea Country Planning Project 2009–2010. Final Report to Department of Sustainability, Environment, Water, Population and Communities, Canberra.

Edyvane KS (2010) *Review of NT Integrated NRM Plan: Update of Asset Descriptions – Coastal and Marine.* Report to the Natural Resource Management Board of the Northern Territory, Charles Darwin University, Darwin.

Edyvane K, Dethmers K (2010) *Identifying Potential Sites for Marine Protected Areas in the Northern Territory.* Final Report on Project CF2007/135 to NT NRM Board. Prepared by Department of Natural Resources, Environment, the Arts and Sport, Darwin.

Fishing World (2011) Territorians rally against no go, no take zones. *Fishing World,* 7 November 2011, <http://www.fishingworld.com.au/news/territorians-rally-against-no-go-no-take-zones>.

Ganter E (1996) Indigenous participation in coastal management in the Northern Territory, Australia: issues and options. *Ocean and Coastal Management* **33**, 193–212. doi:10.1016/S0964-5691(96)00051-8.

Gosford B (2011) 'With monopoly comes responsibility – the NT News, fishos and marine reserves'. *Crikey,* 20 November 2011, <http://www.blogs.crikey.com.au/northern/2011/11/20/with-monopoly-comes-responsibility-the-nt-news-fishos-and-marine-reserves/>.

Govan H, Aalbersberg W, Tawake A, Parks JE (2008) *Locally-managed marine areas: a Guide to Supporting Community-based Adaptive Management.* Locally-Managed Marine Area (LMMA) Network, Suva.

Halpern BS, Walbridge S, Selkoe KA, Kappel CV, Micheli F, D'Agrosa C, *et al.* (2008) A global map of human impact on marine ecosystems. *Science* **319**, 948–952. doi:10.1126/science.1149345.

Harrison L, McGuire L, Ward S, Fisher A, Pavey C, Fegan M, *et al.* (2009) *An Inventory of Sites of International and National Significance for Biodiversity Values in the Northern Territory.* Department of Natural Resources, Environment, The Arts and Sport, Darwin.

Landcare Council of the Northern Territory (LCNT) (2005) *Integrated Natural Resource Management Plan for the Northern Territory. Sustaining Our Resources – People, Country And Enterprises.* Landcare Council of the Northern Territory, Darwin.

Lavers J (2010) *A Review of the Effectiveness of Marine Protected Areas for the Conservation of Biodiversity and Management of Fisheries in Australia.* A report prepared for the National Marine Protected Areas Working Group (NMPAWG) and Commonwealth Scientific Industrial Research Organisation (CSIRO).

Moyses M, Panton W (2008) Indigenous partnerships in Northern Territory protected areas: joint management of national parks and support for Indigenous Protected Areas. In *Protecting Country – Indigenous Governance and Management of Protected Areas* (Eds D Smyth and GK Ward) pp. 9–18. Australian Institute of Aboriginal and Torres Strait Islander Studies, Canberra.

Murdoch L (2011) 'If miners talked to us, they'd know about sacred island's poison tree, say elders'. *Sydney Morning Herald,* 27 June 2011, <http://www.smh.com.au/national/if-miners-talked-to-us-theyd-know-about-sacred-islands-poison-tree-say-elders-20110626–1glsv.html>.

NMPAWG (National Marine Protected Areas Working Group) (2010) *Progress in Implementing the National Representative System of Marine Protected Areas (NRSMPA).* Report to the Natural Resource Management Ministerial Council by the National Marine Protected Areas Working Group, Canberra.

Northern Territory Government (2005) *Draft Northern Territory Parks and Conservation Masterplan.* Department of Natural Resources Environment and the Arts, Darwin.

NRETA (2008) *Northern Territory Marine Protected Areas Strategy (Draft).* Final Report on Project 2005/063 to NT NRM Board. Prepared by the Department of Natural Resources, Environment, and the Arts, Darwin.

Nursey-Bray M, Pedder A (Eds) (2004) *Key Conservation Issues in the Northern Marine Planning Area.* A Report to the National Oceans Office. Prepared by the Northern Australian Environment Alliance, Darwin.

Protect Arnhem Land (2015) *Protect Arnhem Land.* Maningrida, Northern Territory, <http://www.protectarnhemland.org/#!background/cgsp>.

Rose B (2012) Indigenous Protected Areas – Innovation beyond the boundaries. In *Innovation for 21st Century Conservation* (Eds P Figgis, J Fitzsimons and J Irving) pp. 50–55. Australian Committee for IUCN, Sydney.

Scrymgour M (2007) 'NT Marine Parks Strategy'. Media release, Minister for Natural Resources. 9 March 2007, <http://www.territorystories.nt.gov.au/bitstream/10070/82712/2/scrymgour.0903.marinepark.pdf>.

Smit N, Billyard R, Ferns L (2000) *Beagle Gulf Benthic Survey: Characterisation of Soft Substrates.* Technical Report No. 66. Parks and Wildlife Commission of the Northern Territory, Darwin.

Smyth D (2008) Just add water? Taking Indigenous Protected Areas into Sea Country. In *Protecting Country: Indigenous Governance and Management of Protected Areas* (Eds D Smyth and GK Ward) pp. 95–109. Australian Institute of Aboriginal and Torres Strait Islander Studies, Canberra.

Smyth D, O'Leary P (2007) *Protecting Sea Country: Opportunities for Indigenous Engagement in the Northern Territory Marine Protected Area Strategy.* A discussion paper commissioned by the Northern Territory Department of Natural Resources, Environment and the Arts, Darwin.

Ward S, Harrison L (2009) *Recognising Sites of Conservation Significance for Biodiversity Values in the Northern Territory.* Department of Natural Resources, Environment, The Arts and Sport, Darwin.

Ward TJ, Edyvane KS, Larson H, Marsh H, Bustamente R (2007) *Scientific Design and Implementation Principles for a Network of Comprehensive, Adequate and Representative Marine Protected Areas in the Northern Territory.* Department of Natural Resources, Environment, the Arts and Sport, Darwin.

PART 3

Key aspects of Australian marine protected areas

Previous page: Twelve Apostles Marine National Park. (Photo Dean Ingwersen)

Chapter 14

A review of marine protected area legislation in Australia

Erika J. Techera

Introduction

Although Indigenous people have applied traditional marine management and customary laws since ancient times (see also Chapter 18), contemporary marine protected area (MPA) legislation first emerged in Australia just over 50 years ago. International recognition of the need to establish MPAs was documented as early as 1962 (McNeill 1994). In Australia, marine extensions of terrestrial parks were declared in Victoria in 1898 and Western Australia in 1910 (McNeill 1994), but Green Island in Queensland, declared in 1938, was the first stand-alone MPA. Since 1998, Australia's Commonwealth, state and territory governments have worked cooperatively to establish a National Representative System of Marine Protected Areas (NRSMPA) (ANZECC 1998).

The focus on MPAs in Australia partly reflects the country's commitments under the Convention on Biological Diversity (CBD) and the associated protected area coverage targets, as well as their inclusion in Australia's Oceans Policy (Commonwealth of Australia 1998a,b). Kearney *et al.* (2012) also point to the call for more MPAs as a reaction to beliefs that fisheries management was failing worldwide. In response, fisheries management regulations were strengthened in the 1990s and this is clear from a survey of legislation in existence today, most of which dates to this period. Although several area-based tools were used, the focus was upon fauna and fish in particular. In some cases, 'habitats', 'biodiversity' and 'conservation' were included as objectives but rarely recreational uses and tourism. This led to the development of specific, conservation-focused marine parks statutes, independent of fisheries legislation.

Today all states and territories (apart from the ACT) have some form of legislation providing for MPAs (Table 14.1), but there is no consistency between these regulatory regimes. Although there is a relative wealth of literature on the scientific benefits of MPAs, and economic analyses of their efficacy, very little has been written on the law and policy. Two of the few studies that have been written are Kuo (2011) and Gullett (2003). Further detailed analysis is necessary if advances are to be made in the legal regimes. In addition, much can be learned from exploring other jurisdictions, and this adds to the pressing need for comparative legal research. In the sections that follow, the legislative frameworks will be explored.

Canvassing MPA legislation around Australia

Australia is a signatory to various treaties including the CBD, which give rise to obligations in relation to marine protection. The Australian Government has power to pass legislation to implement those treaties and has done so through the *Environment Protection and Biodiversity Conservation Act 1999*. In addition, Australia has ratified the UN Convention on the Law of the Sea (establishing the territorial sea and Exclusive Economic Zone (EEZ)) and implemented the Offshore Constitutional Settlement, which divides jurisdiction for oceans governance; with states and the Northern Territory controlling the first 3 nautical miles and the Commonwealth 3–200 nautical miles (see also Chapter 3). The only substantial exception to this is the Commonwealth *Great Barrier Reef Marine Park Act 1975*, which does regulate some of Queensland's coastal waters (see also Chapters 5, 12). Thus the regulatory environment is complex involving multiple jurisdictions (Table 14.1).

Commonwealth

At the federal level, the *Environment Protection and Biodiversity Conservation Act 1999* (EPBCA) implements Australia's international obligations under several treaties as well as creating a framework for the protection of Commonwealth marine areas. The EPBCA implements commitments under three conventions that can provide directly for the protection of the marine environment: the World Heritage Convention, the Convention on Wetlands of International Importance (the Ramsar Convention) and the International Convention on the Regulation of Whales. Other protection categories include biosphere reserves, Commonwealth reserves and conservation zones (s.390D), which contribute to commitments under the CBD. Most relevantly, the EPBCA provides for Commonwealth reserves to be declared in Commonwealth marine areas (e.g. EEZ) (s.344). These marine reserves are to be managed in accordance with IUCN protected area categories and principles (s.347–348). Thereafter, restrictions are set out on what activities may take place in the reserve and penalties for breaches (s.354–354A). Multiple provisions protect the Great Barrier Reef and link the EPBCA to the *Great Barrier Reef Marine Park Act 1975*.

The object of the *Great Barrier Reef Marine Park Act 1975* is to provide for the conservation and protection of the Great Barrier Reef Region. The Act establishes the Great Barrier Reef Marine Park (ss.30–31) and the Great Barrier Reef Marine Park Authority whose functions include making recommendations and providing advice to the relevant Minister, undertaking investigations and making zoning and management plans (s.7) (see also Chapter 5). Membership of the authority includes one full-time member and two part-time members of which one is nominated by the Queensland Government (s.10); at least one member must be Indigenous and one have knowledge of the tourism industry. The *Great Barrier Reef Marine Park Act 1975* focuses on zoning plans that are to protect, *inter alia*, the World Heritage values of the park (s.32) and permits. The zoning provisions provide for public consultation (s.32C) and a statement of the environmental, economic and social values for any given area (s.35). Relevantly, the Act applies the principles of ecologically sustainable use and ecosystem-based management (ss.3AA and 3AB). The Act includes duties on entrants to the park to minimise or prevent harm (s.37AA), and provides for offences for breaches of zoning plans (s.38BA–BD), certain conduct in unzoned areas (s.38CA-CB), as well as undertaking mining or geological storage operations (s.38AA–AB).

Queensland

Queensland had one of the earliest pieces of specific MPA legislation – the *Marine Parks Act 1982* – now replaced by the *Marine Parks Act 2004* (see also Chapter 12). Under s.5

Table 14.1. Statues and regulations relating to marine protected areas in Australian jurisdictions

Jurisdiction	Statutes	Regulations
Commonwealth	*Environment Protection and Biodiversity Conservation Act 1999*	*Environment Protection and Biodiversity Conservation Regulations 2000*
	Great Barrier Reef Marine Park Act 1975	*Great Barrier Reef Marine Park Regulations 1983*
Queensland	*Marine Parks Act 2004*	*Marine Parks Regulation 2006* *Marine Parks (Declaration) Regulation 2006*
	Fisheries Act 1994	*Fisheries Regulation 2008*
	Nature Conservation Act 1992	*Nature Conservation (Protected Areas) Regulation 1994* *Nature Conservation (Protected Areas Management) Regulation 2006* *Nature Conservation Regulation 1994*
New South Wales	*Marine Estate Management Act 2014* (Replacing the former *Marine Parks Act 1997*)	*Marine Estate Management Regulation 2009* *Marine Estate Management (Management Rules) Regulation 1999* *Aquatic Reserves Notification 2013* (Replacing the former *Marine Parks (Zoning Plans) Regulation 1999* and *Marine Parks Regulation 2009*)
	National Parks and Wildlife Act 1974	*National Parks and Wildlife Regulation 2009*
	Fisheries Management Act 1994	*Fisheries Management (General) Regulation 2010*
South Australia	*Marine Parks Act 2007*	*Marine Parks Regulations 2008* *Marine Parks (Zoning) Regulations 2012*
	Fisheries Management Act 2007	*Fisheries Management (General) Regulations 2007*
	National Parks and Wildlife Act 1972	*National Parks and Wildlife (National Parks) Regulations 2001*
Victoria	*National Parks Act 1975*	*National Parks Regulations 2013*
	Fisheries Act 1995	*Fisheries Regulations 2009*
Western Australia	*Conservation and Land Management Act 1984*	*Conservation and Land Management Regulations 2002*
	Fish Resources Management Act 1994	*Fish Resources Management Regulations 1995*
Tasmania	*National Parks and Reserves Management Act 2002*	*National Parks and Reserved Land Regulations 2009*
	Living Marine Resources Management Act 1995	*Fisheries Rules 2009*
Northern Territory	*Territory Parks and Wildlife Conservation Act 2006*	*Territory Parks and Wildlife Conservation By-Laws 1984* *Territory Parks and Wildlife Conservation Regulations 2006*
	Fisheries Act 1988	*Fisheries Regulation 1992*
	Parks and Wildlife Commission Act 2000	*Parks and Reserves (Framework for the Future) Regulations 2003*

'conservation of the marine environment' is the main purpose of the legislation, which is to be achieved through a 'comprehensive and integrated strategy' including the declaration of marine parks, zoning and management plans. Reference is made to cooperation through the involvement of public authorities, interested persons and groups, and Indigenous people as well as cooperative implementation of international and intergovernmental responsibilities (s.6). A 'coordinated and integrated approach with other environment conservation legislation' is recognised and although reference is made to the *Nature Conservation Act 1992,* fisheries laws are not mentioned. Application of the precautionary principle in decision making is also included. Although intrinsic values are not mentioned, there are two sub-sections referencing 'public appreciation, understanding and enjoyment' as well as 'cultural, economic, environmental and social relationships between marine parks and other areas'. This comprehensiveness is significant and more explicit than in other Australian MPA legislation.

Areas for declaration are broadly defined in s.8(2) as including various waters as well as land. Declarations are made by regulation under s.8(1) and no process or other details regarding who can propose a marine park and how it might be declared, are set out in the Act. The relevant Minister is responsible for the preparation of the draft and final zoning and management plans and must give public notice inviting submissions, which must be taken into account (ss.21 and 23). Under s.29 the Governor in Council approves the management plan for a marine park. Again the Minister is required to give public notice of the plan and invite submissions (s.31), and must prepare the final management plan having considered each submission made (s.32). Unlike other MPA legislation in Australia, there is no advisory body to give assistance to the Minister. If there is any inconsistency between a regulation in the Act and zoning plan the latter prevails (s.151).

The Act provides for enforcement including the appointment of inspectors under s.52 with a broad range of powers (ss.59–74). Offences are provided for in Part 4 including penalties ranging from imprisonment for unlawful serious environmental harm to fines for other wilful actions, as well as non-compliance with conditions and providing false or misleading information or documents.

A range of compliance notices (s.93), directions (s.99) and orders (s.109) are provided for, as well as temporary restricted area declarations (s.95), which prevail over a regulation, zoning plan or authorisation under the Act (s.98). The chief executive of the relevant department or an inspector can bring proceedings in court (s.111).

In Queensland, other protected areas may be declared under the *Nature Conservation Act 1992.* Under this Act, a range of protected areas may be declared including a national park, conservation park, resources reserve, nature refuge, coordinated conservation area, wilderness area, World Heritage management area and international agreement area (s.14); various management principles apply to each (ss.15–26). Land is not defined in this Act but under the *Land Act 1994* includes marine areas (s.5). Other relevant Queensland legislation includes the *Fisheries Act 1994* under which fish habitat areas may be declared (s.120) in which fisheries resources are protected (s.122).

New South Wales

Up until November 2014, NSW also had a specific MPA statute, the *Marine Parks Act 1997* (see also Chapter 11). The new legislation is the *Marine Estate Management Act 2014.* The *Marine Parks Act 1997* aimed at conserving marine biological diversity, maintaining ecological processes as well as, where consistent with these goals, providing for ecologically sustainable use of fish and marine vegetation and 'opportunities for public appreciation,

Fig. 14.1. Signage outlining the allowable uses within zones of the Batemans Marine Park, southern New South Wales. (Photo: James Fitzsimons)

understanding and enjoyment' (s.3). Although coordination and integration were not mentioned, the above objects were articulated broadly, particularly the reference to fisheries and flora. Provision was made for interactions between this Act and fisheries regulations (s. 10). Similarly, coordination (to a certain extent) was achieved through the Advisory Council (discussed further below).

The Act provided for the declaration of marine parks; such areas included waters, but also included any land adjacent to those waters (s.6). The Act spelt out what the regulations may contain, including regulating the use and enjoyment of a park, prohibiting certain activities, taking of flora and fauna, entry, removal of certain people, mooring of vessels, protection of cultural heritage, and fees for use (s.17) (Fig. 14.1).

Three significant bodies were included in the *Marine Parks Act 1997*: first, the Marine Parks Authority, which included members from several government departments (s.29). The functions of the Authority included: investigating, assessing and considering proposals for marine parks; making recommendations as to classifications of areas within the parks; preparation of operational plans; managing and controlling activities that may have an adverse effect on the park's objectives; regulating ecologically sustainable use; disseminating information; and encouraging public appreciation, and understanding and enjoyment of marine parks (s.30).

Second, the Marine Parks Advisory Council, which also consisted of members from various government departments as well as members appointed to represent the Australian Government. In addition, the Council included representatives of the interests of marine conservation, and a peak conservation group, an expert in marine science and members to represent Aboriginal people, the tourism industry and scuba divers, as well as commercial and recreational fishers (s.32(2)). The functions of the Council included advising on proposals, conservation of marine biological diversity, ecologically sustainable use, public use and enjoyment (s.33).

Third, advisory committees, which were created for each park (s.35). Membership of the committees included members appointed by the relevant Ministers to represent the interests of marine conservation, marine science, Aboriginal people, the tourism industry, commercial fishers, recreational fishers, scuba divers and local councils, and were appointed following a public call for nominations. An advisory committee had similar functions to the Marine Parks Advisory Council under s.35AA but included advice on the application of marine park closures (it is not clear why that was not an included function of the Council).

The Marine Parks Authority, in consultation with an advisory committee, drafted the zoning plan (s.17C). Zoning plans included classifications of uses within, and management of, a zone (s.17B(2)). The Authority submitted the plan to the relevant Ministers for approval who then considered submissions made, as well as comments from the relevant advisory committee (s.17C(5)). Under s.23 of the former NSW *Marine Parks Act 1997*, operational plans (rather than management plans) were prepared by the Authority with the objective of identifying and defining a 'scheme of strategies, actions or activities' for the marine park (s.24). The Authority was to refer the draft operational plan to the relevant advisory committee for consideration and advice (s.25), which must be considered by the Authority before adoption.

The Act was enforced by marine park rangers (defined under s.35A and s.36 and included fisheries officers) and penalties were included for breaches. All fees and charges under the Act (e.g. for the use of a park or for a permit to carry out activities) were deposited into a Marine Parks Fund (s.42) to cover costs and expenses of administration.

The statute linked back to the *Environmental Planning and Assessment Act 1979* (EPAA), providing that before any approval can be given under the EPAA regard must be had to the objects of the *Marine Parks Act 1997* (s.19). The *Marine Parks Act 1997* included other provisions that integrated this legislation with the *Fisheries Management Act 1994* and the *National Parks and Wildlife Act 1974* (e.g. s.37). The *Fisheries Management Act 1994* remains relevant as it includes provisions for the declaration of aquatic reserves to 'conserve the biodiversity of fish and marine vegetation' by providing for the protection of fish habitat and threatened species, as well as species management, educational activities and scientific research (s.194). Mining in reserves is prohibited (s.197B) and development applications for other activities are made under the EPAA (s.197C and D). The *National Parks and Wildlife Act 1974* refers to 'national parks legislation' defined as including that Act as well as, *inter alia*, the *Marine Parks Act 1997.* The objects of the *National Parks and Wildlife Act 1974* include the conservation of nature, including the protection of animals, which are defined as including marine mammals and aquatic amphibians and reptiles, but specifically excludes fish (s.5). In relation to marine areas, these may be protected as extensions of land-based protected areas under any of the various classifications under the *National Parks and Wildlife Act 1974* (e.g. nature reserve, national park and state conservation area).

In November 2014, the *Marine Parks Act 1997* was replaced by the *Marine Estate Management Act 2014.* The objects of the new legislation are broader than the former Act, providing for management of the NSW marine estate, the coordination of the public authorities involved in the marine estate, and 'declaration and management of a comprehensive system of marine parks and aquatic reserves' (s.3). Significantly, this Act defines the marine estate as including coastal waters and adjacent lands, as well as estuaries and islands (s.6); it therefore, has the potential of achieving integrated coastal zone management. It provides for the declaration of both marine parks and aquatic reserves (sections 23 and 34). The *Marine Estate Management Act 2014* is administered by the Minister for Primary Industries and the Minister for the Environment. A new authority has been established, the Marine Estate Management Authority, whose functions are more holistic than the Marine Parks Authority thereby facilitating integration. The *Marine Estate Management Regulation 2009* makes further provision for integration by providing for consent from 'relevant ministers'. Reference is made to the preparation of a management strategy for the whole marine estate and promotion of 'collaboration and co-ordination between public authorities', but also to new matters including the assessment of threats and risks (s.8). In addition, the Act creates the Marine Estate Expert Knowledge Panel to provide advice to the Marine Estate Management Authority (s.9). The Panel membership will include 'membership includes independent experts across the environmental, economic and social sciences' (NSW Government 2014). It seems that advisory committees for marine parks are to be created on a non-statutory basis and not under the Act.

The Act is very new and it remains to be seen whether it will achieve its objects and precisely how the Authority and Panel will operate on the ground.

South Australia

The objects of South Australia's *Marine Parks Act 2007* include protection and conservation of marine biological diversity and marine habitats with reference to a 'comprehensive, adequate and representative system of marine parks' (s.8) (see also Chapter 8). These objects are to assist with maintenance of ecological processes, adaptation to climate change, protecting and conserving features of natural or cultural heritage, ecologically sustainable development and use, and providing opportunities for public appreciation, education, understanding and enjoyment. Furthermore, s.8 expands upon ecologically sustainable development and sets out 11 principles to be taken into account in South Australia's MPAs.

Integration is achieved through definitions, and specific provisions cover interactions with the *Aquaculture Act 2001, Coast Protection Act 1972* and *Development Act 1993*, with additional links to the *Environment Protection Act 1999, Fisheries Management Act 2007, Harbors and Navigation Act 1993, Historic Shipwrecks Act 1981, Mining Act 1971, Natural Resource Management Act 2004, Offshore Minerals Act 2000, Petroleum Act 2000* and *Petroleum (Submerged Lands) Act 1982*. Integration with other legislation is further provided for in relation to authorised officers who can be designated under this Act, the *Fisheries Management Act 2007*, the *National Parks and Wildlife Act 1972*, or be police officers (s.32).

The South Australian *Marine Parks Act 2007* provides for the proclamation of a marine park on the recommendation of the Minister who in turn must have regard to the advice of the Marine Parks Council (s.10). The Minister must manage the park in accordance with a management plan (s.12), which establishes zones, directs the management of day-to-day issues and provides guidelines on granting permits. The process to make or amend a management plan is inclusive, involving the views of all 'relevant Ministers', the Marine Parks Council, a representative of signatories to any Indigenous Land Use Agreement and native title holders, and such persons or bodies as the Minister determines to be leading representatives of the environment and conservation sector, local government, fishing industry, aquaculture industry, recreational fishing sector, mining and petroleum industries, tourism sector and general business sector (s.14). Interestingly, the draft management plan is subject to scrutiny by the Environment, Resources and Development Committee of the Parliament (s.15). The Committee can either object, not object, or make recommendations for amendment of the plan. If it objects, then the plan must go before both Houses of Parliament.

A zone is defined by its degree of protection and includes four types: 'General Managed Use Zone', 'Habitat Protection Zone', 'Sanctuary Zone' and 'Restricted Access Zone' (s.4), as well as a special purpose area where specified activities, that would otherwise be prohibited or restricted because of the zoning of the area, will be permitted. Temporary closures are permitted under s.18, by which access to an area or certain activities may be restricted in a marine park for a limited time.

The South Australian legislation sets out the functions and powers of the Minister in detail (s.22), which include seeking and assessing community nominations for marine parks, ensuring protections are in place through the regulation of activities, issuing permits, consulting relevant parties, integrating 'the administration of this Act with ... other legislation', keeping the state of marine parks under review, and enforcement. Such a provision does not exist in most of the other jurisdictions.

The Marine Parks Council is established under s.24 and includes 10 members, including three people with knowledge and experience of commercial fishing, aquaculture, and recreational fishing, three marine conservation specialists, two with scientific expertise relevant to the marine environment, a member with extensive involvement in community affairs, and a member with extensive knowledge of Indigenous culture. The Council provides advice to the Minister on community nominations for marine parks, advice on areas to be specified as marine parks, protection orders, boundaries, zones, management and promotion of community participation (s.29).

Broad enforcement powers are given to authorised officers (s.34) and breaches of the South Australian *Marine Parks Act 2007* attract a range of orders (ss.38 and 40) and some of the highest penalties compared with breaches of other MPA legislation.

Of additional relevance is the *Fisheries Management Act 2007*, which provides for the conservation and management of the aquatic resources of the state, including through the declaration of aquatic reserves. The Act requires the preparation of management plans for commercial and recreational fishing and aquatic reserves (ss.42–43) including consideration of matters raised during public consultation (s.44). In addition, regulations may be made under s.128 for matters such as licensing of activities and a quote system for fisheries. As would be expected, there are a range of offences for breaches of the Act (ss.76–77).

The only other relevant law is a specific piece of legislation to establish the Adelaide Dolphin Sanctuary. The *Adelaide Dolphin Sanctuary Act 2005* created the Sanctuary to protect the dolphins of the Port Adelaide River estuary and Barker Inlet and the natural habitat of that population.

Victoria

In Victoria, MPAs may be established under the *National Parks Act 1975* or the *Fisheries Act 1995* (see also Chapter 9). The objects of the *National Parks Act 1975* include the creation of marine national parks and marine sanctuaries for the preservation and protection of the environment, flora and fauna, scientific study and responsible land management (s.4). Marine national parks and marine sanctuaries must be managed in accordance with these objects and in order to preserve and protect the environment and species, to prevent the introduction of and to eradicate already introduced exotic species as well as to promote the use, enjoyment and understanding of the area (s.17D). Power is given to the Secretary (a body corporate established under the *Conservation, Forests and Lands Act 1987*) to issue permits to take fish or bait from a park or sanctuary for research purposes (s.21A). The Act is enforced by authorised officers who may, for example, direct a person to cease activities in a park or sanctuary (s.38). Offences are prescribed under s.45A and include fishing in a marine national park or marine sanctuary. The Act also provides restrictions on permissions that may be granted under *Mineral Resources (Sustainable Development) Act 1990, Geothermal Energy Resources Act 2005, Greenhouse Gas Geological Sequestration Act 2008, Petroleum Act 1998, Offshore Petroleum and Greenhouse Gas Storage Act 2010, Coastal Management Act 1995* or the *Pipelines Act 2005* for activities in a marine national park or marine sanctuary (ss.40 – 40B). The *National Parks Act 1975* also provides for the establishment of 'other parks', several of which are MPAs (marine and coastal parks, marine reserves, marine parks) as well as for more terrestrial-focused categories such as national parks, state parks and coastal parks, which may include relatively small marine components down to the low-water mark or estuaries.

The *Fisheries Act 1995* provides for the regulation, management and conservation of Victorian fisheries and aquatic habitats (s.1). Fisheries reserves may be declared under s.88, including a spawning, hatchery or nursing ground, a critical habitat, or for aquaculture, harvesting, management or monitoring, education, or non-consumptive uses. A management plan for a fisheries reserve must be prepared and may provide for the measurement of fishing capacity, access licences, as well as duration and review of the plan (s.28) and, in addition, specify guidelines regulating or restricting activities in the fisheries reserve and specify standard terms and conditions for permits (s.30). Authorised officers enforce the Act and may direct someone to leave a reserve, cease a prohibited activity or remove a matter or thing (s.89).

Western Australia

There are several pieces of relevant legislation in Western Australia including the *Fish Resources Management Act 1994* (FRMA) and *Conservation and Land Management Act 1984* (CALMA) (see Chapter 7). Any area of land and water in Western Australia may be declared to be a marine park, marine nature reserve or marine management area under the CALMA (s.13). Under s.13B, a marine park may include recreational and commercial activity only to the extent it is consistent with conservation, restoration, protection or preservation (Fig. 14.2). The Minister must classify the park as a wilderness area, prohibited area, limited access area, temporary control area, or a recreation area (under s.62) or divide it into general use area, sanctuary area, recreation area or special purpose area. Aquaculture, commercial or recreational fishing and pearling activities cannot be carried out in a sanctuary area or recreation area, or special purpose area in which such activities have been found to be incompatible. Furthermore this Act takes precedence over the FRMA or the *Pearling Act 1990* if there is inconsistency (s.13B). Marine management areas may be declared for conservation, recreational, scientific and commercial purposes (s.13C). The Act establishes the Marine Parks and Reserves Authority (MPRA) (s.26A) to have marine reserves vested in it and having the functions of developing policies with respect to the objectives of the Act (s.26B). In addition, the Marine Parks and Reserves Scientific Advisory Committee is established to provide advice to the Minister and the MPRA (s.26F and G). A range of offences are provided with respect to enforcement.

The FRMA has the objects of sustainable management of fisheries and aquaculture, and conservation of fish and other aquatic resources and habitats through ecologically sustainable management, including minimising impacts while enabling sustainable use through management of fishing, aquaculture, and tourism (s.3). This Act is integrated

Fig. 14.2. Whale Shark tourism at Ningaloo Marine Park. (Photo: Erika Techera)

with the CALMA and uses the CALMA definitions of marine park, marine nature reserve and marine management area. In addition, the FRMA provides for the establishment of 'fish habitat protection areas' (s.115). Aquaculture licences and leases, which may be issued under the FRMA, may not be given in respect of marine areas protected under the CALMA (ss.92–94 and 98). Similarly, a designated fishing zone may not be declared in a marine nature reserve, marine management area or marine park established under the CALMA and any prior fishing zone is overridden by the protected area classification (s.110). Furthermore, any marine management area, marine park or marine nature reserve cannot be designated a fish habitat protection area (s.116). No other exclusive licences may be issued in relation to a marine management area, park or nature reserve (s.252) under the FRMA.

Tasmania

In Tasmania, the *National Parks and Reserves Management Act 2002* (NPRMA) provides for the declaration of protected areas on land, defined as including land under the sea (s.3) (see also Chapter 10). A 'nature reserve', 'national park', 'state reserve', 'regional reserve' and 'private nature reserve' are given the same meaning as in the *Nature Conservation Act 2002*. Under the NPRMA, the National Parks and Wildlife Council is to advise the relevant Minister, provide a forum for consultation and review management plans (s.14). The Council comprises up to 12 members appointed on the nomination of the Minister, but no specific criteria are set out apart from interest in, ability to contribute to and capacity to offer independent advice (s.13). Provision is made for management plans to be prepared with respect to reserves, following public consultation and including advice from the

Council (ss.19–25). Under s.30, the managing authority manages reserved land. The Minister may establish the Conservation Management Trust as a managing authority for a conservation area, nature recreation area or regional reserve (s.31). The Act is enforced by rangers (s.10) who have various enforcement powers supported by penalty provisions (ss.61–73).

Under the *Nature Conservation Act 2002*, Crown land may be set aside under s.11, and private land under s.12, for conservation purposes (as set out in Schedule 1). Wildlife is defined under the *Nature Conservation Act 2002* but does not include fish (s.3) and therefore, if the conservation of fish is a goal, the *Living Marine Resources Management Act 1995* (LMRMA) must be used in conjunction with the NPRMA and *Nature Conservation Act 2002*.

Under the LMRMA, the Minister may establish a 'marine resources protected area' for the purpose of the protection of representative samples of habitats and ecosystems, protecting vulnerable species, maintenance of fish productivity, species and genetic diversity, protection of sites of ecological significance or fragility, scientific research sites and education (s.105). The Minister may make rules for management plans in relation to fishing, species protection and the regulation of gear and equipment (s.104). Draft management plans must be available to the public and representations may be made, which must be considered (s.106–109). The Minister must approve the plan if it promotes the conservation of nature and restricts introduced fish, and assists in the public enjoyment areas (s.110). These areas are managed by Managing Authorities (ss.116–117). There are penalties for breaches of the conditions of such plans (s.113), as well as broad offences in a marine resources protected area, which are likely to have a detrimental effect on the environment (s.131) as well as special penalties (ss.267–269). The only other relevant provisions are those relating to habitat protection plans, which may be developed for specific areas (s.118), and 'research areas' (s.143).

Northern Territory

Under the *Territory Parks and Wildlife Conservation Act 2006*, land is defined as including sea (s.9) (see also Chapter 13). Power is given to the Administrator to declare a park or reserve (including a wilderness zone within a park or reserve) (s.12) or a sanctuary (s.25A). Similar power is given to revoke a park, reserve or sanctuary (s.13). The Administrator, however, must consider any report by the Parks and Wildlife Commission of the Northern Territory, and in preparing the report the Commission must call for and consider public comments (s.14). Excavation, building and other works are prohibited in a park or reserve unless authorised under section 17 of the *Mineral Titles Act 2010*, Part 2 of the *Petroleum Act 1984* or section 18A of the *Petroleum (Submerged Lands) Act 1982*. S.18 of the *Territory Parks and Wildlife Conservation Act 2006* provides for the development of management plans (which may include zoning) by the Commission following public consultation. The Administrator may approve or alter the plan, which is then brought before the Legislative Assembly by the Minister (s.19). The plan is implemented and enforced by the Territory's Parks and Wildlife Commission and Conservation Land Corporation (s.21). Provisions are also included for the joint management of certain parks with Aboriginal Traditional Owners. With respect to sanctuaries, the Act restricts unauthorised entry, which is subject to penalty (s.25C). Other provisions prohibit the taking, killing or capturing of a plant or animal, including a fish, breach of which results in a penalty (s.25F). The Act includes provisions for classifications of wildlife, including threatened species and the declaration of essential habitat (ss.37–38).

The Act is largely administered by the Commission, which can make by-laws for the carrying out of its functions including providing for the protection or prohibition of certain activities in parks and reserves (s.71). The Commission is established under the *Parks and Wildlife Commission Act 2000* and its functions include promotion of conservation and protection of the environment, management of areas under the *Territory Parks and Wildlife Conservation Act 2006* and enforcement of protection, conservation and sustainable use of wildlife.

Finally, the NT *Fisheries Act 1998* permits the relevant Minister to designate a fishery management area (s.22) with a management plan prepared by the Director (s.23). The Act is enforced by fisheries officers who have various search and seizure powers (ss.30–35) and included are penalties for a variety of offences (ss.37–46).

Common themes and key differences in MPA legislation in Australia

In summary, it is clear that all states and the Northern Territory have legislation that allows for the declaration of MPAs. The statutes fall into three types: specific marine parks Acts, fisheries regulation or more traditional national parks legislation. Although the objects of the various pieces of legislation differ, all provide for the declaration of protected areas managed through plans, often involving zoning. All the legal frameworks provide an administrative structure, although some have dedicated authorities, and others use existing bodies such as committees. The composition of these bodies also differs, with some being much more inclusive than others. In addition, administration is assisted by enforcement officers; again some are dedicated to MPAs but others common to a range of regulatory instruments. All the legislation includes offence provisions, but there are significant differences: for example, the comprehensiveness of regimes varies, as do the level of the penalties.

Challenges and opportunities for MPA legislation

From a legal perspective, the above analysis indicates that the legal landscape is fragmented, within and between jurisdictions. Protected areas created under specific MPA or national park legislation may be distinguished from those created under fisheries management legislation; the first two focus on conservation and the last on fisheries management. As one would expect, the goals and objectives of fisheries regulations, national parks and MPA statutes do differ. This may result in perverse outcomes where choices are made to create a MPA under one, as opposed to another, piece of legislation. For example, choosing to create a MPA under laws with stronger or weaker levels of protection or penalties may alter the effectiveness. Selecting legislation with easier or harder processes for creating MPAs, and the ease with which they may be reversed, can also change outcomes.

Another concern is that poor communication between government agencies may hinder the achievement of goals in a fragmented system (Baelde 2005). Additionally, significant differences between regimes hamper the creation of a uniform enforcement framework (Gullett 2003). Furthermore, different regimes may create confusion in stakeholders who operate across jurisdictions. It has been noted that 'inconsistent legislative obligations can slow or prevent designation' where they creating administrative and regulatory complexity (Jennings 2009, p. 18). Moreover, having multiple laws and policies may be a barrier to the translation of marine scientific research into law and policy; scientists must navigate the complex landscape in order to communicate their research to government agencies and those seeking to promote conservation. These problems may be overcome where integration mechanisms are established, and some states have done this to a greater extent than others; for example, as discussed above, Queensland and South Australia have done this well.

Where broad decision-making powers are contained in the legislation without detailed criteria as to how those powers are to be used, there is the risk that policy objectives will not be achieved; policy is not legally enforceable until it is translated into legislation. The extent of Ministerial power is also relevant, particularly where an independent advisory body is not established. Having an independent body to assess applications to create MPAs may contribute to the success of a MPA regime (Wescott 2006).

Although a single best practice approach to MPA law making is unlikely, the above survey of Australian legislation does demonstrate strengths and weaknesses in the various legal frameworks. It may be possible to draw out the best practice elements from each to create a model law that would tackle some of the fragmentation concerns. Model laws have been more commonly used at the international level to identify and share potential legal tools and measures to assist states in developing effective domestic law (UNCITRAL 1985; FAO 2009). However, there is no reason why a similar approach could not be taken in a country such as Australia, where it is clear that there is fragmentation of law across states.

Conclusions

It is clear that legal protections for marine areas in the states and territories of Australia have come a long way in the last 50 years. Just as scientific knowledge and management of MPAs have evolved, so have laws and policies. MPAs will not provide a solution to every threat in marine waters, and certainly are challenged by cumulative, diffuse and multi-point impacts. Although legislative advances have generally been positive, an analysis reveals many inconsistencies between jurisdictions, and fragmentation is more likely to hamper than assist marine conservation. Nevertheless, marine protection may be achieved through a combination of tools in different pieces of legislation. Given the existing web of laws and policies, the way forward perhaps lies in better integration: ensuring there is a more comprehensive and coordinated approach. What is needed is a coherent framework if both fisheries resources and biodiversity outcomes are to be achieved for the benefit of overall marine health.

Biography

Erika Techera is Professor and Dean of Law, Faculty of Law, The University of Western Australia. She is also a member of the Oceans Institute at the University of Western Australia.

References

ANZECC (1998) *Guidelines for Establishing the National Representative System of Marine Protected Areas.* ANZECC Task Force on Marine Protected Areas. Environment Australia, Canberra.

Baelde P (2005) Interactions between the implementation of marine protected areas and right-based fisheries management in Australia. *Fisheries Management and Ecology* **12**, 9–18. doi:10.1111/j.1365-2400.2004.00413.x.

Commonwealth of Australia (1998a) *Australia's Oceans Policy Volume 1: Caring, Understanding, Using Wisely.* Commonwealth of Australia, Canberra.

Commonwealth of Australia (1998b) *Australia's Oceans Policy Volume 2: Specific Sectoral Measures.* Commonwealth of Australia, Canberra.

FAO (2009) *Agreement on Port State Measures to Prevent, Deter and Eliminate Illegal, Unreported and Unregulated Fishing.* Food and Agriculture Organization of the United Nations, Rome, <http://www.fao.org/fishery/topic/166283/en>.

Gullett W (2003) Legal framework and enforcement experience of marine protected areas in Tasmania, New South Wales and Commonwealth waters. In *Aquatic Protected Areas: What Works Best and How Do We Know?* (Eds A Beumer, A Grant and D Smith) pp. 388–397. Australian Society for Fish Biology, Perth.

Jennings S (2009) The role of marine protected areas in environmental management. *ICES Journal of Marine Science* **66**, 16–21. doi:10.1093/icesjms/fsn163.

Kearney R, Buxton CD, Farebrother G (2012) Australia's no-take marine protected areas: Appropriate conservation or inappropriate management of fishing? *Marine Policy* **36**, 1064–1071. doi:10.1016/j.marpol.2012.02.024.

Kuo WL (2011) The adequacy of the regulatory regime of marine protected areas in New South Wales for addressing threats to marine biodiversity. PhD thesis. University of Wollongong, Wollongong, <http://ro.uow.edu.au/theses/3372>.

McNeill SE (1994) The selection and design of marine protected areas: Australia as a case study. *Biodiversity and Conservation* **3**, 586–605. doi:10.1007/BF00114203.

NSW Government (2014) Marine Estate Management Act 2014: Frequently Asked Questions, NSW Government, Sydney, <http://www.marine.nsw.gov.au/__data/assets/pdf_file/0008/533078/Marine-Estate-Mgmt-Act-2014-FAQs.pdf>.

UNCITRAL (1985) *Model Law on International Commercial Arbitration.* United Nations Commission on International Trade Law, Vienna, <http://www.uncitral.org/uncitral/en/uncitral_texts/arbitration/1985Model_arbitration.html>.

Wescott G (2006) The long and winding road: the development of a comprehensive, adequate and representative system of highly protected marine protected areas in Victoria, Australia. *Ocean and Coastal Management* **49**, 905–922. doi:10.1016/j.ocecoaman.2006.08.001.

Chapter 15

The economic case for marine protected areas

Caroline Hoisington

Introduction

Some aspects of marine protected areas (MPAs) can be valued in economic terms relatively easily when the existence of MPAs causes an increase in tourism and benefit local businesses. The Great Barrier Reef Marine Park is the classic example, and many smaller MPAs have had positive effects on local economies as well. However, MPAs are primarily created for their important ecological benefits, which are more difficult to assess in economic terms, but are important to us.

The idea of putting economic values on natural areas, including marine areas, has evolved greatly in recent years. It remains controversial for several reasons. Some people consider the idea offensive: a commodification of things that are essential to life on this planet that should not be treated as if they were marketable goods. Others consider that there are simply too many things that will be omitted from any attempt to value nature, from the intrinsic value of natural systems, their aesthetic and spiritual values, to the importance of experiencing oneself in a natural environment, feeling connected, exercising and more (see review by Juniper 2012). Others also doubt that economic valuations have been helpful in preserving natural areas (McCauley 2006) or mistrust the blending of economics with environmental issues (Sagoff 2012). Some authors list economic benefits along with ecological, cultural and social benefits of natural areas, quantifying some and not others (Mulongoy and Gidda 2008).

Many analysts acknowledge these concerns but argue, often from experience, that when policy and investment decisions are being made, whether by government agencies, private businesses or other groups, calculations involving money, often cost–benefit estimations, become the basis for decisions. Environmental factors may be acknowledged verbally or as an appendix to a report, but in practice, if they are not given a monetary value, they are left out of the calculations, making their value equal to zero in the actual decision-making process. This amounts to a market failure, where prices do not represent the full social values.

Ecosystem services valuation attempts to remedy at least part of this problem by pricing services that natural systems provide for free by relating these services to human needs. This approach can be particularly useful when trying to analyse the effects of a potential policy change or investment that may impact the environment: a risk assessment, multicriteria analysis and other decision-making frameworks (DEFRA 2007). Other uses include simply drawing attention to the extent to which our economies function within the larger context of our natural environment.

Different groups of researchers and authors have constructed different categories to organise lists of the services ecosystems provide. They all relate functions of ecosystems to human needs and uses – some directly (e.g. food, materials) and some more indirectly (e.g. water filtering, carbon sequestration, shoreline protection). See, for example, de Groot's seminal work (de Groot 1992) and many subsequent papers, the publications by The Economics of Ecosystem Services and Biodiversity (TEEB) including the Synthesis Report (Sukdev *et al.* 2010), and many more.

Many now use the four types of categories for ecosystem services used by the Millennium Ecosystem Assessment (2005):

- supporting services (nutrient cycle, soil formation, primary production, etc.)
- provisioning services (food, water, wood, fuel, etc.)
- regulating services (climate regulation, water purification, etc.)
- cultural services (aesthetics, spirituality, education, recreation, etc.).

One issue impeding ecosystem services valuation is that we do not understand fully as yet all the biological relationships that make up ecosystems generally, how they function and how they benefit us. A good example of our evolving understanding is the contrast between the values for ecosystem services in two large studies by groups headed by Robert Costanza (Costanza *et al.* 1997, 2014) that attempted to value ecosystems of the entire globe. The first was a landmark study, based on an analysis of very large datasets, that resulted in an estimate for the value of ecosystem services for the Earth at about US$33 trillion per year in 1995 (equivalent to US$46 trillion in 2007 terms). This represented about twice the annual value of the GDP of all the world's countries at that time (Zimmer 2014). The 2014 study re-evaluated the totals in light of a great deal more data and estimated a value of US$145 trillion per year for the same areas as the earlier study and US$125 trillion (both in 2007 dollars for comparison) if the loss of significant areas of different ecosystems since 1995 is taken into account (Costanza *et al.* 2014). So the basic value is now estimated to be more than three times as much. Despite the fact that a great deal has been researched and written in recent years about ecosystems valuation, this is a relatively new field, and it is still evolving.

The basis for an economic assessment of MPAs

The economic value of MPAs is based on the idea that declaring an area to be a MPA will make a difference. Regulating what uses are allowed and not allowed can cause changes in marine ecosystems that will lead to what humans consider to be a gain (or the avoidance of a loss) of value.

Economic valuation in general is based on the idea that value is measured ultimately in terms of human wants and needs. For environmental values, researchers often use a system of total economic value (TEV), which encompasses all sorts of values by summing direct-use (both extractive and non-extractive) value, indirect-use value, non-use value and intrinsic value (DEFRA 2007). Thus, a direct-use extractive value for marine areas could be fishing and, if zoning part of a MPA as a sanctuary zone (an area protected from fishing) led to an increase in productivity of neighbouring fishing grounds, the benefit of this protection could be measured by the value of the incremental catch of fish (which would have to be greater than the value of the catch sacrificed from the closed area to be positive).

A direct use non-extractive value could be swimming, snorkelling, scuba or boating. An increase in revenues from tourists could come from direct uses, both extractive (e.g. fishing) and non-extractive (e.g. renting boats or scuba gear).

The concept of ecosystem services helps to value indirect use. Protecting a marine area may have the effect of increasing (or not losing) biological systems that provide ecosystem services, such as oyster beds and seagrass meadows that filter water, shield and stabilise coastlines or act as nurseries for fish. Protecting marine ecosystems is now also understood to lead to their increased robustness and resistance to threats (IUCN/WCPA 2008; McCook *et al.* 2010). Because of our growing understanding of the importance of marine ecosystem services to human communities, and the knowledge that marine ecosystems are often facing multiple stresses, increased marine ecosystem resilience has an economic value. Ecosystem services have become a large part of the discussion about the value of MPAs and particularly fully protected, no-take sanctuary zones because these areas are where the biggest changes are seen.

Non-use values include option value (for possible future use), bequest value (use of future generations) and existence value. MPAs may be considered to have an existence value if we believe they will protect something we personally feel is important and value, even if we are not planning on making any direct use of it. These non-use values are seldom estimated, but sometimes they can be monetised. An example is a campaigns to 'save the whales,' supported by people who may never expect to see whales but nevertheless feel they are important enough to contribute money. Surveys asking people if they want MPAs have been done in New South Wales (NSW), showing very large support, with responders listing reasons such as 'allowing future generations to enjoy the park' (42% of respondents in one example, see McGregor Tan Research 2008).

Willingness to pay (WTP) estimates, where surveys are taken to ask people how much they would be willing to pay for MPAs are less common. One example, focusing on proposed new MPAs in Commonwealth waters of South-west Australia found that 'The Australian community would appear to have a positive WTP for the establishment of new MPAs in the South-West Marine Region. Aggregating the mean estimated WTP per household to 50% of the population of Australian households gives an aggregate WTP of $400M' (Gillespie 2010, p. 23).

So, the question of determining an economic value of MPAs really rests on two things:

1 Scientific research to show that creating a MPA can be effective in either increasing – or avoiding the loss of – values we care about.
2 Finding indications of the extent to which the community, collectively, cares about such things. Actual monetary values are the easiest, but not always available. Estimations based on scientific research into biological relationships and ecosystem functions and social science surveys into human values are often involved,[1] as well as more straightforward financial data on benefits to local businesses from increased numbers of visitors.

Demonstrating both of the above has been somewhat difficult. The first was not well researched until recent years, but now there is an extensive amount of research (e.g. Toropova *et al.* 2010) that – given the circumstances necessary to make them effective – MPAs can make a significant difference, particularly the no-take sanctuary zones. For example, a summary of a meta-analysis of data is shown in Table 15.1.

The second requirement, assessing relevant human values, is complicated. Simple financial data on businesses may be lacking, although surveys showing increases in tourism revenues and reports of recreational fishing improvement associated with MPAs exist

Table 15.1. Summary of some findings of studies of ecological changes within marine protected areas

(Adapted from Toropova *et al.* 2010)

Indicator	Main findings	Region	Number of MPAs	Source
Biomass Density Size Richness	446% increase 166% increase 28% increase 21% increase	Global (for fish, invertebrates, algae)	124	Lester *et al.* (2009)
Fish density	1.66× higher	Global	33	Molloy *et al.* (2009)
Density Biomass Species richness	1.4–1.92× higher 0.107–3.67× higher 1.27–1.68× higher	Global, temperate only (for fish, invertebrates, algae)	30	Stewart *et al.* (2009)
Biomass Density Size Richness	352% increase 151% increase 29% increase 25% increase	Global (for fish, invertebrates, algae)	81	Halpern (2003)
Fish density Species richness	1.25–3.7× higher 1.11× higher	Global	19	Mosquera *et al.* (2000), Côté *et al.* (2001)
Fish density Biomass	1.2× higher 2.1× greater	Mediterranean	12	Guidetti and Sala (2007)
Fish density Species richness	2.46× higher No effect	Mediterranean	12	Claudet *et al.* (2008)
Fish density	1.64× higher (2.5× for exploited species)	Philippines	19	Maliao *et al.* (2009)

in some cases (Hoisington 2013). Indirect values, such as willingness to pay surveys and travel cost surveys, have been done in some cases. For example, travel cost methods have been used to estimate the value of MPAs to users in NSW (Gillespie Economics 2007). The value of ecosystem services can be very high without the need for people to actually know the underpinning relationships. For example, seagrasses and other ecosystems can be essential as nurseries for a valued commodity such as fish, but people who value fishing and/or eating fish highly may not know anything about these ecosystems or care about them. Estimations of social values are rare.

To assess the value of MPAs, economists generally must return to the cost–benefit framework, albeit with a wider scope for estimating economic benefits. The benefits of creating a MPA are based on projections of future situations: the benefits are the difference between what the future situation would be with a MPA compared with what it would be without one. Costs are generally simpler to project. They include initial investments made into infrastructure, equipment, setting up areas with signage and the like; the ongoing costs may include monitoring or policing areas, surveys, and so on; and finally the value of other uses that are sacrificed, some of which may be included in set-up costs, such as paying out commercial fishing licenses.

Cost–benefit calculations are based on looking at 'income streams' over projected future years, where costs are mostly incurred in the early years as investments are made to put something in place, and benefits generally begin to accrue with some time lag. The timeframe for the projection of costs and benefits is usually based on the projected life of an investment. The present value calculations used for cost–benefit analyses require a discount factor, used to represent the opportunity cost of money, meaning roughly what return could be earned on alternative investments instead of the one being analysed.[2] It is known that the effects on recreational fishers may be rapid with rezoning. These are generally losses in areas where sanctuary zones have been declared in places that may have been used by recreational fishers and gains (in the generally larger areas) where commercial fishing has been banned or reduced in scale and recreational fishing improves.

Biological effects of MPAs increase over time and may show very little effect for the early years. The positive effects of MPAs of increasing fish size and density, ecosystem complexity and resilience and associated ecosystem benefits start in a few years for faster growing species, but can take longer to become evident, easily up to 15 years for the slower growing species. There is also evidence that some benefits can keep increasing for 40 years or more (see below and Molloy *et al.* 2009). Cost–benefits are designed for financial investments and they are calculated over the estimated 'lifetime' of the investment. However, the discount factors used mean that calculations of value for benefits after ~25 years are effectively almost zero. Because of this, the continuing, long-term benefits from MPAs tend to be underestimated in cost–benefit calculations.

Cost–benefit calculations are always based on estimates of future conditions with and without an investment of some sort. In some investment cases, the projection is that without the investment nothing would change in future conditions. This assumption is generally not considered to be correct for MPAs, particularly in areas of heavy usage, because the future situation without any MPA may be deterioration. The point is that looking at the economics of MPAs involves projecting future situations but this has to be done by researching what has happened in MPAs that have been in existence for some time; recently declared MPAs will not provide answers.

Benefits and cost of MPAs in Australia and elsewhere

The use of estimates for ecosystem services as part of valuation of natural areas, including MPAs, is increasing worldwide, but Australia has not been a leader in this field. Although socio-economic assessments are required for new MPA proposals (e.g. NSW Government 2006), most do not include ecosystem service estimations and can be therefore considered incomplete. Nevertheless, researchers continue to expand and develop the use of ecosystem services for marine areas in Australia and internationally.

One of the largest, and the most financially valuable in terms of revenue generated, MPA in Australia is the Great Barrier Reef Marine Park (GBRMP) (see Chapter 5). The economic benefit was summarised by McCook *et al.* (2010) who looked at the effects of rezoning that took place in 2004 and significantly enlarged the area of sanctuary zones. The economic value to Australia of a healthy Great Barrier Reef is more than A$5.5 billion per year in benefits to local economies from actual expenditures. These estimates only include use value (not ecosystem service values and the other types of values mentioned above as part of TEV) and so underestimate the total economic value (McCook *et al.* 2010). Employment is estimated at 53 800 full time jobs. Annual revenue from tourism is estimated to be ~36 times the revenue from commercial fishing (Fig. 15.1). The costs of annual management are estimated to be less than 1% of the annual economic returns, although McCook *et al.* (2010) acknowledge that those costs should be compared with net value rather than gross output values, but net measures are not available. Even the costs of

Fig. 15.1. SCUBA divers on the Great Barrier Reef Marine Park. Tourism in this MPA provides a significant source of income for the region. (Photo: James Fitzsimons)

structural adjustment to increase the highly protected areas amounted to less than 4% of the economic returns in a single year (McCook *et al.* 2010).

The expansion of the GBRMP no-take zones had a large benefit in biological terms (e.g. the widespread recovery of depleted fish stocks in the expanded no-take areas and even diffusion across into other zones; McCook *et al.* 2010) but not all MPAs have such scale nor the possibility of generating such a large financial benefit. Even so, studies of areas around NSW MPAs have shown economic benefits to surrounding communities (Hoisington 2013).

Western Australia's Ningaloo Reef is the closest to the GBRMP in its potential for both developing an international reputation as a good place to see spectacular natural attractions, from Whale Sharks to extensive reefs, and a source of tourism revenues. The fully protected area of the Ningaloo Marine Park was expanded in 2004 to 34% of the MPA (Jones *et al.* 2009). The draw of Ningaloo has changed over the years, from commercial whaling and fishing in the 1950–60s to recreational fishing in the 1970–80s, to more passive, nature-based activities as the adjoining protected areas in Western Australian and Commonwealth waters have evolved. That trend is expected to continue, while recreational fishing is projected to grow and commercial fishing remains in some areas (Marine Parks and Reserves Authority 2005), nature-based and wilderness tourism is now the main source of income in both Exmouth and Coral Bay. Research into data for modelling the effects of Ningaloo (Jones *et al.* 2009) revealed that the estimated tourist expenditure in these regions for the year ending September 2008 was A$141 million, with 179 352 visitors staying for an average of 9.92 nights. Average nightly expenditure was A$79.46. This is consistent with a 2004 study (described in Jones *et al.* 2009) in these regions that estimated annual expenditure to be A$127 million. Although reliable statistics are not available for the early 1990s, it is believed locally that visitor numbers have increased markedly since the MPAs were put in place and protection expanded (Jones *et al.* 2009). Ecosystem service values have not yet been estimated.

MPAs can be created with different approaches. In South Australia, the decision was made to protect a representative set of different ecological systems, mostly in good condition, on the grounds that they are rich with marine species, special habitats and economic, social, Aboriginal and cultural values, which could be damaged without good management (see Chapter 8). The network of MPAs covers 44% of South Australia's waters, and the fully protected sanctuary zones are 6% of the state waters (Bailey *et al.* 2012). The value of the sanctuary zones in ecosystem services terms is estimated at a minimum of A$22.8 million per year, using methodology developed by Centre for Policy Development, Sydney, based on Australian and international research, primarily that for TEEB by the United Nations (Eadie and Hoisington 2011; Hoisington 2012) (Fig. 15.2). The predicted losses to commercial fishing due to exclusion from these areas was important for local remote communities, where the impact on 19 fisheries was calculated at A$5.8 million per year, but at a state-wide level this constituted only about 0.02% of the gross product (Bailey *et al.* 2012). In this assessment, the environmental and expected ecological benefits of protection were covered thoroughly, but their economic value was not estimated. Bailey *et al.* (2012) discussed the value of the ecosystem services and tourism, but the projections of future conditions, with and without the MPAs, was considered to be beyond current knowledge.

Fig. 15.2. Light River Delta Sanctuary Zone in the Upper Gulf St Vincent Marine Park, South Australia. Mangroves and seagrass provide important ecosystem services, including shoreline protection and fish nursery grounds. (Photo: James Fitzsimons).

Therefore the net present value of costs, projected over 20 years became a threshold value of about A$64 million that Bailey *et al.* (2012) projected future benefits should exceed.

In other places, including in NSW, MPAs have been created in areas that are heavily used (see Chapter 11). The largest part of these areas remains open to recreational fishing and some commercial fishing, but (relatively small, generally less than 25%) parts of the MPAs are sanctuary zones, which prohibit fishing and other extractive uses. The direct benefits from these MPAs would be increases in value for visitors and businesses that benefit from tourism. People coming to MPAs, as to parks on land, expect infrastructure and good recreational possibilities. Recreational fishing in general use zones where commercial fishing has been limited can improve immediately. Sanctuary zones are particularly valued by snorkelers, divers and others knowing that there will be something there to see, and by all those who say they value the natural environment being protected. It has been noted, however, that where use of an area has been heavy – and the resulting biological condition is relatively degraded, for instance by high fishing pressure – the positive changes in sanctuary zones can take longer to develop. Current research reveals that older reserves are much more effective than young ones for causing increases in numbers of fish and more productive ecosystem functioning, with the conclusion that reserves should not be judged before at least 15 years (Molloy *et al.* 2009). The NSW marine parks, however, are all relatively new, with the oldest two being zoned in 2002 and the most recent one in 2007. Nevertheless, Solitary Island and Jervis Bay Marine Parks in particular have showed increases in businesses associated with tourism, and Lord Howe Island is largely dependent on tourism associated with its Marine Park (Hoisington 2013).

The creation of MPAs in Australia is mostly fairly recent. Older international examples can offer further insights. One MPA that has worked out very successfully is found in New Zealand. The story was told succinctly in the American *National Geographic* special on 'Saving the Sea's Bounty' (Warne 2007). The director of the University of Auckland Marine Laboratory succeeded in 1977 (after 12 years of trying) in having a 2-mile square section of marine coast – that was not particularly remarkable – fully protected as a no-take zone. The local media proclaimed 'Nothing to do at Goat Island Bay any more'. Almost immediately when fishing for snapper and spiny lobster, the main predators of sea urchins stopped, the kelp forests began to regrow and both predator species gradually became abundant. Spiny rock lobsters within the protected areas are now ~15 times higher in density than populations outside. The local lobster fishermen are reported to be pleased to get their catch with less effort and therefore less cost, because now they can set their traps around the edges of the fully protected zone to catch lobsters as spill over from what they now call 'our reserve'. The area now has a paved road leading to a marine museum that is widely used by schools, and 100 000 visitors per year come to 'look at fish' by snorkelling, or in in glass-bottomed boat tours, and students come to look and learn in school trips by the hundreds (Warne 2007).

Concluding comments

Estimations of benefits are often complex, but MPAs can have large benefits in both financial and economic terms. Although economic evaluation of proposed MPAs is required ahead of declaration, it is complex due to a lack of science and economic data, difficulties in measurement and estimation, and time to show ecological effects.

Although the economic valuation of environmental benefits and ecosystem values for marine areas and the importance of fully protected areas is still young, several interesting conclusions are coming to light as follows:

- MPAs do not always have to be set in spectacular or unique places to be successful in drawing visitors and becoming an economic asset for local communities (Warne 2007).
- Some MPAs have been proved to serve local commercial fishers well, even improving their catch from spill-over leading to increased numbers and size of fish, or reduced costs of catching fish, thereby maintaining or improving fishers' net incomes (Eadie and Hoisington 2011).
- Even when there is strong opposition to the creation of a MPA, local opinion can change and even become positive, as economic benefits become visible. However, the economic returns are based on ecological changes, and these changes can take time. Initially, there are costs to setting up a MPA, and the most important of these may be the foregone fishing opportunities, particularly for recreational fishers who can feel ownership rights in areas that become sanctuary zones, but typically they are not compensated, unlike commercial fishers who typically are.
- If MPAs are not zoned effectively, and particularly if the sanctuary zones are too small or not enforced, the biological changes, and therefore the economic benefits, may not accrue.
- The time factor is crucial. Overall fish densities (size and numbers) can continue to increase for 40 years or more (e.g. Molloy *et al.* 2009), but most do not start immediately when MPAs are declared. This is very significant for economic assessment, because the timing of benefits is a crucial part of valuation.
- Factors affecting surrounding areas are critical too. For example, the Great Barrier Reef Marine Park is under threat from various causes, from sedimentation and pollution to possible effects from dredging for expanding coal ports and therefore potentially under threat of being delisted as a World Heritage area (see Chapter 5), which would most likely damage tourism revenues. This is not the focus of this chapter, but it is highly relevant to economic valuation simply because it demonstrates that, although declaring a MPA may be necessary, it will generally not be sufficient in itself to ensure the changes necessary to get the economic benefits that MPAs can bring.

Overall, MPAs have many economic benefits. Multiple use zones in MPAs can draw visitors because they expect to find things to do in interesting places, generally with access to useful infrastructure such as parking, camping, boating, fishing, designated swimming areas, and so forth. The sanctuary zones in MPAs help to ensure that snorkellers and divers will find enough to see (or fishers to catch just outside the boundaries) and that those people who appreciate natural systems for their existence, aesthetic and spiritual values (or simply feel glad that healthy natural systems exist and can persist) will benefit also.

Increases in visitor numbers, and related increase in economic activities, are very frequently found around MPAs, but – as in most investments – the initial economic effects are the costs. There are initial set-up costs and then ongoing management costs. In the case of establishing sanctuary zones where no fishing or other extractive activities are allowed, some of the costs are in the loss of use of these areas by those who were previously using them. Commercial fishers may be compensated, but recreational fishers will not. On the other hand, commercial fishing removed or restricted from multiple use zones in MPAs can result in improved catch for recreational fishers, both immediately and over time.

All of the benefits from MPAs generally increase with time, if MPAs are managed effectively. Increases in tourism revenues are not immediate, but grow over time as the information about and reputation of a MPA increases. The biological effects of MPAs – the most basic reason for creating them, to recover ecosystem losses or ensure against such losses in the future – develop the most slowly. These are also the most difficult scientific benefits to measure, although much progress has been made. They can also be the most difficult to estimate economically because the economic valuation depends first on biological understanding and data and second on relating observed changes to human values. Perhaps the most important element to understand in the valuation process is that of time – both to see the ecological effects and human understanding of these effects.

Acknowledgements

The author wishes to thank colleagues at Centre for Policy Development for enlarging her understanding of economic and environmental issues in Australia and providing the chance to research and write on marine protected areas, and colleagues at Sydney Institute of Marine Science who have given her some understanding of scientific background on her much-loved home environment of Sydney Harbour.

Biography

Caroline Hoisington is an agricultural and natural resources economist, and an advocate for sustainable development and environmental management. She is the author of *Rough Trade: How Australia's Trade Policies Contribute to Illegal Logging in the Pacific Region* (The Australia Institute), author and co-author of six papers on Australian marine management (Centre for Policy Development) and most recently *Our Harbour Our Asset* (Sydney Institute of Marine Science). She worked primarily in international development for The World Bank and other agencies in 25 countries. She writes on a range of policy areas, focusing now on the economic values of forests, oceans and estuaries.

Endnotes

1. Much of the recent work on economic valuation of ecosystem services has focused on how to value services that are not marketed (Pereira *et al.* 2005).
2. In Australian financial analyses, 6% or 7% is often used as a discount factor for cost–benefit calculations, although there are strong arguments in Australia and internationally for using a much lower discount rate – or even a positive rate – for natural resources investments for several reasons, most particularly because they are seen as increasing in value over the years. This is a complicated issue and not resolved, nor further considered here.

References

Bailey H, Bryars S, Spoehr J, Morison J, Brook J, Barnett K, *et al.* (2012) *Marine Park Regional Impact Statements, Main Report.* Report prepared for the South Australian Department of Environment, Water and Natural Resources. Econsearch, Adelaide, <http://www.environment.sa.gov.au/files/5fb6d22b-39ed-4a6b-9cf7-a0b700edccf0/mp-gen-impactstatement-statewide.pdf>.

Claudet J, Osenburg CW, Benedetti-Cecchi L, Domenici P, Garcia-Charton JA, Perez-Ruzafa A, *et al.* (2008) Marine reserves: size and age do matter. *Ecology Letters* **11**, 481–489. doi:10.1111/j.1461-0248.2008.01166.x.

Costanza R, d'Arge R, de Groot R, Farberk S, Grasso M, Hannon B, *et al.* (1997) The value of the world's ecosystem services and natural capital. *Nature* **387**, 253–260. doi:10.1038/387253a0.

Costanza R, de Groot R, Sutton P, van der Ploeg S, Andersond SJ, Kubiszewski I, *et al.* (2014) Changes in the global value of ecosystem services. *Global Environmental Change* **26**, 152–158. doi:10.1016/j.gloenvcha.2014.04.002.

Côté IM, Mosquera I, Reynolds JD (2001) Effects of marine reserve characteristics on the protection of fish populations: a meta-analysis. *Journal of Fish Biology* **59** (suppl. A), 178–189. doi:10.1111/j.1095-8649.2001.tb01385.x

de Groot RS (1992) *Functions of Nature: Evaluation of Nature in Environmental Planning, Management and Decision Making.* Wolters Noordhoff, Groningen, The Netherlands.

DEFRA (2007) *An Introductory Guide to Valuing Ecosystem Services.* Department for Environment, Food and Rural Affairs, London, UK.

Eadie L, Hoisington C (2011) *Stocking Up: Securing our Marine Economy.* CDP Occasional Paper 14. Centre for Policy Development, Sydney.

Gillespie R (2010) *Non Use Economic Values of Marine Protected Areas in the South-West Marine Region.* Paper presented at the Economics and the Environment Network Symposium, 22–24 November 2010, Australian National University, Canberra, <https://taxpolicy.crawford.anu.edu.au/research_units/eerh/workshops/network_symposium/pdf/presentations/day_two/robert_gillespie_paper.pdf>.

Gillespie Economics (2007) *The Recreation Use Value of NSW Marine Parks.* Prepared for the NSW Department of Environment and Climate Change. Gillespie Economics, Sydney, <http://www.environment.nsw.gov.au/resources/research/RecreationUseValueNSWMarineParks.pdf>.

Guidetti P, Sala E (2007) Community-wide effects of marine reserves in the Mediterranean Sea. *Marine Ecology Progress Series* **335**, 43–56. doi:10.3354/meps335043.

Halpern BS (2003) The effects of marine reserves: do reserves work and does reserve size matter? *Ecological Applications* **13**, 117–137. doi:10.1890/1051-0761(2003)013[0117:TIOMRD]2.0.CO;2.

Hoisington C (2012) *Zoning South Australia's Marine Parks, an Economic Evaluation.* CPD Occasional Paper 22. Centre for Policy Development, Sydney.

Hoisington C (2013) *Marine Protection Dividend: NSW Marine Parks Deliver More Over Time.* CPD Occasional Paper 34. Centre for Policy Development, Sydney.

IUCN/WCPA (2008) *Establishing Resilient Marine Protected Area Networks – Making it Happen.* IUCN World Commission on Protected Areas, IUCN/WCPA, National Oceanic and Atmospheric Administration, and The Nature Conservancy, Washington DC.

Jones T, Hughes M, Wood D, Lewis A, Chandler P (2009) *Ningaloo Coast Region Visitor Statistics Collected for the Ningaloo Destination Modelling Project.* Sustainable Tourism Cooperative Research Centre, Gold Coast, <http://www.crctourism.com.au/wms/upload/resources/90067b%20Jones%20Ningaloo%20Visitor%20Survey_WEB.pdf>.

Juniper T (2012) We must put a price on nature if we are going to save it. *The Guardian* 10 August 2012, <http://www.theguardian.com/environment/2012/aug/10/nature-economic-value-campaign>.

Lester SE, Halpern BS, Grorud-Colvert K, Lubchenco J, Ruttenberg BI, Gaines SD, *et al.* (2009) Biological effects within no-take marine reserves: a global synthesis. *Marine Ecology Progress Series* **384**, 33–46. doi:10.3354/meps08029.

Maliao RJ, White AT, Maypa AP, Turingan RG (2009) Trajectories and magnitude of change in coral reef fish populations in Philippine marine reserves: a meta-analysis. *Coral Reefs* **28**, 809–822. doi:10.1007/s00338-009-0532-6.

Marine Parks and Reserves Authority (2005) *Management Plan for the Ningaloo Marine Park and Muiron Islands Marine Management Area 2005–2015*. Department of Conservation and Land Management, Perth.

McCauley DJ (2006) Selling out on nature. *Nature* **443**, 27–28. doi:10.1038/443027a.

McCook LJ, Ayling T, Cappo M, Choat H, Evans RD, De Freitas DM, *et al.* (2010) Adaptive management of the Great Barrier Reef: a globally significant demonstration of the benefits of networks of marine reserves. *Proceedings of the National Academy of Sciences of the United States of America* **107**, 18278–18285. doi:10.1073/pnas.0909335107.

McGregor Tan Research (2008) *Jervis Bay Marine Park Community Survey Final Report*. NSW Marine Parks Authority, Sydney.

Millennium Ecosystem Assessment (2005) *Ecosystems and Human Well-being: Synthesis*. Island Press, Washington DC.

Molloy PP, McLean IB, Côté IM (2009) Effects of marine reserve age on fish populations: a global meta-analysis. *Journal of Applied Ecology* **46**, 743–751. doi:10.1111/j.1365-2664.2009.01662.x.

Mosquera I, Côté IM, Jennings S, Reynolds JD (2000) Conservation benefits of marine reserves for fish populations. *Animal Conservation* **3**, 321–332. doi:10.1111/j.1469-1795.2000.tb00117.x.

Mulongoy KJ, Gidda SB (2008) *The Value of Nature: Ecological, Economic, Cultural and Social Benefits of Protected Areas*. Secretariat of the Convention on Biological Diversity, Montreal.

NSW Government (2006) *Socio-Economic Assessment of the Batemans Marine Park*. NSW Government, Sydney, <http://www.mpa.nsw.gov.au/pdf/bmp-socio-economic-report.pdf>.

Pereira HM, Reyers B, Watanabe M, Bohensky E, Foale S, Palm C, *et al.* (2005) Condition and trends of ecosystem services and biodiversity. In *Millenium Ecosystem Assessment. Ecosystems and Human Well-being, Volume 4 Multiscale Assessments*. (Eds D Capistrano, CK Samper, MJ Lee and C Raudsepp-Hearne) pp 171–203. Island Press, Washington DC.

Sagoff M (2012) The rise and fall of ecological economics: a cautionary tale. *Breakthrough Journal* **2**, <http://thebreakthrough.org/index.php/journal/past-issues/issue-2/the-rise-and-fall-of-ecological-economics>.

Stewart GB, Kaiser MJ, Côté IM, Halpern BS, Bayliss HR, Pullen AS (2009) Temperate marine reserves: global ecological effects and guidelines for future networks. *Conservation Letters* **2**, 243–253. doi:10.1111/j.1755-263X.2009.00074.x.

Sukdev P, Wittmer H, Schröter-Schlaack C, Nesshöver C, Bishop J, ten Brink P, *et al.* (2010) *The Economics of Ecosystems and Biodiversity – Mainstreaming the Economics of Nature: A Synthesis of the Approach, Conclusions and Recommendations of TEEB*. TEEB, Geneva, Switzerland.

Toropova C, Meliane I, Laffoley D, Matthews E, Spalding M (Eds) (2010) *Global Ocean Protection: Present Status and Future Possibilities*. Agence des aires marines protégées, Brest, France; IUCN WCPA, Gland, Switzerland; UNEP-WCMC, Cambridge, UK; TNC Arlington VA; UNU Tokyo, Japan; WCS New York. <https://portals.iucn.org/library/efiles/edocs/2010-053.pdf>.

Warne K (2007) Blue haven. *National Geographic* **April 2007**, 70–89.

Zimmer C (2014) Putting a price tag on nature's defences. *The New York Times*, 5 June 2014, <http://www.nytimes.com/2014/06/05/science/earth/putting-a-price-tag-on-natures-defenses.html?_r=0>.

Chapter 16

The science in Australia's marine protected areas

Trevor J. Ward and Romola R. Stewart

Introduction

In Australia, as elsewhere, ocean biodiversity is under pressure and being degraded by a variety of human-derived activities such as pollution, habitat damage and excessive fishing (SEC 2011). Marine protected areas (MPAs) offer the potential for assisting with the protection of ocean biodiversity. Together with other specific measures, well-designed MPAs may provide a buffer against the impacts from several threats, to enable healthy marine biodiversity and ecosystems to persist in an uncertain future.

The benefits of MPAs are well articulated in the scientific literature, particularly the potential for 'insurance' against the cumulative effects of a poorly known suite of pressures. For example, no-take marine sanctuaries (IUCN protected area category I or II; Day *et al.* 2012) that exclude all forms of fishing and mining are uniquely capable of protecting and supporting the resilience of biodiversity and habitats. By supporting large-bodied and long-lived individuals that can contribute disproportionately to reproductive output from a population (e.g. Barrett *et al.* 2007), they provide a measure of insurance against management uncertainties, and a benchmark for evaluating the effects of both human activities and natural stresses outside the boundaries of a MPA (McLeod *et al.* 2005; Emslie *et al.* 2015).

Australia's commitment to develop a National Representative System of Marine Protected Areas (NRSMPA) was formalised through the Intergovernmental Agreement on the Environment (1992) and fulfils important and long-standing international obligations as set out in the 1992 Convention of Biological Diversity (CBD) and its Jakarta Mandate. More recently, the Aichi Target 11 formalised the commitment of signatories to conserve 10% of the global oceans in 'effective and equitably managed, ecologically representative and well connected systems of protected areas' by 2020 (CBD 2010). However, this target is now seen by many as being inadequate – the international conservation community at the 2014 IUCN Sydney World Parks Congress called for 30% of the ocean to be protected within no-take MPAs to more confidently provide for conservation of the structure and function of ocean habitats and ecosystems (IUCN World Parks Congress 2014).

Australia's efforts towards achieving MPA commitments have earned it the reputation as world leader in the science of marine conservation planning, and yet we still fall short of the Aichi Target 11 (CBD 2010) and further still from the emerging scientific best practice of at least 30% in no-take MPAs. In this chapter we examine the role of science in supporting implementation of Australia's NRSMPA program, with examples to highlight where good practice has delivered good MPA planning outcomes. We consider where the gaps lie between the theory and practice of MPA design and point to further aspects of science needed to address these challenges.

Australia's NRSMPA program

Australia's commitment to develop the NRSMPA came at a time when the science of conservation planning was undergoing a renaissance. The early 1990s saw the advancement of a new theory of conservation planning, which emerged from widespread recognition of the failure of the terrestrial national parks estate to effectively conserve the full suite of biodiversity due largely to the opportunistic and piecemeal approach in which areas were being added to the reserve estate (Pressey *et al.* 1993). Analysis had shown that such *ad hoc* approaches were highly biased towards certain species/habitats, while failing to protect many others (Kirkpatrick 1983). Ecological theory was shifting away from single species approaches in favour of reserve networks, with the primary objective of nature conservation and the maintenance of biodiversity more broadly (Margules 1989; Pressey 1994; Margules and Pressey 2000).

The NRSMPA's inception therefore came at a time where there was only limited theory to guide the selection of areas for biodiversity conservation, but foreshadowed a new era in developing a theory of conservation planning in which Australian scientists and politicians were to play a leading role. Implementation initially focused on developing the NRSMPA policy and governance frameworks, together with the introduction of the Interim Marine and Coastal Regionalisation for Australia (IMCRA; later the Integrated Marine and Coastal Regionalisation of Australia), a spatial framework for classification of Australia's marine environment into bioregions at a scale useful for regional planning (Commonwealth of Australia 2006). Two types of bioregions were established by the IMCRA process: 41 provincial-scale regions covering all waters of the Exclusive Economic Zone, and 60 smaller meso-scale regions covering the mainland continental shelf (Commonwealth of Australia 2006). Significantly, IMCRA offers a practical framework for identifying the 'broad-scale' components of a representative system of MPAs, at the scale of both types of bioregions. It underpins the goal of the NRSMPA to include examples of all 'representative' ecosystem types within MPAs, as well as 'rare' ecosystems/habitat types, and other sites of biological and ecological importance, within each bioregion. The principles 'comprehensive, adequate and representative' (CAR), and the IMCRA bioregionalisation continue today to guide the implementation of protected area systems in all Australian jurisdictions (Department of the Environment 2014).

Marine bioregionalisations have been mapped, using a variety of approaches, for all Australian jurisdictions. Physical datasets used to inform bioregional classifications include: oceanographic data (temperature, salinity, currents, nutrients and upwelling zones); bathymetry; sediments and substrate geology; coastal geomorphological features; and climate (wind patterns, tides, wave patterns, rainfall, etc.). Biological datasets include: distribution of 'key species', fish and 'megafauna'; dominant and 'indicator' species for ecosystems; ecosystem/habitat maps (often delineated using aerial photography and remote sensed data in addition to survey data); and 'communities' or major 'assemblage' types. Numerous national and state-level organisations have contributed data to the bioregionalisation process in Australia (e.g. Muldoon 1995; IMCRA Technical Group 1998; Connell Wagner 2000).

Although bioregionalisation is a suitable technique for classifying and mapping large spatial scale units, it cannot realistically be used to define the smaller spatial scale variation in habitats and their component flora and fauna across space and time (Thackway and McRae 1995). Yet it is at the small spatial scale of habitats and species distributions that decisions about MPA placement and zoning are made. Recognition of the multiple scales at which planning and spatial information is required was a pivotal crossroad for the implementation of the NRSMPA and the theory of conservation science more broadly, as decisions on the selection and placement of MPAs locally were understood to have ramifications for the performance of the national MPA network as a whole. It became evident that knowledge of the finer scale patterns of marine biodiversity, together with information on the local social and economic activities, must be framed within a regional context if the objectives of the NRSMPA system as a whole were to be achieved.

The launch of Australia's Oceans Policy in 1998 (Commonwealth of Australia 1998a,b) further advanced the framework for delivery of the NRSMPA by committing to implement an integrated and ecosystems-based oceans planning and management system for all Australian jurisdictions (see also Chapter 19). This shifted the focus towards large-scale marine plans that would have carriage of the NRSMPA and facilitate the selection of MPA networks that were integrated with a range of other uses, so forming an integrated network targeting both local and regional-level protection. For this purpose, the information requirements to support MPA planning expanded beyond the biophysical elements to encompass social, cultural and economic data for the purposes of characterising the assets, values and threats in the region.

As the Oceans Policy began to take shape, it was recognised that there was significant resistance to implementing the policy in state waters, and the process contracted to mainly focus on the Commonwealth jurisdiction (see also Chapter 3). Each Marine Planning Region was the subject of intensive scientific studies to assemble and profile the available information about system attributes and provide a basis for management (see Box 16.1). Informed by activities within the Oceans Policy development process, the first large-scale bioregional approach to MPA system planning was carried out in the Great Barrier Reef (GBR) region, implemented by the Commonwealth's Great Barrier Reef Marine Park Authority (GBRMPA) under the mandate of its implementing Act (Chapter 5). This was conducted independently of the Oceans Policy but resulted in the extensive rezoning of ~33% of the GBR region into no-take marine sanctuaries (Fernandes *et al.* 2005; Chapter 5). Importantly, the GBR process was able to harness concurrent Australian advances in conservation planning theory and new spatial optimisation tools that allowed, for the first time, the selection of MPA networks to be framed as a systematic problem of decision analysis drawn from the science of operations research (Fernandes *et al.* 2005, 2009; Ball and Possingham 2000).

Below, we summarise experiences in two instances where Australian jurisdictions have applied (with varying success) the science together with systematic planning tools, for MPAs in their regions. We also outline the findings of an independent science-based design for MPAs in the Commonwealth's South-west Marine Region that proposed a network of marine sanctuaries across the region based on optimal trade-offs for specific uses.

Box 16.1. MPAs in the context of Regional and Bioregional Marine Plans

The planning process for each of the six Australian marine regions commenced with a Regional Marine Plan for the South-east Region, and subsequently, Marine Bioregional Plans for four of the five remaining marine regions (leaving out the Coral Sea Region) (Department of the Environment 2015). The *South-east Regional Marine Plan* focused on all marine waters of the region, but the subsequent bioregional plans focused on matters pertaining mainly to the Commonwealth waters, recognising interactions with the contiguous state waters where they occur. Both regional and bioregional plans included aspects of the design and implementation of MPAs, heavily supported by collection, analysis and synthesis of the available data about the natural systems of the relevant region. For each region, an unprecedented effort was devoted to collating the readily available data on species, habitats and environments of the Commonwealth waters. This involved review of empirical data from major university programs, museum data, CSIRO, Australian Institute of Marine Science and, in some instances, data from the private sector. In addition to collating and synthesising the biophysical knowledge base, relevant social and economic information was summarised for the mainland near the Commonwealth marine regions (e.g. Greiner *et al.* 2006; Clifton *et al.* 2007). The result of this was, for each region, the first systematic catalogue of broad-scale knowledge of Australia's mainland ocean waters (Department of the Environment 2015).

The biophysical knowledge base consists of inventories and broad distributions of many species, the geomorphological attributes, and physical ocean features. This knowledge was summarised into a series of bioregional profiles for each region, with an accompanying set of reports and datasets, many in unpublished and privileged client reports and databases for the Australian Government (e.g. McClatchie *et al.* 2006; Rochester *et al.* 2007). Typical products include systematic classification of geomorphic and ecological attributes (such as found in IMCRA V4; Commonwealth of Australia 2006). Although the biological data assembled and assessed included much of the readily available data, the species-level knowledge base was heavily focused on fished species and data derived from fishery-based programs, and on Commonwealth-listed (protected) species. The broader structure and function of the regional biological systems were largely described through the development of proxies derived from biota-environmental models and empirical relationships, informed to some extent by expert elicitation (such as Fig. 16.1).

For planning purposes and as a basis for assessing performance, the regional planning process also provides a 'report card' for each region, describing some of the natural values of each region and assigning qualitative estimates of 'pressures' that apply to those values (although there is no estimate of performance or condition). This includes a subjective and limited set of 'Key Ecological Features', established to represent a sample of a region's broad-scale natural values, with species-level knowledge (as above) derived heavily from fisheries programs (Commonwealth of Australia 2012). The risks of using surrogates for the purpose of regional planning (including MPA design) are well known (e.g. Williams *et al.* 2009), but they are nonetheless widely used as the only readily available form of knowledge. Similarly, the performance assessment of the regions are also problematic, based on model-dependent performance structures using indicators based on presumed effects of known pressures (including fishing) on the assets (Dambacher *et al.* 2012). While this is a data-rich approach, it is accompanied with a high risk of Type III error (Bark *et al.* 2013). In the context of assessing the contributions of MPAs to regional biodiversity conservation outcomes, such indicators – derived mainly from assumptions and models of existing threats and impacts – may not provide an accurate assessment of the achievements of MPA networks in the marine regions. This is because many pressures have impacts that have not been well studied, and unpredicted impacts (the surprises) and off-reserve pressures typically remain outside the performance evaluation structure.

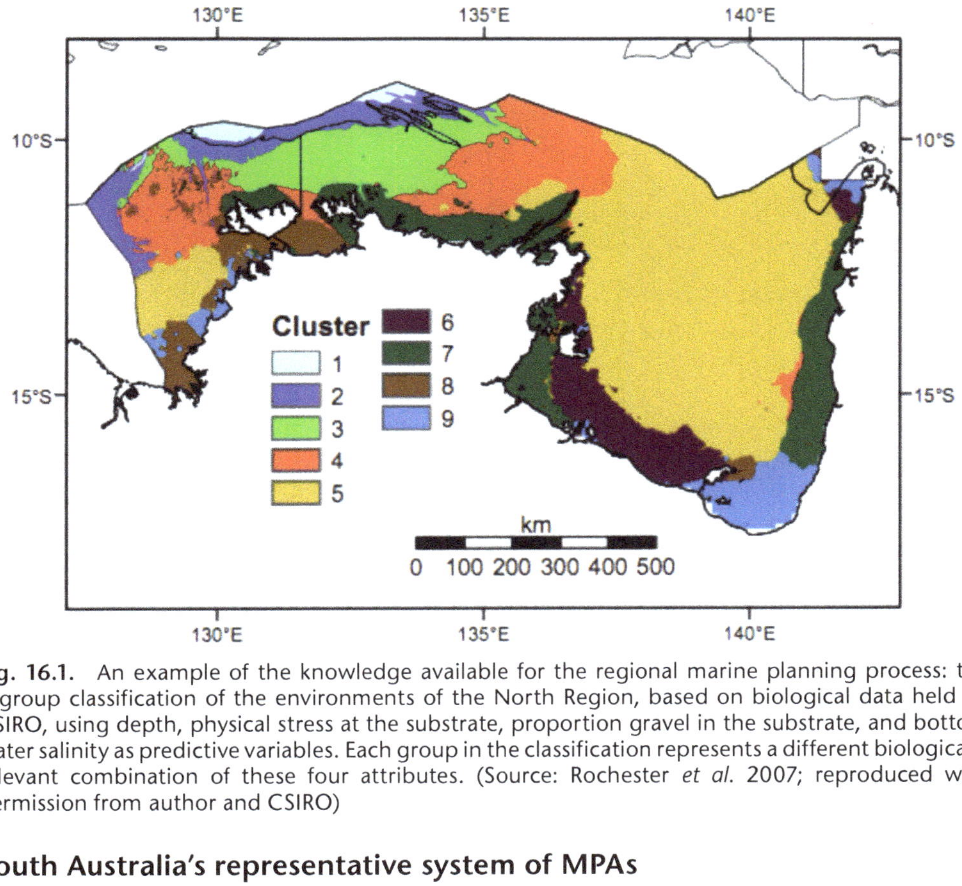

Fig. 16.1. An example of the knowledge available for the regional marine planning process: the 9-group classification of the environments of the North Region, based on biological data held by CSIRO, using depth, physical stress at the substrate, proportion gravel in the substrate, and bottom water salinity as predictive variables. Each group in the classification represents a different biologically relevant combination of these four attributes. (Source: Rochester *et al.* 2007; reproduced with permission from author and CSIRO)

South Australia's representative system of MPAs

In October 2014, the South Australian (SA) Government announced a MPA network comprising 19 new marine parks encompassing ~45% of the state waters, of which 6% was designated as no-take marine sanctuaries (Fig. 16.2) (see Chapter 8). This achievement resulted from more than 20 years of extensive research and planning across the state and followed a rigorous process underpinned by science.

The process commenced in the early 1990s, with an information-gathering phase involving a series of state-wide biophysical benthic surveys. These surveys were targeted towards habitat mapping, which then informed an ecosystem-based hierarchical classification at a scale useful for conservation planning. The improved understanding of the

Fig. 16.2. Rocky coastline on the northern coast of Kangaroo Island, Southern Spencer Gulf Marine Park, South Australia. (Photo: Kris Carlyon)

spatial pattern of biodiversity in the state's marine waters highlighted deficiencies within the existing SA MPA system and prompted a comprehensive policy and legislative overhaul. This resulted in the South Australian *Marine Parks Act 2007* that signalled a departure from the piecemeal, issue-based MPAs in favour of a scientifically principled and systematic planning framework, strongly focused on biodiversity conservation for the whole of SA waters as its primary objective. With the policy and legislative frameworks then in place, selection of the preliminary candidate MPA network was guided by best practice scientific principles together with world-class software (Marxan: Kirkman 2013) and clearly defined objectives and design criteria. Notably, these initial design criteria were ecological and biophysical, with social, cultural and economic values considered later. This two-stage process, as in the Commonwealth design process (below), was problematic and may have contributed to a lengthy and, arguably more protracted and contentious, public consultation phase than would have been needed if these other criteria had been incorporated together with the initial ecological objectives. It was not until public consultation that economic and social design constraints were formally considered and, at this second stage of the process, there was significant pressure to provide for specific user interests in places that would be affected by zoning, with a consequent erosion of the extent of achievement of the biodiversity conservation objectives (Kirkman and Shepherd 2015). Nonetheless, the design held widespread scientific grounding, and was further iteratively refined and developed over an extensive period of consultation with stakeholders, community and scientists, eventually maintaining biodiversity conservation as the primary design driver.

Conservation planning software was applied to audit how well the outer boundaries of the SA MPA network performed against the design criteria. Generally, the finally adopted network achieved appropriate representation targets, although with just 6% of the region represented within no-take marine sanctuary zones the SA system may not be able to achieve conservation objectives in SA waters, because effective no-take MPAs are likely to require minimum sizes and proportions (Edgar *et al.* 2014), and high levels of complementarity to meet CAR objectives.

The 2014 Commonwealth MPA network

Australia, as a signatory to the Convention on Biological Diversity, had agreed to previous protected area targets of 10% reservation of marine waters by 2012, and more recently the Aichi Target 11 (CBD 2010) of 10% in an 'ecologically and well-connected' MPA network by 2020. National policy also emphasised the need for a representative system with goals to ensure bioregional representation (including the need for high protection – IUCN category I and II reserves – in each bioregion), including different depth ranges, benthic/demersal biological systems, and types of seafloor (e.g. ANZECC TFMPA 1998; Commonwealth of Australia 2008). In 2004, the NRSMPA covered just 7% of Australia's marine jurisdiction and, although this expanded to nearly 9.5% of Australia's marine waters by 2008, much of this increased area of MPAs was neither on the continental shelf nor within the nearshore state waters, where much of the highly valued biodiversity is located. Furthermore, only 4.5% was protected in no-take marine sanctuaries (IUCN Category I or II).

The zonings for the MPAs designed within the planning process for Commonwealth waters were to have come into effect in 2014 (South-west, North-west, North, Temperate East, Coral Sea regions), but, at the time of writing, the promulgation of management zones in these MPAs has been suspended pending a full review (see Chapter 4). If implemented as designed, they were to have increased the MPA network to ~3.1 million km^2, which is 36.5% of the Commonwealth waters, with 13.3% proposed as no-take sanctuaries (IUCN I or II). Yet despite this intended significant increase in area, these initial zoning plans would still fail to fulfil the most fundamental CAR objective of representativeness – the capture at least 10% of each bioregion in a highly protected MPA (IUCN I or II). Even

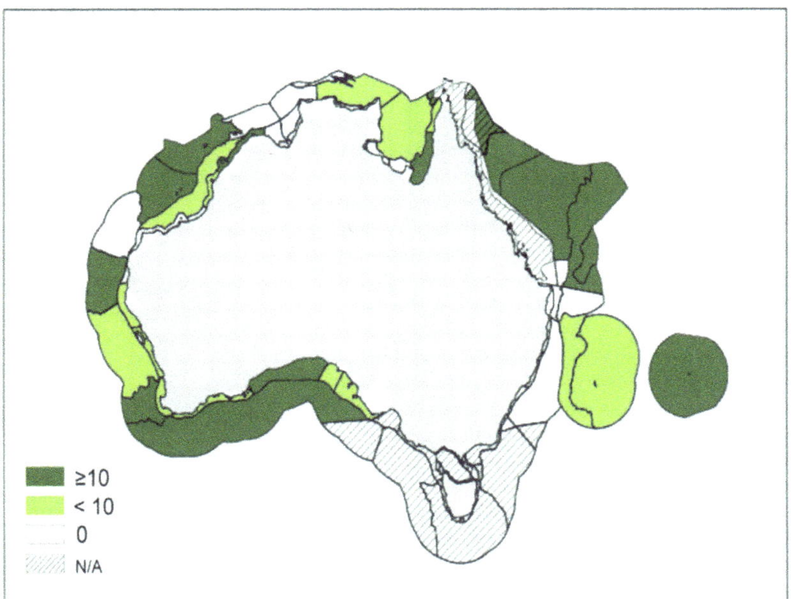

Fig. 16.3. Achievement reported as the proportion of no-take marine sanctuaries (IUCN I or II) in individual bioregions, for the proposed 2014 Commonwealth MPA network. NA = MPAs designed and implemented under earlier programs. (Source: McGowan and Possingham (2015); reproduced with permission of authors)

at the coarse scale of bioregions (Fig. 16.3), Australia has failed to achieve at least 10% of each marine bioregion in a no-take marine sanctuary (IUCN I or II) (McGowan and Possingham 2015) in nearly half (27 of the 58) bioregions over which the Regional Marine Zoning Plans apply (Barr and Possingham 2013; see also Chapter 17). The majority of unrepresented bioregions are the shelf areas in the Temperate East, North and North-west regions (McGowan and Possingham 2015). The no-take marine sanctuaries that are proposed occupy just 2.3% of the shelf area, compared with 16.4% of the deeper water off-shelf area (Devillers *et al.* 2015).

The 2014 Commonwealth MPA network design has also been labelled by Devillers *et al.* (2015) as 'residual', because of the predominance of offshore areas in the MPA network that are well away from most specific current threats to biodiversity. In all categories of protection, the MPAs occupy 16.9% of shelf area and 36.2% of non-shelf area. There is no doubt that the offshore areas have an important complement of biodiversity that will now be better protected when these MPAs have sanctuary zones that are agreed and declared. However, in contrast, the biodiversity in most urgent need of present-day protection is predominantly on the shelf and in shallow coastal waters, which are poorly represented in the overall network, and contain very small marine sanctuary areas (Devillers *et al.* 2015). This outcome appears to result from the lack of an effective application of a suitable decision framework to the complete problem of MPA design. In particular, the design process seems to suffer from a lack of assessment of conservation impact on the targets for MPA protection, and may also be the outcome of a two-stage design process that gave preference to ease of MPA establishment over achievement of conservation outcomes (Devillers *et al.* 2015).

The science of MPA decision making

Decision analysis is a branch of operations research used in many fields of human endeavour (Goodwin and Wright 2009) to cope with complex problems that have weakly defined attributes, uncertain quality of information and uncertain levels of risk. A decision-analysis framework is typically invoked to provide efficient and effective pathways for decision making that maintain a focus on achieving the intended (often multiple and competing) objectives within an uncertain policy and knowledge environment (Possingham 2001).

The basic problem of MPA design is to identify a complement or network of areas that fulfil a specified set of design criteria as completely as possible (Cocks and Baird 1989). Conceptually, there are constraints to achieving the specific design criteria, and correctly expressing these constraints is central to the problem statement. They should reflect limitations to achieving the objectives or goals, whether that is the number or coverage of the areas or, say, the cost of the MPA network. As a result, efficiency and complementarity are key principles underpinning conservation planning (Pressey and Nicholls 1989; Bedward *et al.* 1992).

Designing efficient MPA networks is best achieved by selecting combinations of areas that complement each other well. Doing so contributes to the goal of comprehensiveness in the most efficient way. There are several different ways in which the selection of areas may proceed, but in modern designs, systematic conservation planning tools are commonly used to provide an efficient minimum-area solution to a defined MPA design problem (Pressey 1994; Pressey and Tully 1994). When applied within a decision-theory framework (Table 16.1), these tools can efficiently support informed decision making that accounts for social, economic and conservation objectives, and can robustly explore different assumptions and scenarios that reflect a range of uncertainties, including present and projected competing uses.

Table 16.1. A typical generic recursive decision-analytic model, with examples of a MPA application, derived from principles of decision theory/analysis

Generic decision analysis flow	Example of MPA-specific action
MPA design phase	
Identify and bound the decision situation and understand the objectives	Determine the spatial boundaries for the decisions; define assets/values and their target occurrence within the MPA system; determine the types of user activities where consistent with conservation and their interactions with other resource users to define the constraints of the MPA problem
Identify alternatives	Choose alternative sets of MPAs that may be consistent with the objectives
Deconstruct and model the problem	Explore the costs of different arrangements of MPAs and different zones of use, in terms of biodiversity conserved; iterate alternatives steps
Choose the best alternative	Assess/report on performance of alternative MPA designs in achieving conservation objectives and in minimising trade-offs.
Conduct sensitivity analysis	Test the choice against risks and uncertainties, including with stakeholders
Iterate if needed	Revisit the cost sensitivity and trade-off analyses as needed
Combine options if needed, iterate	Revisit the candidate arrangements
MPA declaration and management phase	
Implement chosen option	Legislate, declare
Monitor and report on indicators	Evaluate success and progress using unbiased indicators testing achievement of objectives
Iterate/revise as needed	Revisit final set arrangements, adapt/revise within agreed schedule/timelines

In typical systematic planning implementation, a decision-theory approach is used to frame the selection problem at each step, to examine alternative choices, and to support decisions about trade-offs that may need to be considered. Systematic conservation planning software complements this approach by providing the opportunity for low-risk exploration of the costs and benefits of various options, and examines different types of trade-offs that might need to be further considered to ensure that biodiversity goals are achieved at the minimum possible cost to users. Used correctly, such software also provides a gap analysis (known as '*post-hoc* accounting') by making explicit the type and extent of biodiversity that will not be represented in the existing or proposed MPA network (highly protected or otherwise). In this role, decision-analytic science provides a basis for considering alternative proposals from different users by evaluating the likely impact of their proposal, the interactions and acceptability. Each alternative can be benchmarked against the outcomes for biodiversity protection within the MPA network, the effectiveness of different types of management zones in relation to specific assets and values to be protected, and the type and cost-effectiveness of performance assessment structures required to monitor the efficiency and effectiveness of the MPA once established.

Twenty-five years after the inception of the NRSMPA, best scientific practice is now accepted as structuring the problem of MPA selection within a decision-theory framework, including biophysical, social and economic objectives, as pioneered by GBRMPA and the Australian science community for the GBR region (Fernandes *et al.* 2005, 2009). The challenge for conservation planners is to properly articulate the design constraints so that conservation outcomes can be achieved with minimal trade-offs to existing users. Plans for MPAs developed in this way can also inform management arrangements that will continue to meet the needs of intersectoral users and issues, as well as provide a basis for effective ongoing MPA management.

Using the decision sciences to support MPA design: an independent example for the South-west Region

To help guide and inform the Commonwealth's Regional Marine Planning program, an independent analysis of the Commonwealth's South-west Marine Region (SWMR) was performed to determine a candidate set of no-take marine sanctuaries for the region (Fig. 16.4). The analysis examined the extent of interaction between competing uses for the same areas (Fig. 16.5), and considered spatial options for implementation of marine sanctuaries that provide for the continuation of a range of existing and likely future uses (Possingham *et al.* 2009). This analysis found that a network of marine sanctuaries could be efficiently designed for the region with small impacts on the existing uses. The study

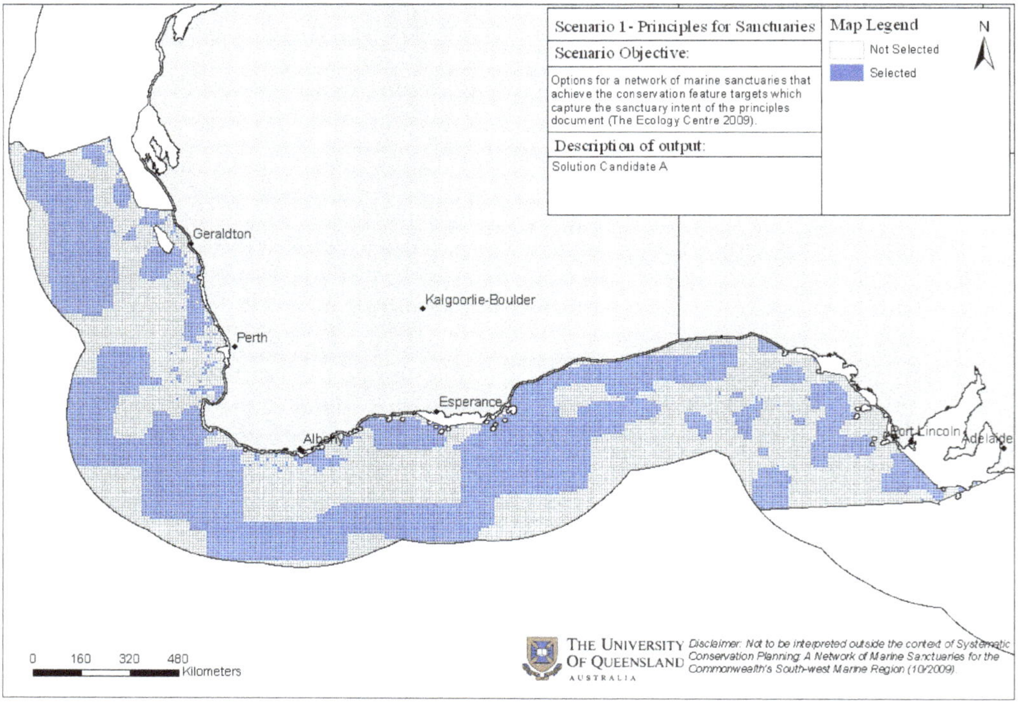

Fig. 16.4. A potential network of marine sanctuaries for the South-west Marine Region derived from a systematic analysis of available knowledge to provide high-level protection for a comprehensive and representative sample of 1712 biodiversity conservation features across the region. Areas shaded in blue have been located to efficiently minimise both the proportion of the region required for the marine sanctuaries network and the spatial displacement of 46 known user activities, while simultaneously fully meeting the conservation targets for biodiversity. The scenario represented here allocates costs equally to all users proportional to the area that their activities occupy (as represented by proxies for spatial 'cost' for each user type; see Fig. 16.5). Note that marine sanctuaries were the only MPA category modelled here: a multiple use set of MPAs would require a different arrangement of areas. (Source: Possingham *et al.* 2009; reproduced with permission of the authors)

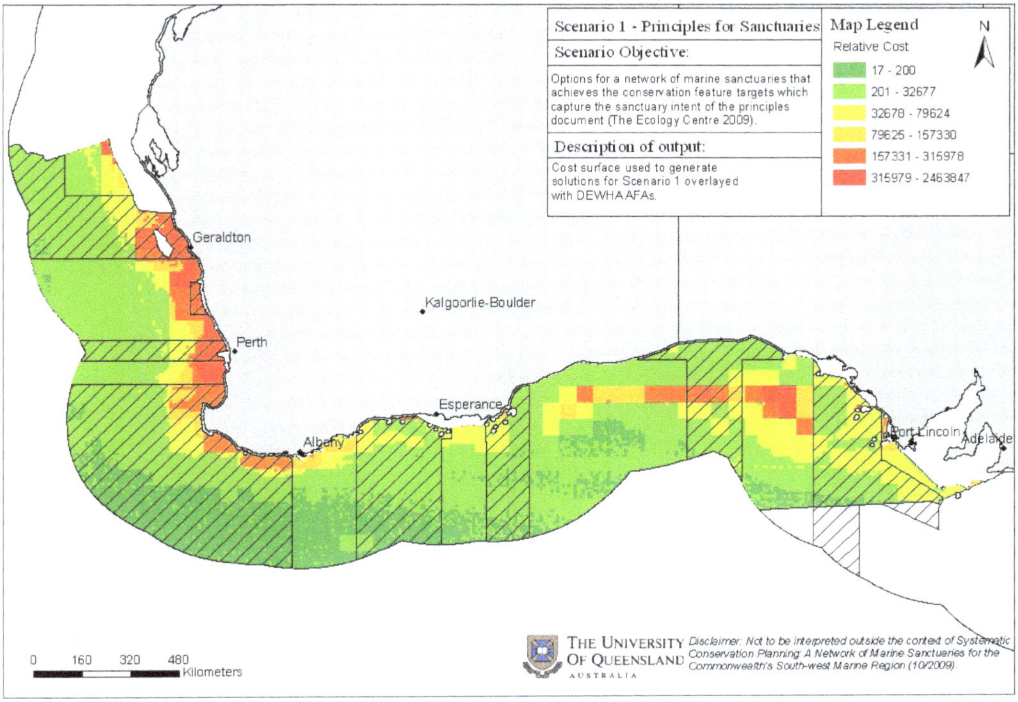

Fig. 16.5. The cost surface, representing a combination of user activities of different types and levels of resolution, used to minimise the effects of sanctuary location on the aggregate of uses across the region. The map represents the intensity of combined existing uses, increasing from lowest in the light green to highest in the red coloured areas, used to avoid (as far as possible) overlay of MPAs with existing uses. Individual uses are equally weighted in this cost surface. (Source: Possingham *et al.* 2009; reproduced with permission of the authors)

also showed that it is important to assign trade-offs explicitly in specifying the decision problem, because with only implicit weighting of the various current or putative uses (or deference to a second stage of stakeholder consultations) the outcomes for both MPAs and users can be affected and in a way that can be inequitable (Table 16.2). Using a systematic conservation planning framework, the principles of decision analysis could be implemented and the interests of users in specific areas examined for their impact on both other users and the achievement of biodiversity outcomes. The need for additional science input can then be made explicit (such as high-resolution data on a specific set of uses), consistent with the now standard approach in Australia (Fernandes *et al.* 2005; Kirkman 2013).

The analysis (using Marxan: Possingham *et al.* 2009) compared the cost (in terms of area lost to a use) that would be incurred by preferencing one user over the others, while fully achieving the biodiversity objectives (Table 16.2). As an example of comparative trade-off impacts, the analysis showed that if areas of value to the petroleum sector were given preference via a high weighting (i.e. to avoid the placement of MPAs in areas important to petroleum interests), then:

- a greater area of the region was needed to meet the biodiversity objectives in sanctuaries (an additional 7%)
- the impact of any resulting set of sanctuaries on the Western Rock Lobster (WRL) fishery would be increased by ~32% more than that imposed by an equal weighting of all uses that included petroleum.

Table 16.2. Impacts of 'user-preferencing' on other resource users in the design of marine sanctuaries for the Commonwealth South-west Marine Region

(Source: Possingham *et al.* 2009)

Model scenarios – achieving all conservation targets but minimising the impact of marine sanctuary declaration	Protection as no-take marine sanctuaries (proportion of the total region; %)	Trawl Fisheries (proportion of area of current activities affected; %)	Non-trawl Fisheries (proportion of area of current activities affected; %)	Recreational Fishing (proportion of area of current activities affected; %)	Western Rock Lobster Fishing (proportion of area of current activities affected; %)
1. Modelled base case: total area (km^2) of sanctuaries network (no user preferences applied)	50	18	25	32	28
2. Areas of value to petroleum sector excluded from marine sanctuaries network	57	41	32	43	37
3. Areas of value to recreational fishing preferentially excluded from marine sanctuaries network	53	41	40	18	39
4. Minimising the effect of the sanctuaries network on fisheries yield $ value (gross value product, fisheries production, all forms, 30 × 30 nautical mile grid)	51	33	27	32	26

Similarly, by avoiding placement of sanctuaries in areas where recreational fishing occurs, a slight (3%) increase in the size of the marine sanctuary network would be needed across the region to meet the biodiversity objectives. However, this would increase the cost to the WRL fishery from 28% to 39% of the area currently WRL-fished: an increase in total cost to the WRL fishery of ~40% of currently fished area.

In this study of the South-west, the modelled costs are upper estimates because of embedded heterogeneity in the input data, and the real costs to specific uses would likely have been much less if high-spatial resolution information had been available. Even so, the trade-off analyses highlighted specific areas where high-resolution data were needed, and where further investments in scientific analysis would likely yield more robust long-term outcomes for MPAs. Overall, this work demonstrates that a systematic decision-analytic approach can enable a high level of achievement of biodiversity conservation objectives to be accompanied with (potentially) only minor costs to users (in terms of area lost to a specific type of use). This form of decision-support structure can also provide the opportunity for assessment of the trade-offs between total area occupied and costs to specific users, the impacts of preferencing one user over other users, and the implications of alternative arrangements, all of which can explored as an integral part of the MPA design process.

Conclusions

1 *Australia's present and proposed MPAs are an important contribution to conservation, but are limited by a lack of decision-making science.*

Australia's network of MPAs makes an important contribution to biodiversity conservation in specific places across the jurisdiction. The knowledge-base developed for the Great Barrier Reef, Ningaloo Reef and the Commonwealth Marine Regions to underpin the systematic processes for design and management of MPAs is globally unparalleled in MPA practice at the national scale. However, although scientific knowledge and science-based tools have contributed significantly to the design and implementation of these and other MPAs in Australia, no amount of high-quality but narrowly derived or discipline-level science can substitute for a poorly structured decision process that allows issues of only peripheral importance to prevent achievement of good biodiversity conservation outcomes. Despite the wealth of world-wide evidence that, given appropriate design, highly protected MPAs are effective in providing biodiversity outcomes (Ballantine 2014; Edgar *et al.* 2014), and the exceptional track record and capacity in Australian science, MPAs that generate limited benefits for biodiversity continue to be declared in Australian waters because of a failure to fully apply a systematic decision-theory framework to the problem of MPA design and management.

2 *The two-stage process of MPA design truncates the role for the sciences and results in weaker MPA outcomes.*

The current approach favoured by many Australian governments is to adopt a two-stage process to MPA planning – an initial design phase that is primarily concerned with identification of outer boundaries of the MPA network, followed by a public consultation process that then leads to a detailed zoning plan with permitted uses. This two-stage process is, effectively, a decision structure that marginalises the role of science, constraining it mainly to the initial phase. This serves to weaken MPA biodiversity outcomes by allowing issues to be introduced in the second stage that may not have been well articulated in the initial design problem, and are subsequently considered outside the science-based framework. Without the framework provided by a systematic decision structure that is propagated into the second stage (and thence to management), vested interests can circumvent science-based designs and trade-off high-level conservation outcomes for specific resource-use benefits. Independent analyses have demonstrated that such benefits may be able to be largely realised, together with the targeted conservation outcomes, if they are correctly articulated and formulated as part of the initial design problem (Possingham *et al.* 2009). Yet a decision-analytic approach (Possingham 2001) is not yet being used for MPAs by Australian institutions, resulting in detrimental trade-offs to MPA conservation objectives and an inefficient use of limited public and private resources. This raises questions about the broader motivations and capacity of governments to implement long-standing and science-supported national policy and capacity relating to MPAs.

3 *Australia has the skills, experience and capacity to deploy the full range of sciences to improve MPA design and management.*
The GBRMP is singularly the most successful large-scale zoned MPA derived from a spatial planning framework and is universally recognised as setting the MPA global benchmark for multiple-use forms of MPA. The GBR design process, based heavily on science, demonstrated the ability to deliver good conservation outcomes with minimised trade-offs when explicit design targets are defined and carried throughout all phases of a MPA design program. In addition to design, performance assessment of MPAs is problematic without design criteria and quantitative objectives that are embedded in a fully scoped decision-analysis framework. A key benefit of designing a MPA network within a systematic decision-analysis framework is that a robust basis for quantitative performance assessment is also provided, including the capacity to evaluate off-reserve pressures. In the GBR network, for example, key conservation benefits continue to be assessed and evaluated against objectives for the network (Emslie *et al.* 2015). The systematic and science-based approach, as developed in the design and implementation of MPAs for the GBR, and the associated science capacity, offers an important guide to how to design well-structured MPA networks that provide for biodiversity conservation and for retention of substantial components of user requirements. Designing and implementing a system of

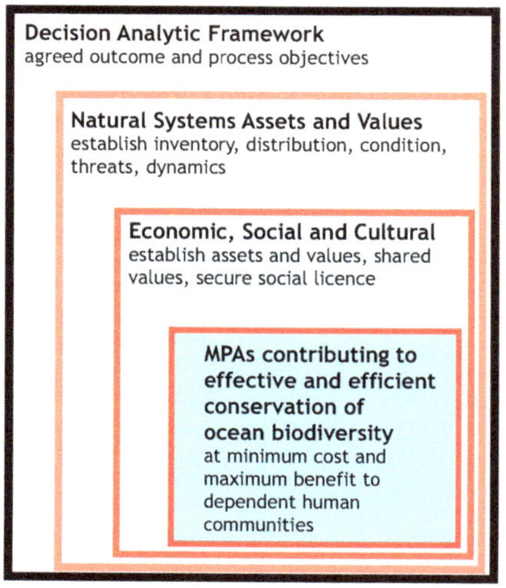

Fig. 16.6. Conceptual model for the relationship of the sciences in MPA design and implementation that will more confidently provide for effective and efficient networks of MPAs for Australia.

MPAs that only weakly considers national objectives for biodiversity conservation is neither efficient nor effective. It is likely to result in a greater burden on the public and industry purse in the future as repeated revisions of the MPAs and development of other mechanisms will be needed to better conserve species and ecosystems, albeit in future terms of government. Future deployment of the full breadth of Australia's science capacity for the purposes of reviewing and refining the design of the national MPA network will ensure that excessive future costs are avoided, and provide for more effective MPA outcomes.

4 *Future MPA networks will need to include explicit and substantive roles for decision analysis and social science.*
There has been a substantive focus on biophysical sciences in establishing Australian MPAs, and this has provided a strong basis for identifying the natural values that are in need of protection. However, the last two decades of experience in the design and implementation of MPAs at national and state level has shown that the design and delivery of such broad-reaching regulatory reform as MPA networks requires a much broader science base than has typically been applied. In particular, the science of decision analysis should frame the overall process (as discussed above), and tools from the social sciences dealing with the issues of cultural, social and economic values should be included as an integral component of MPA design. We conceptualise this as an 'integrated role for the decision, biophysical and social

Box 16.2. Securing the social licence for MPAs

Securing public trust and confidence (the 'social licence') to implement the public policy and regulatory reform required for a comprehensive, adequate and representative system of MPAs in Australia's waters is paramount. Achieving a network of MPAs that is likely to achieve objectives in nature conservation (as opposed to securing a set of boundaries around waters and seabed that may or may not contribute to achieving conservation outcomes) constitutes a major program of regulatory reform. Establishing an inventory of the structure and function of natural systems, and the associated values and pressures acting to degrade these attributes, will not, of itself, necessarily lead to public acceptance of either the need for or the cost of a set of such a set of reforms.

The challenge directly addressed by elements of social science is to establish the need and to map out a pathway for developing shared values and higher levels of trust in the institutions and actors engaged in such major policy programs. This challenge is identical to that of major business sectors, and there are established norms for how to approach the question of establishing and securing the 'social licence' among the major actors for such reforms (Sapp *et al.* 2009).

The design and effective implementation of MPAs depends on forms of science that span a range of disciplines. Beyond the natural systems inventory stage, tools from the social sciences (including political sciences) are needed, with specific attention to issues of developing and maintaining public trust and confidence in such major regulatory reform programs. This is especially important in areas where land and seabed title is held by Indigenous owners, and where rights to use of resources are already allocated for other purposes. Establishing the inventory of structure, function, connectivity and biophysical values of the natural systems is an essential but insufficient role for the sciences in MPA initiatives, and the role of natural sciences needs to be supported by tools from social science, set within an effective decision-analytic framework and decision-analysis tools.

sciences' to support the whole process of MPA design, implementation and management to simultaneously achieve objectives of conservation and the 'social licence' for the regulatory reform that constitutes a network of MPAs (Box 16.2, Fig. 16.6). Without such a broad science-based framework spanning at least these three major science disciplines, and supported with appropriately designed performance assessment systems, it seems that 'effective' (achieving their goals and objectives) and 'efficient' (designed and implemented in a cost-effective manner) MPAs will remain an illusion, and will limit Australia's capacity to achieve its national and international objectives in ocean conservation for many years.

Biographies

Trevor Ward is a marine ecologist, specialising in performance and risk-assessment systems for marine biodiversity and ecosystems. He is internationally recognised and has published more than 140 technical publications in international journals, books and reports, including two university texts on ecolabelling in marine fisheries. His career includes 20 years with CSIRO, and in 1996 he was jointly awarded the CSIRO Chairman's Medal for Excellence in Science. He is an accredited auditor of wild catch fisheries for the Marine Stewardship Council (MSC) global program, led the marine component of Australia's 2011 State of the Environment report, and is a former member of Western Australia's Marine Parks and Reserves Authority.

Rom Stewart is a marine scientist and conservation planning expert with broad experience across academic, government and private sectors. Her interests are in applying the decision sciences to better support conservation outcomes. She has published widely in the field of marine conservation planning, with her particular research interests focusing on the development of decision-support tools with a reach to real-world conservation problems. Specific research areas include incorporating multiple objectives and socio-economic factors to support efficient conservation decisions that strike a balance between sustainable use and conservation.

References

ANZECC TFMPA (1998) *Guidelines for Establishing the National Representative System of Marine Protected Areas.* Australian and New Zealand Environment and Conservation Council, Task Force on Marine Protected Areas. Environment Australia, Canberra.

Ball IR, Possingham HP (2000) *MARXAN (V1.8.2): Marine Reserve Design Using Spatially Explicit Annealing, a Manual.* University of Queensland, Brisbane.

Ballantine B (2014) Fifty years on: lessons from marine reserves in New Zealand and principles for a worldwide network. *Biological Conservation* **176**, 297–307. doi:10.1016/j.biocon.2014.01.014.

Bark RH, Peeters LJM, Lester RE, Pollino CA, Crossman ND, Kandulu JM (2013) Understanding the sources of uncertainty to reduce the risks of undesirable outcomes in large-scale freshwater ecosystem restoration projects: an example from the Murray–Darling Basin, Australia. *Environmental Science & Policy* **33**, 97–108. doi:10.1016/j.envsci.2013.04.010.

Barr LM, Possingham HP (2013) Are outcomes matching policy commitments in Australian marine conservation planning? *Marine Policy* **42**, 39–48. doi:10.1016/j.marpol.2013.01.012.

Barrett NS, Edgar GJ, Buxton CD, Haddon M (2007) Changes in fish assemblages following 10 years of protection in Tasmanian marine protected areas. *Journal of Experimental Marine Biology and Ecology* **345**, 141–157. doi:10.1016/j.jembe.2007.02.007.

Bedward M, Pressey RL, Keith DA (1992) A new approach for selecting fully representative reserve networks: addressing efficiency, reserve design and land suitability with an iterative analysis. *Biological Conservation* **62**, 115–125. doi:10.1016/0006-3207(92)90932-D.

CBD (2010) *Convention on Biological Diversity. 'Aichi Biodiversity Targets'.* CBD, Montreal, <http://www.cbd.int/sp/targets>.

Clifton J, Olejnik M, Boruff B, Tonts M (2007) 'Patterns of future development in the South-west Marine Region'. Report for Australian Government Department of the Environment, Water, Heritage and the Arts. Institute for Regional Development, School of Earth and Geographical Sciences, University of Western Australia, Perth.

Cocks KD, Baird IA (1989) Using mathematical programming to address the multiple reserve selection problem: an example from the Eyre Peninsula, South Australia. *Biological Conservation* **49**, 113–130. doi:10.1016/0006-3207(89)90083-9.

Commonwealth of Australia (1998a) *Australia's Oceans Policy Vol 1: Caring, Understanding, Using Wisely.* Commonwealth of Australia, Canberra.

Commonwealth of Australia (1998b) *Australia's Oceans Policy Vol 2: Specific Sectoral Measures.* Commonwealth of Australia, Canberra.

Commonwealth of Australia (2006) *A Guide to the Integrated Marine and Coastal Regionalisation of Australia. Version 4.0.* Department of the Environment and Heritage, Canberra.

Commonwealth of Australia (2008) *Goals and principles for the establishment of the National Representative System of Marine Protected Areas in Commonwealth waters.* Australian Government, Canberra.

Commonwealth of Australia (2012) *Commonwealth marine environment report card. Supporting the marine bioregional plan for the North-west Marine Region.* Department of Sustainability, Environment, Water, Population and Communities, Canberra, <http://www.environment.gov.au/system/files/pages/1670366b-988b-4201-94a1-1f29175a4d65/files/north-west-report-card-commonwealth.pdf>.

Connell Wagner (2000) 'NRSMPA strategic plan of action: status of benthic habitat mapping (Action 6 – Ecosystem Mapping)'. A report to the ANZECC Task Force on Marine Protected Areas. Connell Wagner Pty Ltd, Sydney, <http://www.environment.gov.au/system/files/resources/1bb7e9d4-adf3-465e-8f09-10b537f877f4/files/nrsmpa-action6.pdf>.

Dambacher JM, Hayes KR, Hosack GR, Lyne V, Clifford D, Dutra LXC, *et al.* (2012) 'Project Summary: National Marine Ecological Indicators'. A report prepared for the Australian Government Department of Sustainability, Environment, Water, Population and Communities. CSIRO Wealth from Oceans Flagship, Hobart.

Day J, Dudley N, Hockings M, Holmes G, Laffoley D, Stolton S, *et al.* (2012) *Guidelines for Applying the IUCN Protected Area Management Categories to Marine Protected Areas.* IUCN, Gland, Switzerland.

Department of the Environment (2014) *Commonwealth Marine Reserves – Background.* Department of the Environment, Canberra, <http://www.environment.gov.au/topics/marine/marine-reserves/overview/background>.

Department of the Environment (2015) *Marine Bioregional Plans.* Department of the Environment, Canberra, <http://www.environment.gov.au/marine/marine-bioregional-plans>.

Devillers R, Pressey B, Grech A, Kittinger J, Edgar G, Ward T, Watson R (2015) Reinventing residual reserves in the sea: are we favouring ease of establishment over need for protection? *Aquatic Conservation: Marine and Freshwater Ecosystems* **25**, 480–504. doi:10.1002/aqc.2445.

Edgar GJ, Stuart-Smith RD, Willis TJ, Kininmonth S, Baker SC, Banks S, *et al.* (2014) Global conservation outcomes depend on marine protected areas with five key features. *Nature* **506**, 216–220. doi:10.1038/nature13022.

Emslie MJ, Logan M, Williamson DH, Ayling AM, MacNeil MA, Ceccarelli D, *et al.* (2015) Expectations and outcomes of reserve network performance following re-zoning of the Great Barrier Reef Marine Park. *Current Biology* **25**, 983–992. doi:10.1016/j.cub.2015.01.073.

Fernandes L, Day J, Lewis A, Slegers S, Kerrigan B, Breen D, *et al.* (2005) Establishing representative no-take areas in the Great Barrier Reef: large-scale implementation of theory on marine protected areas. *Conservation Biology* **19**, 1733–1744. doi:10.1111/j.1523-1739.2005.00302.x.

Fernandes L, Day J, Kerrigan B, Breen D, De'ath G, Mapstone B, *et al.* (2009) A process to design a network of marine no-take areas: lessons from the Great Barrier Reef. *Ocean and Coastal Management* **52**, 439–447. doi:10.1016/j.ocecoaman.2009.06.004.

Goodwin P, Wright G (2009) *Decision Analysis for Management Judgment*. John Wiley & Sons, London.

Greiner R, Herr A, Larson S (2006) 'Literature review of relevant socio-economic information available for the Northern Planning Area'. Final Report, prepared for the National Oceans Office Branch, Department of the Environment and Heritage. CSIRO Sustainable Ecosystems, Townsville, <http://riverconsulting.com.au/reports/Northern-Planning-Area_Report.pdf>.

IMCRA Technical Group (1998) *Interim Marine and Coastal Regionalisation for Australia: An Ecosystem-based Classification for Marine and Coastal Environments. Version 3.3.* Environment Australia, Canberra.

IUCN World Parks Congress (2014) *A Strategy of Innovative Approaches and Recommendations to Enhance Implementation of Marine Conservation in the Next Decade.* International Union for Conservation of Nature, Gland, Switzerland. <http://www.worldparkscongress.org/downloads/approaches/ThemeM.pdf>.

Kirkman H (2013) Choosing boundaries to marine protected areas and zoning the MPAs for restricted use and management. *Ocean and Coastal Management* **81**, 38–48. doi:10.1016/j.ocecoaman.2012.07.017.

Kirkman H, Shepherd S (2015) Further efforts to protect biodiversity in coastal waters of South Australia. *Ocean and Coastal Management* **104**, 115–123. doi:10.1016/j.ocecoaman.2014.12.013.

Kirkpatrick JB (1983) An iterative method for establishing priorities for the selection of nature reserves: an example from Tasmania. *Biological Conservation* **25**, 127–134. doi:10.1016/0006-3207(83)90056-3.

Margules CR (1989) Introduction to some Australian developments in conservation evaluation. *Biological Conservation* **50**, 1–11. doi:10.1016/0006-3207(89)90002-5.

Margules CR, Pressey RL (2000) Systematic conservation planning. *Nature* **405**, 243–253. doi:10.1038/35012251.

McClatchie S, Middleton J, Pattiaratchi C, Currie D, Hendrick G (Eds) (2006) *The South-west Marine Region: Ecosystems and Key Species Groups.* Department of the Environment and Heritage, Canberra.

McGowan J, Possingham HP (2015) Submission to the Commonwealth Marine Reserves Review, <http://www.researchgate.net/publication/275584041_Submission_to_the_Commonwealth_Marine_Reserves_Review>.

McLeod KL, Lubchenco J, Palumbi SR, Rosenberg AA (2005) 'Scientific consensus statement on marine ecosystem-based management'. Signed by 221 academic scientists and policy experts with relevant expertise. Communication Partnership for Science and the Sea, Portland, OR, <http://www.compassonline.org/sites/all/files/document_files/EBM_Consensus_Statement_v12.pdf>.

Muldoon J (Ed.) (1995) *Towards a Marine Regionalisation for Australia. Proceedings of a Workshop Held in Sydney, New South Wales, 4–6 March 1994*. Ocean Rescue 2000 workshop series, Publication No. 1. Great Barrier Reef Marine Park Authority, Townsville.

Possingham H (2001) *The Business Of Biodiversity: Applying Decision Theory Principles To Nature Conservation*. Australian Conservation Foundation, Melbourne.

Possingham H, Ward TJ, Stewart R, Segan D, Kircher L (2009) 'Systematic conservation planning: a network of marine sanctuaries for the Commonwealth's South-west marine region'. Report prepared for Pew Environment Group, Uniquest Project No: 15621. University of Queensland, Brisbane, <http://www.pewtrusts.org/~/media/Imported-and-Legacy/uploadedfiles/peg/publications/report/Systematic20Conservation20Plannin-g20A20Network20of20Marine20Sanctuariespdf.pdf>.

Pressey RL (1994) Ad hoc reservations: forward or backward steps in developing representative reserve systems. *Conservation Biology* **8**, 662–668. doi:10.1046/j.1523-1739.1994.08030662.x.

Pressey RL, Nicholls AO (1989) Efficiency in conservation evaluation: scoring versus iterative approaches. *Biological Conservation* **50**, 199–218. doi:10.1016/0006-3207(89)90010-4.

Pressey RL, Tully SL (1994) The cost of ad hoc reservation – a case study in western New South Wales. *Australian Journal of Ecology* **19**, 375–384. doi:10.1111/j.1442-9993.1994.tb00503.x.

Pressey RL, Humphries CJ, Margules CR, Vane-Wright RI, Williams PH (1993) Beyond opportunism: key principles for systematic reserve selection. *Trends in Ecology & Evolution* **8**, 124–128. doi:10.1016/0169-5347(93)90023-I.

Rochester WA, Moeseneder CH, Miller MJ, Milton DA, Fry GC, Griffiths SP, *et al.* (2007) 'The north marine region marine bioregional plan: information and analysis for the regional profile'. Final report to the Australian Government Department of the Environment and Water Resources, CSIRO Marine and Atmospheric Research, Brisbane, <https://publications.csiro.au/rpr/download?pid=procite:196dd54f-636b-4058-ba2c-53f93d69c978&dsid=DS1>.

Sapp SG, Arnot C, Fallon J, Fleck T, Soorholtz D, Sutton-Vermeulen M, *et al.* (2009) Consumer trust in the U.S. food system: an examination of the recreancy theorem. *Rural Sociology* **74**, 525–545. doi:10.1526/003601109789863973.

SEC (State of the Environment 2011 Committee) (2011) *Australia State of the Environment 2011*. Independent report to the Australian Government Minister for Sustainability, Environment, Water, Population and Communities, Canberra, <http://www.environment.gov.au/topics/science-and-research/state-environment-reporting/soe-2011/soe-2011-report/6-marine>.

Thackway R, McRae D (1995) A national system of protected areas. *Conservation Australia* **1**, 7–9.

Williams A, Bax NJ, Kloser RJ (2009) Remarks on "Comment on: Williams *et al.* (2009) Australia's deep-water reserve network: implications of false homogeneity for classifying abiotic surrogates of biodiversity, ICES Journal of Marine Science, 66: 214–224" by Peter T. Harris, Andrew D. Heap, Tara J. Anderson, and Brendan Brooke. *ICES Journal of Marine Science* **66**, 2086–2088. doi:10.1093/icesjms/fsp212.

Chapter 17

Halfway there? Gaps and priorities in the development of Australia's marine protected area network

Daniel Beaver

Introduction

After over 20 years of concerted effort, Australians now find themselves with a tapestry of protection for marine life that spreads across its considerable oceans. Through each jurisdiction, and through the major environments from estuaries to shallow coastal waters to the deep ocean, there are some places that have been set aside for nature. This is a remarkable achievement of which the Australian community is rightfully proud (EMC 2012). Are there still gaps in this tapestry of protection? And if so, which gaps are most in need of filling? These are the two central questions of this chapter. The contents are a personal opinion based on over a decade's involvement in planning processes for the protection of marine life in Australia and Antarctica.

Australia's marine protected area (MPA) network contains both highly protected, no-take marine protected areas (herein termed 'marine sanctuaries') and a very wide variety of other marine protected areas (Fig. 17.1). Marine sanctuaries are places where nature has an opportunity to flourish free from direct human exploitation (i.e. prohibiting fishing, mining, etc.). Over a long period of time, scientists have been gathering evidence in Australia and all around the world that marine sanctuaries are critical to the protection of marine life (e.g. Lubchenco *et al.* 2003; Edgar *et al.* 2014). Australia has a range of other marine protected areas where the constraints on human exploitation vary considerably. Some restrict commercial trawling, such as the new Lalang-garram/Camden Sound Marine Park in the Kimberley that contains a large zone named 'special purpose zone (whale conservation)', but place no other restrictions on commercial or recreational fishing (Department of Parks and Wildlife 2013). Others have almost no restrictions, such as the general use zones of South Australia's marine parks where the only disallowed activity is littering (South Australian Government 2012). The benefits to marine life from these other marine protected areas vary and are currently difficult to predict (Shears *et al.* 2006; Alós and Arlinghaus 2013; Rife *et al.* 2013). This variability and unpredictability in effectiveness led a recent study of 87 protected areas worldwide to conclude that complete protection from human exploitation was one of five key factors for an effective marine protected area (Edgar *et al.* 2014).

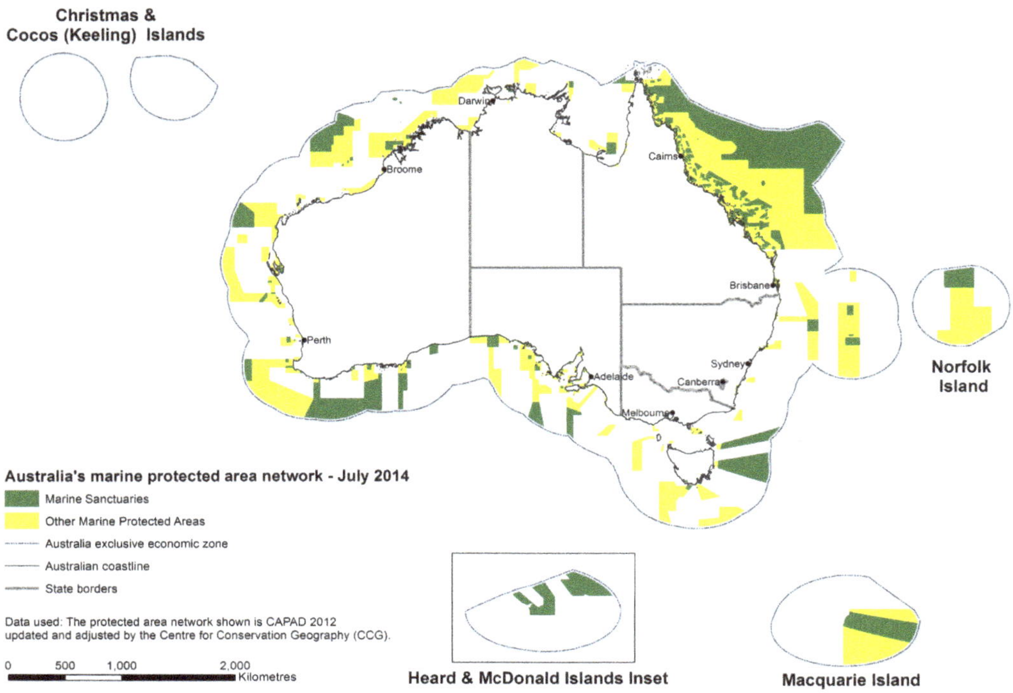

Fig. 17.1. Australia's marine sanctuaries and other marine protected area network in July 2014. Marine sanctuaries are highly protected, no-take MPAs or zones within MPAs.

Reservation levels in Australia's marine waters

Completely protected marine sanctuaries currently extend over 13% of Australia's oceans.[1] Is this enough? To help answer this question and others, Australia's scientific community has developed a guidance statement to inform the development of Australia's National Representative System of Marine Protected Areas (The Ecology Centre, University of Queensland 2009). This guidance statement is a non-binding scientific consensus statement produced by Australian scientists and planners involved in or studying the development of Australia's marine protected area network. In this guidance statement, the suggested minimum level for marine sanctuaries is 30% of each type of habitat within each of Australia's marine bioregions. This is the same standard as that recommended by the 2014 World Parks Congress (IUCN World Parks Congress 2014) and helps to interpret the primary goal of Australia's National Representative System of Marine Protected Areas (NRSMPA) established in 1998, which is:

> *'to establish and manage a comprehensive, adequate and representative system of MPAs to contribute to the long-term ecological viability of marine and estuarine systems, to maintain ecological processes and systems, and to protect Australia's biological diversity at all levels'.* (ANZECC TFMPA 1998, p. 5)

It also helps to interpret the 'highly protected areas' principle of the NRSMPA, which is the 'NRSMPA will aim to include some highly protected areas (IUCN Categories I and II) in each bioregion' (ANZECC TFMPA 1998, p. 6).

As a signatory to the Convention on Biological Diversity, Australia is also committed to helping to achieve a global target of 10% of the world's oceans protected by 2020 (Aichi Target 11; Convention on Biological Diversity 2010).

If Australia's oceans consisted of a single uniform habitat within one large bioregion, Australia's MPA network would be almost half way to a minimum target of 30%. Of course, this is not the case. Australia's ocean are vast, the third largest in the world, and include the world's greatest diversity of marine life (Costello *et al.* 2010). They currently contain 85 described bioregions and a myriad of distinct habitats (Commonwealth of Australia 2006; Douglass *et al.* 2014). These bioregions are the cornerstone of the Australian marine planning process (ANZECC TFMPA 1998).

The extent to which Australia's 85 marine bioregions are reserved within marine sanctuaries ranges from 0% to 60% (Fig. 17.2 and Appendix 1). However, even those bioregions with the highest levels of protection contain some major gaps. For example, the Cape Province bioregion, which straddles the Coral Sea and Great Barrier Reef Marine Parks, has Australia's highest level of marine sanctuaries (60%). However, although these marine sanctuaries provide very high levels of protection for some seafloor types (basins, canyons, valleys, small hills, plateaus, ridges and slopes), others are not protected at all (pinnacles, reefs, saddles, terraces and troughs) (Table 17.1). In this case, the habitats that have no protection in marine sanctuaries include coral reefs and related habitats.

A second example is the Kerguelen Province around Heard and McDonald Islands (see also Chapter 6). Currently, 17% of the bioregion is within marine sanctuaries, but while some seafloor types have high levels of protection within marine sanctuaries, others currently have none (Table 17.1). In the case of Heard and McDonald Islands, this

Table 17.1. Proportion of seafloor types *protected by* marine sanctuaries within the Cape Province and Kerguelen Province bioregions

(Data used is geomorphology of the seafloor described by Harris *et al.* (2005) for the Cape Province and by Douglass *et al.* (2014) for the Kerguelen Province and CAPAD 2012 updated and adjusted by the Centre for Conservation Geography)

Cape Province		Kerguelen Province	
Seafloor type	**Proportion within marine sanctuaries[a]**	**Seafloor type**	**Proportion within marine sanctuaries[b]**
Basins	100%	Deep ocean floor	21%
Canyons	59%	Banks/shoals	76%
Valleys	100%	Canyons	0%
Knolls	100%	Kerguelen Plateau	19%
Pinnacles	0%	Ridges	0%
Plateaus	94%	Seamounts	0%
Reefs	0%	Slopes	4%
Ridges	61%		
Reef saddles	0%		
Slopes	60%		
Terraces	0%		
Queensland Trough	0%		

[a]Almost all areas within the Cape Province outside marine sanctuaries are incorporated into partially protected zones of the Great Barrier Reef Marine Park and the Coral Sea Commonwealth Marine Reserve

[b]The Kerguelen Province contains only highly protected marine sanctuaries (there are no other types of MPAs within it)

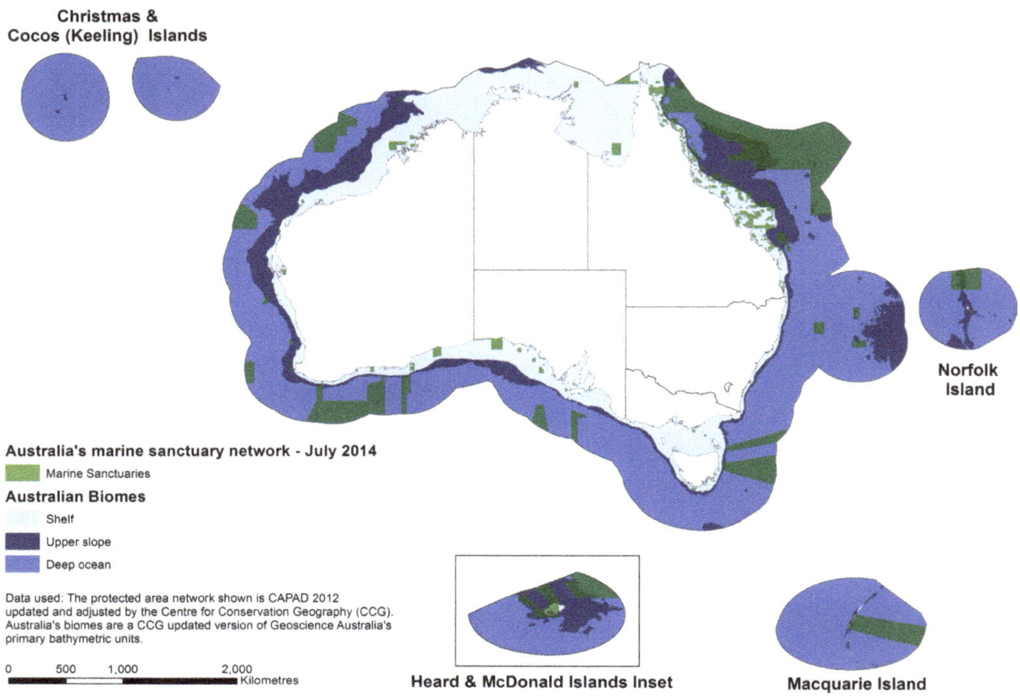

Fig. 17.2. Australia's shelf, upper slope and deep ocean biomes overlaid with Australia's marine sanctuary network as of July 2014.

is because planning was focused in the shallower waters of the plateau where fishing occurs (Meyer *et al.* 2000). At the time when the reserves were designed and implemented, very little was known about the deep ocean to the south where fishing currently does not occur. We now know a lot more about this area (Douglass *et al.* 2014) and it should be a target for new marine sanctuaries to incorporate these unique and unprotected habitats into Australia's NRSMPA.

These two examples are designed to illustrate a key point. Some of Australia's bioregions have extensive marine sanctuary networks but there are still gaps in these networks. This is true across all Australia's bioregions and jurisdictions, from shallow coastal waters to the deep ocean. Australia's oceans extend across the jurisdictions of eight governments (federal, state and territory) from the tropics, through temperate waters to the sub-Antarctic. Attempting to fill all the gaps in the network at once may not be feasible. Up to this point, the development of the marine sanctuary network has involved step changes in protection as different governments engage with planning new marine protected areas (including sanctuaries) at different times. In 2014, Australia has reached the point where the majority of its marine environment has been assessed in a marine protected area selection and designation process at some point, many of these planning processes within the last 10 years. There are also ongoing marine planning processes; for example, the Western Australian Government has established new marine parks at Camden Sound and Eighty Mile Beach and is proposing new marine parks at Horizontal Falls, Roebuck Bay, the North Kimberley and the Dampier Archipelago (see Chapter 7). All other things being equal, opportunities for new marine protected areas are more likely to exist in areas where planning has never occurred, such as Christmas and Cocos (Keeling) Islands, or where planning happened some time ago, such as Ningaloo or Macquarie Island, than in loca-

Table 17.2. Proportion of marine sanctuaries and marine protected areas within the jurisdiction of Australia's governments, July 2014

(Data used is CAPAD 2012 (DSEWPC 2012) updated and adjusted by the Centre for Conservation Geography and the Australian Maritime Boundaries dataset from Geoscience Australia)

		Proportion of jurisdictional waters protected as at July 2014 (%)	
Jurisdiction	**Area (km^2 rounded to nearest 000)**	**Marine sanctuaries**	**All marine protected areas (including marine sanctuaries)**
Commonwealth	8 642 000	13.7	35.9
New South Wales	9000	6.6	36.0
Northern Territory	72 000	1.0	4.1
Queensland	121 000	19.1	73.7
South Australia	60 000	5.3	44.2
Tasmania	23 000	4.3	5.7
Victoria	10 000	5.2	10.3
Western Australia	116 000	3.7	23.3
Total (all Australian waters)	**9 054 000**	**13.4**	**35.9**

tions where planning has occurred very recently, such as the Coral Sea. Possible exceptions include areas where planning occurred relatively recently but where the outcomes were particularly poor, such as the South-east, Temperate East, North and North-west Commonwealth planning regions.

Are there gaps in the comprehensiveness, adequacy and representativeness of the current network that are priorities to fill? For example, the current marine sanctuary network is biased towards tropical marine environments on the Australian east coast, with the 10 bioregions with the highest levels of protection all occurring within the Coral Sea and the Great Barrier Reef off Queensland (Fig. 17.3). Of the 17 bioregions with no marine sanctuaries, all but three occur off Tasmania, the Northern Territory and within the Kimberley in northern Western Australia. With 1% of coastal waters within marine sanctuaries, the Northern Territory has the lowest level of marine sanctuaries of any of Australia's governments (Table 17.2) (see Chapter 13). Although Tasmania appears to have about four times the protection at 4.3%, this conceals the fact that Tasmania has fully protected the ocean it controls around Macquarie Island, leaving the protection of coastal waters around the main island of Tasmania and its adjacent islands much lower at 1%. Western Australia also currently has a lower level of marine sanctuaries, but this is expected to increase in the short term with the implementation of the proposed new marine parks in the Kimberley (see Chapter 7). Queensland, including the Great Barrier Reef Marine Park under federal jurisdiction, although jointly managed, has Australia's best network of marine sanctuaries, with protection approaching 20% of coastal waters.

When considering broader classifications of Australia's oceans, the marine sanctuary network is also currently skewed towards deeper waters, with 13% of the upper slope and 16% of the deep ocean protected within marine sanctuaries, compared with 6% of the shallower waters of the continental shelf (Fig. 17.2, Table 17.3). The shelf is the submerged edge of the Australian continent and in most places extends down to around 200 m (Harris *et al.* 2005). The edge of the Australian continent is a large steep slope that extends down 4000 m or more. The upper slope is defined as that part of this slope that is at depths of 1500 m and shallower, in recognition of the fact that very little fishing or mining in Australia currently takes place at depths greater than 1500 m (Williams *et al.* 2009). The deep

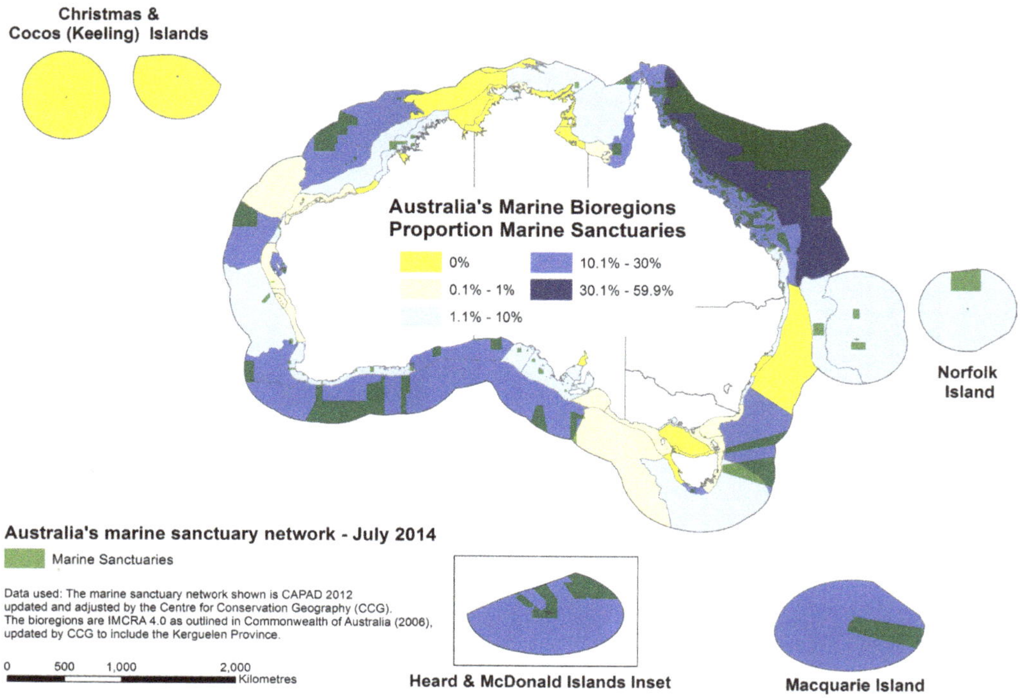

Fig. 17.3. The proportion of each Australian marine bioregion within marine sanctuaries.

Table 17.3. The proportion of marine sanctuaries and marine protected areas within the Australian shelf, upper slope and deep ocean in July 2014

(Data used is CAPAD 2012 (DSEWPC 2012) updated and adjusted by the Centre for Conservation Geography and Geoscience Australia's primary bathymetric units dataset updated by the Centre for Conservation Geography)

Biome	Area (km² rounded to nearest 1000)	Marine sanctuaries	All marine protected areas (including marine sanctuaries)
Shelf	1 975 000	6%	32%
Upper slope	1 386 000	13%	43%
Deep ocean	5 685 000	16%	35%

ocean at depths predominantly deeper than 1500 m contains much less exploitation than the shelf and upper slope. Almost all of Australia's upper slope and deep ocean environments are managed by the Australian Government, while the management of the shelf is split between eight Australian governments. For each of these broad-scale environments the remainder of this chapter focuses on outlining three of the top priorities for the establishment of new marine sanctuaries.

Priorities for the establishment of new marine sanctuaries

1. The shelf

The shelf is home to the highest diversity of marine life, but also some of the highest levels of exploitation (Williams *et al.* 2009). The competing human interests and uses for the waters on the shelf, and a much longer history of those uses, have made marine sanctuaries more difficult to establish on the shelf.

Three of the top priorities for new marine sanctuaries on the shelf are: (i) establishing a network of marine sanctuaries in the Kimberley; (ii) expanding existing marine sanctuaries on the shelf that are currently too small for long-term effectiveness; and (iii) parts of the shelf where marine sanctuary planning and declaration processes are yet to occur.

The Kimberley

The Kimberley is home to the world's largest calving ground for Humpback Whales *Megaptera novaeangliae*, contains some of Australia's most important seabird and shorebird habitats and is a hotspot for endangered species of shark and turtle (Morgan *et al.* 2011; Rogers *et al.* 2011). The Kimberley is a key component of Australia's tropical north: the most intact large tropical shelf ecosystem left in the world (Halpern *et al.* 2008). The Western Australian Government, recognising the importance of the area, is aiming to establish a network of marine sanctuaries to protect the marine life of the Kimberley (Barnett and Faragher 2010) (see also Chapter 7).

Expanding small shelf sanctuaries

The Australian shelf has high numbers of small sanctuaries (Fig. 17.4). Recent research into the effectiveness of marine protected areas has identified size as one of five key factors

Fig. 17.4. Merri Marine Sanctuary, located at the mouth of the Merri River in Warrnambool, Victoria, covers 25 ha and contains a deep canyon with a diverse range of marine life, from kelps and sponges to Giant Cuttlefish, Draughtboard Sharks, parrotfish, schools of Zebra Fish and Southern Rock Lobsters. (Photo: James Fitzsimons)

in determining marine protected area effectiveness (i.e. the larger the size, the more effective: Edgar *et al.* 2014). A high number of these small marine sanctuaries established on the Australian shelf could better protect marine life if they were larger. Expanding small marine sanctuaries into larger marine sanctuaries is likely to be a priority in the next stage of the development of Australia's network of marine sanctuaries on the shelf.

New planning frontiers

Tasmania and the Northern Territory are the only two Australian governments yet to initiate planning and establishment of marine sanctuaries on a large scale (see also Chapters 10, 13). The coastal waters of Tasmania and the Northern Territory are globally important for the establishment of marine sanctuaries: Tasmania for the uniqueness of its marine life (National Oceans Office 2004) and the Northern Territory for the relative intactness of its tropical coastal waters (Halpern *et al.* 2008). Currently only 1% of the coastal waters around the island of Tasmania and the Northern Territory are protected by marine sanctuaries. Consequently, the marine bioregions in these two parts of Australia have very low levels of marine sanctuaries. Of the 20 bioregions with the lowest levels of protection in Australia, more than half are found off the Northern Territory and Tasmania (Fig. 17.2).

Some other parts of Australia's coast are also yet to be examined. The most obvious examples include the south coast of Western Australia, the Torres Strait and Western Cape York off Queensland, the Hawkesbury Shelf bioregion off New South Wales (NSW), and the extension of marine sanctuaries into Commonwealth waters on the continental shelf of the South-east Region off Victoria, Tasmania (including Bass Strait) and South Australia.

2. The upper slope

Many of Australia's upper slope habitats are characterised by high levels of endemic marine life (Last *et al.* 2005) (species that occur nowhere else on Earth). Particular locations such as the Bonney Upwelling, Kangaroo Island Canyons and Perth Canyon support very large aggregations of marine life, including Australia's only known feeding grounds for the world's largest living creature, the Blue Whale *Balaenoptera musculus* (Commonwealth of Australia 2014). Protection of Australia's upper slope is currently unevenly distributed, with 96% of marine sanctuaries on the upper slope occurring within the bioregions associated with the Great Barrier Reef, Coral Sea, Norfolk Island, Macquarie Island and the Kerguelen Plateau. Each of these bioregions contains 13% or more of its upper slope habitats within marine sanctuaries. All other bioregions contain 6% or less of their upper slope habitats within marine sanctuaries, with more than half having no marine sanctuaries at all and are thus priorities for new marine sanctuaries. These bioregions stretch around the Australian continent from Fraser Island in south-east Queensland down around Tasmania, across the Great Australian Bight and all the way up the Western Australian coast to the Arafura Canyons in Australia's far north. There are very large gaps in the protection of the Australia's continental slope. For example, as you head south from the Great Barrier Reef, it is over 1800 km before you reach the marine national park zone of the Flinders Commonwealth Marine Reserve off the east coast of Tasmania. Likewise, heading west from Tasmania it is over 2000 km before you reach the marine national park zone of the Eastern Recherche Commonwealth Marine Reserve in the Commonwealth waters off Esperance, Western Australia.

Three of the top priorities for new marine sanctuaries on the upper slope are the Bonney Upwelling, the Kangaroo Island Canyons and the Arafura Canyons. Other priorities include the Lord Howe Rise, the Ningaloo Canyons, the Abrolhos Canyons, the Perth Canyon and the slope communities of Tasmania and Australia's east coast.

Bonney Upwelling

The Bonney Upwelling along the South Australian/Victorian coast is a seasonal emergence of nutrient-rich water that supports one of just three known Australian feeding grounds for endangered Blue Whales. In 2001, the area was identified by the CSIRO as one of Australia's most unique areas for marine life (Butler *et al.* 2002). As well as Blue Whales, the Bonney Upwelling's distinct and rich marine life, including cold water corals, sponges and bryozoans, is a feeding ground for aggregations of seabirds, fish, whales, seals and penguins (Butler *et al.* 2002). The Bonney Upwelling is located within the West Tasmanian Transition bioregion (0.7% marine sanctuaries). Currently, the Bonney Upwelling contains no marine sanctuaries and the West Tasmanian Transition bioregion contains no marine sanctuaries over any of its upper slope habitats.

Kangaroo Island Canyons

The Kangaroo Island Canyons, also a feeding ground for Blue Whales, are located to the south of Kangaroo Island at the eastern edge of Australia's largest marine bioregion, the Southern Province (20.2% marine sanctuaries). The Southern Province contains one of Australia's largest marine sanctuaries, Marine National Park Zone South within the Southwest Corner Commonwealth Marine Reserve, over the deep sea mountains of the Diamantina Fracture Zone. Unfortunately, the Southern Province's upper slope habitats are less well protected, with 1.1% being within marine sanctuaries. The Kangaroo Island Canyons, while identified as a key ecological feature for Australian marine life, currently contain no marine sanctuaries (Commonwealth of Australia 2012).

Arafura Canyons

Contained within the Timor Transition (0% marine sanctuaries) and Arafura (1.2% marine sanctuaries) bioregions, the Arafura Canyons are Australia's largest canyon system in tropical waters. The Arafura Canyons are isolated by distance from all other Australian canyon systems (the nearest canyons in Australian waters are 900 km away to the east on the Great Barrier Reef and a 1000 km away to the west at Ashmore Reef). In 2012 and 2013, they were identified by successive Australian governments as important for the protection of marine life (Commonwealth of Australia 2013) such that a multiple-use marine protected area, the Arafura Commonwealth Marine Reserve, has been declared over some of the canyons' area. Currently, no marine sanctuaries have been declared over the canyons.

3. The deep ocean

Deep oceans, defined as those areas of the ocean below 1500 m, comprise 62% of Australia's ocean estate, and include some of Australia's largest mountain ranges. With lower levels of human exploitation, the deep ocean contains some of Australia's most intact and little known marine environments. Three top priorities for the expansion of Australia's deep ocean marine sanctuaries are Christmas and Cocos (Keeling) Islands, Australia's sub-Antarctic deep oceans and the deep ocean east of NSW.

Christmas and Cocos (Keeling) Islands

Arguably the number one priority for establishing marine sanctuaries in Australia's deep oceans is in the waters surrounding Christmas and Cocos (Keeling) Islands (Hobbs 2014). The oceans around these two bioregions are highly unique (Brewer *et al.* 2009; Hobbs 2014). Both bioregions currently contain no marine sanctuaries and have never been subject to a marine protected area planning process.

Sub-Antarctic deep oceans

Another top priority is the deep ocean habitats around Australia's sub-Antarctic island territories of Macquarie Island and Heard and McDonald Islands. Although both of these bioregions have relatively high levels of protection – 12% for Macquarie Island and 17% for Heard and McDonald Islands – the marine sanctuaries for Macquarie Island were put in place over 10 years ago, while the deep ocean ecosystems south of Heard and McDonald Islands have so far been excluded from the planning process conducted by the Australian Antarctic Division (Meyer *et al.* 2000). Over the last 10 years, our knowledge of the ecology of these unique areas has greatly increased and the Australian Government, in collaboration with commercial fishers and environment groups, would now be in a good position to plan additional marine sanctuaries for these two bioregions (see also Chapter 6).

New South Wales to Norfolk Island

Of the 10 deep sea bioregions with the lowest levels of protection in Australia, four are found off NSW. Directly off the NSW coast, the Central Eastern Province contains no marine sanctuaries, while the Tasman Basin Province (5.2% in marine sanctuaries), the Lord Howe Province (2.2%) and Norfolk Island Province (9.7%) each contain at least one marine sanctuary. Collectively, these bioregions extend over some of Australia's most ecologically important deep sea habitats, including the Norfolk Island seamounts (Williams *et al.* 2006), the Lord Howe seamount chain and the Tasmantid seamount chain (Commonwealth of Australia 2012). These seamount habitats are currently poorly represented within marine sanctuaries. Currently only four of the 36 seamounts mapped in these bioregions are fully protected within marine sanctuaries.

Conclusions

The current tapestry of marine sanctuaries and other marine protected areas across Australia's seascape (Fig. 17.1) is the result of efforts by all eight jurisdictions, with input and influence from a broad range of the Australian community, including politicians, scientists, advocates, fishers, planners, divers, surfers and the extractives industries. Australia is the only country in the world able to boast such a diverse and extensive network of protection (see IUCN and UNEP-WCMC 2014). The fact that gaps in the network remain and that further work needs to be taken to fill those gaps does not detract from the achievements of the last 40 years, starting with the establishment of the Great Barrier Reef Marine Park in 1975. These achievements deserve to be celebrated and provide the foundation needed to take the next steps towards becoming the first country in the world to comprehensively protect its unique and abundant marine life.

Biography

Daniel Beaver is the director of the Centre for Conservation Geography and an adjunct research fellow at the Centre for Biodiversity and Conservation Science at the University of Queensland. With 15 years of experience in the theory and practice of systematic conservation planning, both on land and in the ocean, Daniel has been engaged in planning for marine protected areas in Australia since 2004.

Endnote

1 As at July 2014. Includes those Commonwealth MPA networks that are undergoing a review (see Chapter 4).

References

Alós J, Arlinghaus R (2013) Impact of partial marine protected areas on coastal fish communities exploited by recreational angling. *Fisheries Research* **137**, 88–96. doi:10.1016/j.fishres.2012.09.007.

ANZECC TFMPA (1998) *Guidelines for Establishing the National Representative System of Marine Protected Areas.* Australian and New Zealand Environment and Conservation Council, Task Force on Marine Protected Areas. Environment Australia, Canberra.

Barnett C, Faragher D (2010) 'Kimberley Wilderness Parks'. Media statement from the Premier, the Hon Colin Barnett MLA and Environment Minister, the Hon Donna Faragher MLC, 22 October 2010, <http://www.mediastatements.wa.gov.au/Pages/StatementDetails.aspx?StatId=3215&listName=StatementsBarnett>.

Brewer DT, Potter A, Skewes TD, Lyne V, Andersen J, Davies C, *et al.* (2009) *Conservation Values in Commonwealth Waters of the Christmas and Cocos (Keeling) Islands Remote Australian Territories.* CSIRO, Brisbane.

Butler A, Althaus F, Furlani D, Ridgeway K (2002) *Assessment of the Conservation Values of the Bonney Upwelling Area.* CSIRO Marine Research, Hobart, Tasmania.

Commonwealth of Australia (2006) *A Guide to the Integrated Marine and Coastal Regionalisation of Australia Version 4.0.* Department of the Environment and Heritage, Canberra, Australia.

Commonwealth of Australia (2012) *Marine Key Ecological Features.* Department of the Environment, Canberra, <http://www.environment.gov.au/fed/catalog/search/resource/details.page?uuid={DEC30942-9AD7-4B4E-AB3C-D8DF9F40F8AC}>.

Commonwealth of Australia (2013) *Environment Protection and Biodiversity Conservation (Commonwealth Marine Reserves) Proclamation 2013.* Commonwealth of Australia, Canberra, <http://www.comlaw.gov.au/Details/F2013L02108>.

Commonwealth of Australia (2014) *Biologically Important Areas of Regionally Significant Marine Species.* Commonwealth of Australia, Canberra, http://www.environment.gov.au/fed/catalog/search/resource/details.page?uuid={2ed86f5a-4598-4ae9-924f-ac821c701003}>.

Convention on Biological Diversity (2010) *Strategic Plan for Biodiversity 2011–2020, The Aichi Biodiversity Targets.* United National Environment Program, Nairobi, <https://www.cbd.int/sp/targets/>.

Costello MJ, Coll M, Danovaro R, Halpin P, Ojaveer H, Miloslavich P (2010) A census of marine biodiversity knowledge, resources, and future challenges. *PLoS ONE* **5**, e12110. doi:10.1371/journal.pone.0012110.

Department of Parks and Wildlife (2013) *Lalang-garram/ Camden Sound Marine Park Management Plan 2013–2023.* Department of Parks and Wildlife, Perth.

Douglass LL, Turner J, Grantham HS, Kaiser S, Constable A, Nicoll R, *et al.* (2014) A hierarchical classification of benthic biodiversity and assessment of protected areas in the Southern Ocean. *PLoS ONE* **9**, e100551. doi:10.1371/journal.pone.0100551.

DSEWPC (2012) *Collaborative Australian Protected Area Database 2012.* Department of Sustainability, Environment, Water, Population and Communities, Canberra, <https://www.environment.gov.au/land/nrs/science/capad/2012>.

Edgar GJ, Stuart-Smith RD, Willis TJ, Kininmonth S, Baker SC, Banks S, *et al.* (2014) Global conservation outcomes depend on marine protected areas with five key features. *Nature* **506**, 216–220. doi:10.1038/nature13022.

EMC (2012) *Approval of Marine Reserves, Essential Report.* Essential Media Communications, Melbourne, Australia, <http://essentialvision.com.au/approval-of-marine-reserves>.

Halpern BS, Walbridge S, Selkoe KA, Kappel CV, Micheli F, D'Agrosa C, *et al.* (2008) A global map of human impact on marine ecosystems. *Science* **319**, 948–952. doi:10.1126/science.1149345.

Harris P, Heap A, Passlow V, Sbaffi L, Fellows M, Porter-Smith R, *et al.* (2005) *Geomorphic Features of the Continental Margin of Australia.* Record 2003/30. Geoscience Australia, Canberra.

Hobbs JPA (2014) A glaring omission in Australia's marine conservation planning. *Marine Policy* **44**, 149–151. doi:10.1016/j.marpol.2013.08.020.

IUCN and UNEP-WCMC (2014) *The World Database on Protected Areas (WDPA).* UNEP-WCMC, Cambridge, UK, <www.protectedplanet.net>.

IUCN World Parks Congress (2014) *A Strategy of Innovative Approaches and Recommendations to Enhance Implementation of Marine Conservation in the Next Decade.* International Union for Conservation of Nature, Gland, Switzerland.

Last P, Lyne V, Yearsley G, Gledhill D, Gomon M, Rees T, *et al.* (2005) *Validation of National Demersal Fish Datasets for the Regionalisation of the Australian Continental Slope and Outer Shelf (>40m depth).* National Oceans Office, Hobart.

Lubchenco J, Palumbi SR, Gaines SD, Andelman S (2003) Plugging a hole in the ocean: the emerging science of marine reserves. *Ecological Applications* **13**, S3–S7. doi:10.1890/1051-0761(2003)013[0003:PAHITO]2.0.CO;2.

Meyer L, Constable A, Williams R (2000) *Conservation of Marine Habitats in the Region of Heard Island and McDonald Islands.* Australian Antarctic Division, Hobart.

Morgan DL, Whitty JM, Phillips NM, Thorburn DC, Chaplin JA, McAuley R (2011) North-western Australian as a hotspot for endangered elasmobranchs with particular reference to sawfishes and the northern river shark. *Journal of the Royal Society of Western Australia* **94**, 345–358.

National Oceans Office (2004) *South-east Regional Marine Plan: Implementing Australia's Oceans Policy in the South-east Marine Region.* National Oceans Office, Hobart.

Rife AN, Aburto-Oropeza O, Hastings PA, Erisman B, Ballantyne F, Wielgus J, *et al.* (2013) Long term effectiveness of a multi-use marine protected area on reef fish assemblages and fisheries landings *Journal of Environmental Management* **117**, 276–283. doi:10.1016/j.jenvman.2012.12.029.

Rogers DI, Hassell CJ, Boyle A, Gosbell K, Minton C, Rogers KG, *et al.* (2011) Shorebirds of the Kimberley coast – populations, key sites, trends and threats. *Journal of the Royal Society of Western Australia* **94**, 377–391.

Shears NT, Grace RV, Usmar NR, Kerr V, Babcock RC (2006) Long-term trends in lobster populations in a partially protected vs. no-take marine park. *Biological Conservation* **132**, 222–231. doi:10.1016/j.biocon.2006.04.001.

South Australian Government (2012) *Marine Parks (Zoning) Regulations 2012.* South Australian Government, Adelaide, <http://www.legislation.sa.gov.au/LZ/C/R/MARINE%20PARKS%20%28ZONING%29%20REGULATIONS%202012/CURRENT/2012.243.UN.PDF>.

The Ecology Centre, University of Queensland (2009) *Scientific Principles for Design of Marine Protected Areas in Australia: A Guidance Statement.* University of Queensland, Brisbane, Australia. <http://www.uq.edu.au/ecology/docs/Scientific_Principles_MPAs.pdf>.

Williams A, Althaus F, Furlani D (2006) *Assessment of the Conservation Values of the Norfolk Seamounts Area: A component of Commonwealth Marine Conservation Assessment Program 2002–2004. Report to the Australian Government Department of the Environment and Heritage.* CSIRO Marine and Atmospheric Research, Hobart.

Williams A, Bax NJ, Kloser RJ, Althaus F, Barker B, Keith G (2009) Australia's deep-water reserve network: implications of false homogeneity for classifying abiotic surrogates of biodiversity. *ICES Journal of Marine Science* **66**, 214–224. doi:10.1093/icesjms/fsn189.

Appendix 1: The level of marine sanctuaries and other marine protected areas within Australia's marine bioregions (as of July 2014)

Domain	Bioregion	Bioregional area (km^2)	Marine sanctuaries	All marine protected areas (including marine sanctuaries)
Off-shelf Provincial Bioregions	Cape Province	109 343	60%	97%
	Central Eastern Province	266 589	0%	6%
	Central Eastern Transition	67 149	13%	70%
	Central Western Province	268 460	1%	23%
	Central Western Transition	162 891	20%	47%
	Christmas Island Province	327 997	0%	0%
	Cocos (Keeling) Island Province	467 115	0%	0%
	Kenn Province	57 419	47%	100%
	Kenn Transition	377 132	44%	77%
	Kerguelen	417 817	17%	17%
	Lord Howe Province	485 348	2%	21%
	Macquarie Island Province	476 289	12%	34%
	Norfolk Island Province	430 793	10%	44%
	North-east Province	442 871	46%	100%
	North-east Transition	148 701	53%	100%
	North-west Province	178 651	0.4%	16%
	North-west Transition	184 424	19%	53%
	South-east Transition	241 910	25%	27%
	Southern Province	770 270	20%	40%
	South-west Transition	101 055	12%	49%
	Tasman Basin Province	156 416	5%	50%
	Tasmania Province	299 560	8%	31%
	Timor Province	217 092	13%	25%
	Timor Transition	24 090	0%	49%
	West Tasmania Transition	289 809	1%	12%
Shelf Mesoscale Bioregions	Abrolhos Islands	6628	1%	66%
	Anson Beagle	17 534	0%	16%
	Arafura	140 371	1%	14%

Domain	Bioregion	Bioregional area (km^2)	Marine sanctuaries	All marine protected areas (including marine sanctuaries)
Shelf Mesoscale Bioregions (continued)	Arnhem Wessel	22 698	0%	12%
	Batemans Shelf	6962	3%	15%
	Boags	8077	0%	3%
	Bonaparte Gulf	57 589	0%	2%
	Bruny	7278	0.1%	11%
	Cambridge-Bonaparte	10 523	0%	42%
	Canning	20 012	2%	37%
	Carpentaria	228 959	1%	4%
	Central Bass Strait	49 311	0%	1%
	Central Reef	31 389	26%	100%
	Central Victoria	4457	2%	2%
	Central West Coast	26 181	1%	24%
	Cobourg	8446	0.2%	16%
	Coorong	31 991	0.5%	22%
	Davey	6784	10%	32%
	East Cape York	14 428	29%	89%
	Eighty Mile Beach	9725	0%	96%
	Eucla	111 072	13%	27%
	Eyre	72 051	4%	45%
	Flinders	20 780	1%	13%
	Franklin	10 362	0%	1%
	Freycinet	7948	0.1%	14%
	Groote	16 587	0%	0%
	Hawkesbury Shelf	9026	0.03%	0.2%
	Karumba-Nassau	56 133	13%	26%
	Kimberley	59 652	8%	53%
	King Sound	4166	0%	0%
	Leeuwin-Naturaliste	26 547	3%	32%
	Lucinda-Mackay Coast	15 136	14%	99%

Domain	Bioregion	Bioregional area (km^2)	Marine sanctuaries	All marine protected areas (including marine sanctuaries)
Shelf Mesoscale Bioregions (continued)	Mackay-Capricorn	54 917	29%	100%
	Manning Shelf	9196	1%	23%
	Murat	35 475	3%	28%
	Ningaloo	9698	9%	68%
	North Spencer Gulf	4445	0.01%	37%
	North-west Shelf	153 987	2%	25%
	Oceanic Shoals	153 880	0%	48%
	Otway	37 272	0.4%	7%
	Pellew	21 268	0%	10%
	Pilbarra (nearshore)	14 568	1%	4%
	Pilbarra (offshore)	40 466	1%	22%
	Pompey-Swains	56 372	26%	100%
	Ribbons	47 805	30%	89%
	Shark Bay	13 919	11%	58%
	Shoalwater Coast	20 277	25%	99%
	Spencer Gulf	11 846	1%	23%
	St Vincent Gulf	12 827	3%	42%
	Tiwi	5102	0%	0.2%
	Torres Strait	35 627	13%	19%
	Tweed-Moreton	43 034	5%	35%
	Twofold Shelf	31 336	1%	4%
	Van Diemens Gulf	12 749	5%	10%
	Victorian Embayments	3147	3%	17%
	WA South Coast	47 225	5%	26%
	Wellesley	26 708	0.2%	27%
	West Cape York	21 845	3%	27%
	Wet Tropic Coast	5823	13%	100%
	Zuytdorp	36 597	1%	39%

PART 4

Differing perspectives on Australian marine protected areas

Previous page: Macquarie Island Nature Reserve. (Photo: Bob Zuur)

Chapter 18

Protecting sea country: Indigenous peoples and marine protected areas in Australia

Dermot Smyth and Miya Isherwood

Introduction and background to sea country

What is sea country?

Although the several hundred coastal and island Indigenous groups around Australia have different languages and their own unique belief systems, ceremonies and relationships with their local environments, they all regard the estuaries, beaches, bays, and marine areas (collectively known as 'sea country' or 'saltwater country') as integral parts of their traditional estates. Sea country is not additional to a clan estate on land: it is part of an integrated terrestrial and marine domain (Smyth 1997, 2001).

According to local cultures, sea country, as on land, contains evidence of the ancient mystical events by which all geographic features, animals, plants and people were created. Sea country contains sacred sites, often related to these creation events, and it contains tracks (or Songlines) along which mythological beings travelled during the creation period. The sea, like the land, is integral to the identity of each clan, and clan members have a kin relationship to the important marine animals, plants, tides and currents.

Today, most Indigenous people with marine clan estates live in coastal regions of the mainland and Tasmania, the larger islands off the Northern Territory coast, and islands in Torres Strait and Bass Strait. However, in the past, many Indigenous people also lived permanently or periodically on smaller offshore islands, particularly off the Queensland, Northern Territory and Kimberley coasts. Island dwellers were, and continue to be, particularly dependent on the subsistence resources of the sea and they maintain a strong interest in the management of large marine estates radiating out from their island homes.

Aboriginal and Torres Strait Islander people's relationship to their sea country brings with it a complexity of cultural rights and responsibilities, including the right to access, use and distribute resources, and the responsibility to manage those resources from generation to generation. Clan members are owners of their country, they belong to their country, they identify with their country and they are stewards of their country, including their sea country (Smyth 1994).

In the past, and continuing in many areas today, marine environments were managed through a variety of strategies and cultural practices, including:

- conduct of ceremonies, songs, dances, storytelling and other rituals with the purpose of nurturing the wellbeing of particular places, species and habitats
- control of entry into marine clan estates by outsiders – restricting resource use to clan members and others by agreement
- seasonal exploitation of particular marine resources
- the opening and closure of seasonal exploitation of particular resources marked by ecological events, such as the flowering of particular plants or the arrival of migratory birds
- restriction on the harvesting of particular species based on age, gender, reproductive conditions, health, fat content, and so on, of individual animals
- restrictions on resource use and distribution by clan members and others based on age, gender, initiation status, marital status and other factors
- restrictions on the use of particular animals and plants of totemic significance to individual clans – each clan usually identified closely with at least one natural element (usually animal or plant), the use of which was often highly restricted or prohibited
- prohibition of entry to certain areas on land and sea; entry and/or hunting and fishing in these areas was believed to cause severe storms or other forms of danger, not only to the intruders but also to other people in the region.

Together, these strategies and practices resulted in a system of marine protection, management and exploitation that was conservative and that enabled Indigenous societies to live within the carrying capacity of local environments.

The geographic extent of pre-colonial use of Australia's oceans by coastal Aboriginal groups varied through time and between regions. Aboriginal occupation of Australia extends at least 50 000 years, and possibly considerably longer (Hiscock 2008). During this time, sea levels have risen over 100 m, resulting in inundation of extensive areas of coastal lands previously occupied by Indigenous groups, particularly around northern Australia with a low gradient shoreline and extensive continental shelf. Land bridges between Australia and Papua New Guinea, and between Tasmania and the Australian mainland were also inundated by the rising seas. Following stabilisation of the sea level at its present height, ~6000 years ago, Indigenous patterns of marine use observed at the time of British colonisation began to be established. Around northern Australia, this included the use of tidal fish traps and extended sea voyages by canoe to exploit resources and manage clan sea country estates, in some places out of sight of the mainland. Off the Kimberley and north Queensland coasts, journeys to outlying reefs and islands could be achieved by stopping off at numerous islands along the way. Marine sacred sites have been recorded up to 80 km off the Northern Territory coast (Peterson and Rigsby 1998).

Before British colonisation in the late 1700s, Indigenous people had established a comprehensive network of sustainably managed and protected maritime estates, including culturally endorsed and enforced 'no-take zones', which had many of the characteristics of contemporary marine protected areas (MPAs) and which had been adapted to dramatic climate and sea level changes over millennia. The effectiveness of this network of marine estate management was enhanced by the integration of land and sea management (which is not often the case with modern MPAs) and the protection and management regime embraced the entirety of Australia's coastal marine environments (in comparison with the modest patchwork of MPAs today).

Recognition of Indigenous sea country rights and interests

The formal recognition of Indigenous people's rights and interests in sea country lags considerably behind recognition of similar rights and interests on land (NAILSMA 2012). This is because of fundamental different perspectives on human interactions with the sea between Indigenous cultures and the broader Australian community. Whereas Indigenous saltwater people regard the sea as an inseparable extension of coastal land estates and subject to the same characteristics of traditional ownership, custodianship, exclusive resources use and customary law as on land, the broader community views the marine environment as open commons, owned and managed by governments on behalf of all Australians. This view can be traced back to Roman jurist Marcianus in the second century AD, who stated that the sea and the shores are 'common to all men' (Cordell 1991; Allen 1994) and subsequently imported into Australia as part of the British colonial legal framework.

On the other hand, although a great many Aboriginal groups were displaced and prevented from accessing their traditional lands as a result of British colonisation and its aftermath, their relationship with the sea could often be maintained, and was sometimes encouraged by government policy. Many early Aboriginal missions and reserves were deliberately established on the coast to encourage self-sufficiency in food. Some of these communities became involved in commercial fisheries and other marine industries, such as the production of Dugong (*Dugong dugon*) oil, collection of pearl and trochus shell and harvesting of bêche-de-mer (or trepang) in northern Australia, and whaling in some southern areas (Smyth 1993).

The first land rights legislation in Australia, the *Aboriginal Land Rights (Northern Territory) Act 1976* (Cth), applies to the whole Northern Territory, including intertidal land, which in many parts of the Northern Territory extends up to 1 km or more from the high tide mark. In 2008, the High Court of Australia found that ownership of intertidal land under the Act includes the right to control access to tidal waters over Aboriginal-owned land between the high- and low-water marks (Brennan 2008).

In other jurisdictions, however, statutory land rights legislation does not extend to marine areas, including intertidal areas, thereby limiting Indigenous people's legal recognition of their cultural connections to sea country; this, in turn, limits their formal engagement in the governance and management of MPAs. However, fisheries legislation in all Australian jurisdictions does recognise, to greater and lesser effect, the existence of distinct Indigenous fisheries to support ongoing access to traditional marine resources by Indigenous people.

Following the Mabo High Court decision of 1992 (Sharp 1996), Indigenous people's customary rights to land and marine areas are now recognised as part of Australian common law in all jurisdictions. The *Native Title Act 1993* (Cth) provided a mechanism for statutory recognition of those rights, as well as procedures for determining their continued existence through native title claims, mediation with other interested parties by the National Native Title Tribunal, and formal determination by the Federal Court. However, the High Court has also determined that where other rights exist – such as the right of commercial fishers, the declaration of a MPA, the right of free passage by mariners, and the right of public access to beaches and the sea – native title rights and interests must 'yield' to those other legal rights. Native title rights and interests in sea country, therefore, are generally not as comprehensively recognised as on land and must coexist with the rights of others (Australian Human Rights Commission 2000).

Indigenous engagement in marine protected areas

Until ~40 years ago, all of Australia's protected areas were managed almost exclusively for their biodiversity and scenic values, with some recognition of archaeological values (and for marine areas, fisheries management purposes), but little regard to other Indigenous cultural values. Indigenous people were excluded from living in and using traditional resources within protected areas, and they played no part in managing these areas, which had been in their care for millennia. In this respect, protected areas were part of the broader colonial project that denied Indigenous Australians ownership of, cultural relationships with, and economic benefit from their traditional estates (Smyth 1995).

Since ~1975, in parallel with, and as a consequence of, the emergence of statutory Indigenous rights to land and subsequently the recognition of native title, various mechanisms have been developed for the involvement of Indigenous Australians in the management of protected areas, including the transfer of ownership of some national parks to Indigenous groups and the development of formal co-management arrangements (usually referred to in Australia as 'joint management') (Smyth 2001).

Indigenous engagement in the governance and management of MPAs is far less advanced than has occurred in terrestrial protected areas, reflecting the more limited legal recognition of Indigenous rights over marine environments as noted above. As a result, there are no MPAs for which Indigenous people have ownership rights equivalent to the co-managed national parks referred to above. Nevertheless, there are examples (summarised below) of Indigenous engagement in MPAs that demonstrate attempts to reconcile the contrasting Indigenous and non-Indigenous perspectives on sea country (Bauman *et al.* 2013).

Northern Territory

Gurig National Park, located 200 km north-east of Darwin in the Northern Territory, became Australia's first jointly managed terrestrial protected area in 1981, and subsequently became the first co-managed integrated terrestrial and marine protected area when the adjacent Coburg Marine Park was merged with the national park to become the 450 000 ha Garig Gunak Barlu National Park (see also Chapter 13). The key features of the joint management of Garig Gunak Barlu National Park are:

- declaration of the park under its own legislation by the Northern Territory Parliament (rather than existing protected area legislation)
- Aboriginal ownership of the terrestrial and intertidal components of the park (but not ownership of the subtidal component of the park)
- a management board comprising 8 members, of whom 4 are Aboriginal Traditional Owners and 4 are representatives of the Northern Territory Government; the Board is chaired by one of the Traditional Owner members who also has a casting vote
- the payment of an annual fee by the Northern Territory Government to Traditional Owners for use of their land as a National Park
- day-to-day management by the Parks and Wildlife Commission of the Northern Territory, including the training and employment of local Traditional Owners as rangers
- recognition of the rights of Traditional Owners to use and occupy the Park
- inclusion of terrestrial and marine areas in a single protected area – reflecting local Traditional Owners' holistic view of country.

Marine areas also form part of Kakadu National Park in the Northern Territory and Booderee National Park in Jervis Bay Territory (a small coastal Commonwealth territory excised from New South Wales – originally envisaged as a port for Australia's inland capital, Canberra), both of which are jointly managed by Parks Australia and local Aboriginal people under Commonwealth legislation. Once again, only the terrestrial components of the parks are Aboriginal-owned, while the joint management boards, on which local Aboriginal people are in a majority, have responsibility for both the terrestrial and marine components of the park.

Queensland

Australia's largest MPA, the Great Barrier Reef Marine Park (GBRMP) off the Queensland coast, was established during the 1970s (see Chapter 5), at a time when there was less recognition of Indigenous people's maritime interests than there is today. Provision was made to permit a continuation of Aboriginal traditional hunting, and efforts were made to document traditional knowledge of the marine park. However, there was no formal provision for recognition of the wider Indigenous interests associated with ownership, use, governance and management rights and responsibilities for the many clan estates that lie within the marine park.

Over recent years there has been a broadening of Aboriginal involvement in management of the GBRMP as a result of lobbying by coastal Indigenous groups, several research projects that documented Aboriginal maritime culture and management aspirations in the area of the marine park (e.g. Smyth 1994; Ross *et al.* 2005), and the prospect that continuing native title may exist in parts of the park.

Initially the Australian Government's *Great Barrier Reef Marine Park Act 1975* made no reference to Aboriginal interests in the marine park. Although subsequent changes fall short of full recognition of Indigenous ownership and management of Indigenous sea country estates, substantial efforts have been made by legislators, the Great Barrier Reef Marine Park Authority (GBRMPA) and by coastal Indigenous groups themselves to put in place measures that recognise the long-standing and continuing relationship saltwater Indigenous groups have with their sea country, while maintaining the GBRMP as a multiple use MPA managed by a government agency on behalf of all Australians. These measures now include:

- amendments to the *Great Barrier Reef Marine Park Act 1975* providing for Indigenous representation on the GBRMPA board and for the accreditation of Traditional Use of Marine Resource Agreements (TUMRAs) (see below)
- the establishment of an Indigenous Reef Advisory Committee comprising Traditional Owners and others with expertise in Indigenous sea country rights and interests
- appointment of Indigenous employees, some of whom are assigned to GBRMPA's dedicated Indigenous Partnerships Unit
- accreditation of, and financial support for the implementation of, TUMRAs, which are negotiated by and among Traditional Owner groups to determine how they will sustainably use and manage their traditional marine resources (e.g. Dugongs, marine turtles, fish) within their own sea country estates
- recognition and support for the dedication of sea country estates as Indigenous Protected Areas (IPAs) within the GBRMP (see below for further discussion on IPAs)
- compliance training for Indigenous sea country rangers located in coastal communities adjoining the GBRMP
- science and management workshops for Traditional Owners to share knowledge and understanding about sea country within the GBRMP
- funding to coastal Aboriginal groups to support their sea country planning and management initiatives, including documenting Indigenous knowledge of the

GBRMP, through GBRMPA's Reef Rescue Land and Sea Country Indigenous Partnerships Program
- special consultative procedures, such as community visits and newsletters, aimed at improving communication between GBRMPA and Indigenous groups
- assessment of potential impacts on Indigenous cultural values of any proposed development within the marine park, such as the construction of pontoons for tourists on islands and outer barrier reefs.

Indigenous Land Use Agreements (ILUAs) provide another mechanism to recognise Indigenous interests within the GBRMP. For example, an ILUA between the Kuuku Ya'u people on eastern Cape York Peninsula, the state of Queensland and the GBRMPA includes recognition of a previous commitment to train and authorise Kuuku Ya'u people as state marine park inspectors with certain compliance and enforcement powers, as well as in-kind support provided by the GBRMPA. The ILUA also provides for information exchange, a publication protocol and wildlife protection measures (ATNS 2011).

There are several other informal agreements between the Queensland and/or Commonwealth governments and Traditional Owner groups, including memoranda of understanding, which set out agreed practices for hunting and fishing, the use of marine resources, law enforcement training and the delegation of powers to manage aspects of MPAs.

A key aspect of these developments in Queensland is the recognition that particular groups of Traditional Owners have cultural associations with, as well as native title rights and interests for, particular sea country estates within the GBRMP. This contrasts with earlier more generic recognition of hunting and fishing rights for all Aboriginal people across the GBRMP.

Torres Strait

Although the Torres Strait region is part of the state of Queensland, it is dealt with separately here because of its unique characteristics: it is primarily a marine region, its population is predominantly Indigenous people of Melanesian origin, and its closest international neighbours are Papua New Guinea and Indonesia. There are ~247 islands in the Torres Strait, scattered over a geographic area of 4 800 000 ha, with 18 communities established on 17 inhabited islands, and two additional communities comprising mainly of Torres Strait Islanders located on northern Cape York Peninsula (known locally as Northern Peninsula Area communities).

The Torres Strait is renowned for its ecological complexity and biodiversity, providing a multitude of habitats and niches for the highly diverse Indo-Pacific marine flora and fauna, including Dugongs and marine turtles. The Torres Strait is also of enormous significance from an Indigenous cultural resource management perspective. Marine and island resources traditionally have been, and continue to be, vital to Torres Strait Islanders from a subsistence and cultural viewpoint. Torres Strait Islanders have a strong and abiding connection with their islands and sea country, governed by their unique *Ailan Kastom* (Island Custom).

These relationships are now legally recognised, with over 20 Registered Native Title Bodies Corporate (RNTBCs) established following native title consent determinations on all inhabited islands and several uninhabited islands throughout the region, as well as the recent High Court determination in the Regional Sea Claim (*Akiba on behalf of the Torres Strait Regional Seas Claim Group v Commonwealth of Australia* [2013] HCA 33 7 August 2013 B58/2012).

In 2001, a native title claim, known as the Torres Strait Regional Sea Claim, was lodged over most of the sea and seabed in Torres Strait. The High Court decision in 2013 in *Akiba* was a significant milestone for all Torres Strait Islanders in terms of their continued exercise of rights and responsibilities over sea country. The High Court, in upholding the decision of Justice Finn in 2010 in the Federal Court, recognised the long history of Torres Strait Islander people and their tradition of trading and using marine resources, and affirmed their ongoing customary rights to take any resources from the waters and use them according to traditional laws and customs, including for livelihood, community and commercial purposes. Importantly, the Court rejected the arguments of the state and Commonwealth that Torres Strait Islander people's traditional rights only covered small areas around the islands, and found the rights covered a continuous area between the maritime estates and shared areas of the communities. The Court also recognised that different parts of the sea were owned by the people of different island communities, and some parts were shared between multiple communities. Many of these findings were underpinned by the Court's recognition of the traditional and cultural significance of *Ailan Kastom*, and the relationships between Torres Strait people and communities that included rights and obligations in relation to marine resources. The High Court's determination only related to the parts of the claim area (Part A) that were not also claimed by the Kaurareg and Gudang people (the Aboriginal Traditional Owners of the continental islands in the south Torres Strait adjacent to Cape York Peninsula). Due to the overlapping claims between the Kaurareg and Gudang people in Part B, the claim to these areas is still to be determined.

Malu Lamar (meaning 'deep water spirit') Torres Strait Islander Corporation was appointed in 2014 as the RNTBC to hold the native title rights and interests in Part A of the Sea Claim area, covering approximately 40,000 km^2 of Torres Strait waters. In 2012, Gur A Baradharaw Kod Torres Strait Sea and Land Council (GBK) was formed by Traditional Owners in the region to act as a peak body to promote the collective interests of native title holders and to develop a culture of governance that aligns with *Ailan Kastom*. The Directors of GBK comprise the Chairs of the RNTBCs in the region. GBK aspires to play a strong future leadership role in relation to land and sea management arrangements.

The Torres Strait Treaty between Australia and Papua New Guinea (PNG), ratified in 1985, has laid important foundations for Torres Strait Islander engagement in protected area management. The Treaty establishes maritime boundaries between the two countries, and recognises the sovereign rights of both countries in respect to the shared management and conservation of fisheries and marine resources. The Treaty recognises the long history of traditional movement, trade and subsistence fishing by both Torres Strait Islanders and PNG people in the waters of Torres Strait, and seeks to minimise interference with their lives, livelihoods and traditional practices. The Treaty establishes a Protected Zone within which traditional movements can occur (Fig. 18.1). A further purpose of the Protected Zone is to protect and preserve the marine environment and endemic flora and fauna (though the Protected Zone is not a recognised MPA). Under the Treaty, several consultative mechanisms have been established to progress the implementation of the Treaty, including a Traditional Inhabitants Meeting, Treaty Liaison Meeting, Environmental Management Committee, Fisheries Advisory Committee and Joint Advisory Council. Figure 18.1 shows the Protected Zone, and fisheries and seabed jurisdiction lines established under the Treaty. Note also the limits of coastal and Australian territorial waters, and the northern extent of the Great Barrier Reef Marine Park boundary.

In terms of formal statutory protection for marine areas within Torres Strait, it is important to note that the Great Barrier Reef Marine Park does not extend up into the

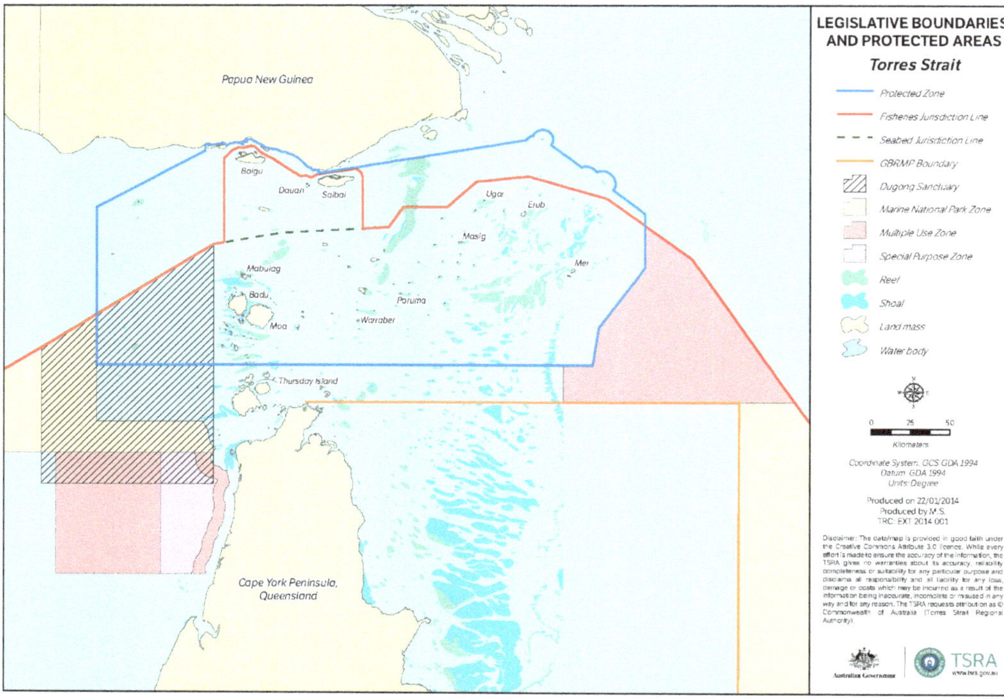

Fig. 18.1. Legislative boundaries and protected areas in the Torres Strait. (Source: Torres Strait Regional Authority)

Torres Strait, although the reefs of Torres Strait are ecologically the northern extension the Great Barrier Reef, and several Torres Strait Islander groups regard the northern section of the GBRMP as part of their traditional fishing grounds. Islander involvement in consultative processes regarding the management of the Great Barrier Reef has been limited to date, though there have been some recent attempts to facilitate Torres Strait Islander Traditional Owner involvement in joint monitoring activities on islands in the northern section of the GBRMP, as well as in collaborative research and ranger compliance training initiatives delivered through GBRMPA (see also Chapter 5).

Perhaps the most significant developments in terms of Islander engagement in marine protected area management have occurred since the development of the *Land and Sea Management Strategy for Torres Strait* in 2005 (Torres Strait NRM Ltd 2005), and the subsequent establishment of the Land and Sea Management Unit (LSMU) within the Torres Strait Regional Authority (TSRA), with funding under the Natural Heritage Trust. Before then, the region lacked a framework and institutional capacity for coordinating strategic and integrated environmental management approaches, covering both terrestrial and marine areas.

Since its inception in 2006, the LSMU has advocated for, established and implemented a community-based management approach as the most appropriate long-term strategy for the sustainable management of the unique environmental values of the Torres Strait, in line with cultural priorities and customary knowledge held by Torres Strait Islander and Aboriginal people living in the region. The LSMU acts as the regional natural resource management body for Torres Strait, and administers a range of projects through a combination of funding sources, in collaboration with Traditional Owners, communities, researchers and all levels of government.

One of the many projects administered by the LSMU is the IPA project, delivered through a funding agreement with the Australian Government. The first IPA for the Torres Strait region was declared in 2000 at Warul Kawa (Deliverance Island) in the top western area of the Torres Strait Protected Zone. Due to its remoteness and the lack of resourcing and institutional support, IPA management activities were very limited on Warul Kawa until quite recently. In 2009, a second IPA in the region, at Pulu Islet, adjacent to Mabuiag Island, was declared. In 2014, the Warraberalgal and Porumalgal IPA, was also declared with the support of Traditional Owners. Plans of Management, developed with Traditional Owner input and based on the best available scientific information, are in place for all three IPAs. At present, the focus for management activities is primarily land based, but the plans make reference to Traditional Owner aspirations for management regimes to eventually extend over traditional maritime estates also, in line with the Sea Country IPA model discussed below.

Indigenous rangers play an integral role in implementing these IPA management plans, including through enforcing protocols for visitation, maintaining cultural heritage sites, carrying out beach surveys and marine debris clean ups, and managing weeds and feral animal incursions, among other things.

With funding under the Australian Government's Working on Country initiative, the LSMU has supported the establishment of ranger groups in 14 Torres Strait communities, employing over 40 local Indigenous staff on the outer islands. Rangers are responsible for carrying out a variety of cultural and natural resource management activities on their islands and in surrounding marine areas, in accordance with the priorities identified in community-endorsed Working on Country Plans. These plans embed and reflect management activities identified in IPA Plans of Management, as well as community-based Dugong and Turtle Management Plans.

The Torres Strait Dugong and turtle fisheries are classified as traditional subsistence fisheries under the *Torres Strait Fisheries Act 1984* (Cth), and are limited to Traditional Inhabitants of the Torres Strait. Dugong and turtle may only be taken in the course of traditional fishing and used for traditional purposes. The legislated management arrangements for these traditional fisheries are outlined in Fisheries Management Notices No. 65 and 66, and are complemented by the community-based Dugong and Turtle Management Plans being implemented on a voluntary basis by communities throughout the region with the support of the TSRA.

The Torres Strait Dugong and Turtle Management Project, which commenced in 2005, aims to support Torres Strait Islanders to sustainably manage these endangered and culturally iconic species through a combination of western scientific research approaches and contemporary management measures, and traditional knowledge and customary management practices. Non-statutory community-based Dugong and Turtle Management Plans have been developed and endorsed in all Torres Strait outer island communities. Whole-of-government support has been negotiated for the implementation of plans through the Protected Zone Joint Authority. Rangers are actively implementing priority activities under the plans, including protection of turtle nesting and foraging areas, Dugong surveys, turtle tagging, seagrass monitoring, marine debris management, culturally appropriate management of harvest, catch monitoring, seasonal and area-based closures, and community education and awareness-raising about the conservation of these species.

Under the Dugong and Turtle Management Project, a consultative process is currently underway regarding the proposed expansion of the Dugong Sanctuary (a large no hunting zone in western Torres Strait; Fig. 18.1) and the inclusion of marine turtles in the protective regime established under the Sanctuary. The Dugong Sanctuary is a protected area established in 2003 under Fisheries Management Notice No. 65 in accordance with the *Torres Strait Fisheries Act 1984*.

Results from Dugong aerial surveys show that the Torres Strait is the most important Dugong habitat in Queensland, Australia and probably the world, supporting a healthy and relatively stable population of more than 12 000 Dugongs (Marsh *et al.* 2011). The region contains 54% of the high and very high density Dugong habitat in north-east Australia. Almost one-quarter of this area occurs in the Dugong Sanctuary, which illustrates the potential value of this spatial closure. The extension of the Sanctuary northwards into an area that is rarely frequented by hunters, and the protection of turtles as well as Dugongs within the Sanctuary, could complement and build upon existing community-based management and legislative measures to manage the species sustainably into the future.

With recognition of native title rights and interests over islands and sea areas in Torres Strait, strong institutional foundations and support for a community-based environmental management approach in the region, and effective arrangements in place for ongoing collaboration between communities, scientists and government agencies, there is considerable scope for the expansion and enhancement of existing protected area regimes over Torres Strait islands and marine areas.

Gur A Baradharaw Kod Torres Strait Sea and Land Council and Malu Lamar (Torres Strait Islander) Corporation, will no doubt be key driving forces in consolidating Traditional Owner-led management approaches across marine areas in the region, and in fulfilling long-held Islander aspirations for greater control over marine resources. Locally dedicated and managed MPAs are likely to feature strongly in the raft of management measures and options that are considered in the efforts to progress this broader agenda.

Western Australia

In Western Australia, the Roebuck Bay Marine Park, near Broome in the Kimberley region, is part of the multi-tenure Yawuru conservation estate that comprises a 100 km long coastal (terrestrial and marine) park complex. The complex is coordinated by the Yawuru Park Council, made up of representatives of Yawuru Traditional Owners, the Western Australian Department of Parks and Wildlife and the Shire of Broome. A comprehensive Yawuru Cultural Management Plan sets out the cultural values and principles to be embedded in the four management plans being developed jointly for the separate but interconnected components of the conservation estate. The imminent dedication of the Yawuru IPA across all components of the Yawuru conservation estate will further strengthen the integration of the land and sea country management within the protected area complex.

A similar collaboration between Traditional Owners and the Western Australian Government has been achieved with respect to 80 Mile Beach Marine Park, located between Broome and Port Hedland, the intertidal portion of which has recently been incorporated into the Nyangumart Warrarn IPA. The policy implications of IPAs in marine areas are discussed further below.

Other MPAs established by the Western Australian Government are currently not co-managed with local Traditional Owners but there is scope for joint management of these marine areas under the *Conservation and Land Management Act 1984* (WA), and/or through the establishment of sea country IPAs over the same marine areas.

In October 2015, the Western Australian Government released a draft Plan of Management for establishing a network of jointly managed marine protected areas and national parks on Dambimangari Country along the Kimberley coast (Department of Parks and Wildlife 2015). This proposal, which would require the consent of Traditional Owners who hold exclusive native title over some of the area, envisages legislated, jointly managed protected areas to be established over the existing Dambimangari IPA and sea country that Traditional Owners intend to include in their IPA. The Western Australian Government proposal contains many of the elements of the joint management arrangements that emerged elsewhere in Australia during the 1980s, in contrast to the Indigenous-led, collaborative arrangements that have evolved in recent years through country-based IPAs (Kerins 2015; see also further discussion on IPAs below).

New South Wales

In New South Wales (NSW), the *Marine Estate Management Act 2014* (NSW) makes no provision for co-management of MPAs with Indigenous people, but allows for Indigenous cultural resource use within marine parks where the activity is consistent with biodiversity conservation, ecosystem integrity and ecosystem function.

Aboriginal involvement in the management of marine parks is facilitated by the Marine Estate Management Authority and implemented in accordance with the Aboriginal Engagement and Cultural Use of Fisheries Resource Policy, which encourages Aboriginal people to nominate for membership of local marine park advisory committees, provides for the establishment, on request by Aboriginal people, of an Aboriginal Advisory Group for each MPA and the convening of community consultative meetings as required (Marine Parks Authority 2010) (see also Chapters 11, 14).

South Australia

In South Australia, the *Marine Parks Act 2007* (SA) makes no provision for formal Aboriginal involvement in MPA governance or management but requires that consideration should be given to Aboriginal heritage values. A step towards greater Indigenous engagement has been made in the Great Australian Bight Marine Park where there is an agreement with the local Yalata Community, through their land management group, to advance cooperation in relation to visitor management, research and monitoring, surveillance and reporting, collection of marine debris and cliff rescue assistance.

Victoria

In Victoria there are no provisions for co-management of MPAs under the *National Parks Act 1975* (Vic). However, under the *Traditional Owner Settlement Act 2010* (Vic) there may be opportunities for agreements to be reached regarding Aboriginal management of MPAs, including the possibility of the granting of Aboriginal Title in respect of a marine national park or marine sanctuary, though this has not yet occurred. Meanwhile, the Management Strategy for Victoria's Marine National Parks and Marine Sanctuaries (Parks Victoria 2003) provides the foundations for joint management partnerships, including:

- committing to working in partnership with Indigenous communities towards the long-term protection and conservation of marine national parks and sanctuaries
- acknowledging traditional ownership of marine areas
- committing to improved consultation and involvement
- developing an Indigenous Cultural Awareness Program

- helping Indigenous communities to build capacity
- increasing Indigenous employment opportunities within Parks Victoria.

Tasmania

In Tasmania there are no legislative provisions for Indigenous co-management of MPAs, but the *Tasmanian Marine Protected Areas Strategy* (Marine and Marine Industries Council 2001) aims to cater for the management of marine areas and species in partnership with Indigenous communities, to recognise the interests of Australia's Indigenous people and to incorporate Indigenous people in decision making.

Indigenous Protected Areas

IPAs are areas of land and/or sea that are voluntarily dedicated as protected areas by the Traditional Owners associated with the area, recognised by the Australian Government and all state and territory governments as part of Australia's National Reserve System (NRS) of protected areas and managed according the protected area management guidelines of the International Union for Conservation of Nature (IUCN). Currently only the terrestrial components of IPAs are recorded in the Collaborative Australian Protected Area Database (CAPAD) maintained by the Australian Department of the Environment. This is because there are currently two separate reserve systems – the NRS (for terrestrial protected areas) and the National Representative System of Marine Protected Areas (NRSMPA) for marine areas. Although the NRS recognises all terrestrial protected areas that meet the IUCN protected area definition, the NRSMPA only recognises MPAs established under legislation. The NRSMPA criteria were developed in the early 1990s before the development of IPAs and before the recognition of native title in Australia and have not been subsequently reviewed (Smyth 2008). Because it is increasingly difficult to make precise distinctions between terrestrial and marine protected areas (e.g. IPAs on coastal Aboriginal land in the Northern Territory extend to the low-water mark and therefore include extensive marine areas), there is an argument for establishing a single reserve system to include all of Australia's terrestrial and marine protected areas.

The first IPA was established in South Australia in 1998 and, as of May 2015, there were 72 IPAs across Australia, together comprising over 60 million ha and contributing over 40% of the total area of the NRS. A further three IPAs were expected to be declared by the end of 2015. Most IPAs are established on Indigenous-owned land, but IPAs have also been established over multiple tenures, including existing national parks and marine parks, in collaboration with other parties, based on the concept of 'country' (traditional land and sea estates) rather than tenure (Rose 2013).

IPAs are established and managed by Indigenous people independently of government legislation, consistent with the IUCN protected area definition, which refers to protected area management by 'legal or other effective means'. IPAs are not government protected areas, but they are supported by the Australian Government's IPA Program and most IPAs are also supported through partnerships with state and territory agencies, natural resource management bodies, business enterprises and research institutions.

The first IPA to include a significant area of sea country was the 101 000 ha Dhimurru IPA on the north-east coast of Arnhem Land (Fig. 18.2). It was declared in 2000 with the inclusion of 9000 ha of sea country that had been registered as a marine sacred site under the Northern Territory's *Aboriginal Sacred Sites Act 1989* (Fig. 18.3). However, limiting the marine component of the IPA to registered marine sacred sites excluded a large area of sea

Fig. 18.2. Land and sea country in Dhimurru Indigenous Protected Area. (Photo: Dhimurru Aboriginal Corporation)

country for which there is currently no legal recognition (Smyth 2008). Dhimurru Aboriginal Corporation released a strategic Sea Country Plan in 2006, which outlined their vision for integrated land and sea management in partnership with government agencies, research institutions and others (Dhimurru Aboriginal Corporation 2006). Following a collaborative planning process over several years, Yolngu Traditional Owners dedicated an additional 400 000 ha of sea country as part of the Dhimurru IPA in April 2013. The inclusion of this additional sea country into the IPA was formally recognised by the Australian and Northern Territory governments at a ceremony during the World Indigenous Network Conference in Darwin in May 2013. The expanded Dhimurru IPA includes a Terrestrial Zone (Aboriginal Land) and a Sea Country Zone (collaborative management) in recognition that management of the land is the sole responsibility of Traditional Owners, while management of the sea country is undertaken in collaboration with government agencies and resource users (Dhimurru Aboriginal Corporation 2015). Dhimurru IPA's Sea Country Zone overlaps with part of the Wessel Commonwealth Marine Reserve – one of a network of offshore MPAs recently established by the Commonwealth Government around Australia (see Chapter 4). Integration of land and sea management across the IPA is coordinated through the Dhimurru IPA Advisory Group chaired by Dhimurru and including representatives of all IPA collaborative agencies.

The concept of this collaborative approach to managing the sea country component of the IPA is that all government and non-government partners bring to the IPA table their respective commitments to achieving the goals of the IPA through the shared application

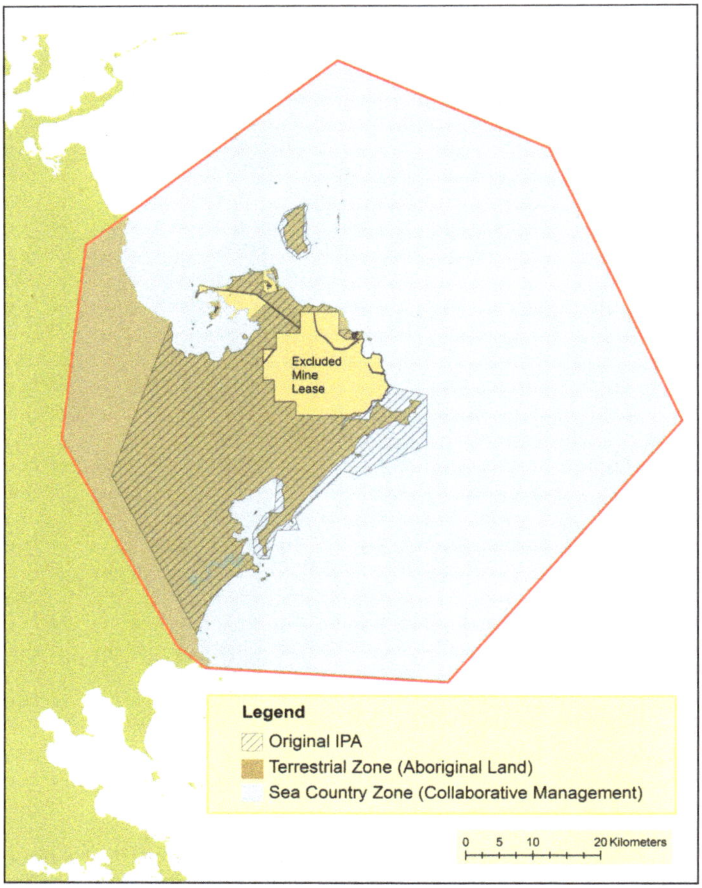

Fig. 18.3. Map of Dhimurru Indigenous Protected Area (IPA) (area within the red boundary) on the north-east coast of Arnhem Land, Northern Territory. (Source: Dhimurru Aboriginal Corporation)

of their respective authorities, resources and capabilities. Further details of the IPA governance and management arrangements are provided in the Dhimurru IPA Management Plan (Dhimurru Aboriginal Corporation 2015).

As noted above, IPAs can overlap and coexist with government MPAs, as a voluntary, non-legal, Indigenous-led mechanism that complements existing MPA management arrangements in the MPA. The first of these in Australia was the Mandingalbay Yidinji IPA near Cairns in north Queensland, which includes a portion of the Queensland Government's Great Barrier Reef Coast Marine Park (Chapter 7), as well as a national park, forest reserve and conservation park – thus reuniting land and sea country that had been fragmented into several terrestrial and marine protected areas managed by different agencies.

Several other IPAs have subsequently been established over parts of the GBRMP and adjacent terrestrial protected areas, including the Girringun Regional IPAs between Innisfail and Ingham, and Eastern KukuYalanji IPA between Daintree and Cooktown (see also Chapter 5).

This country-based approach (Smyth 2011) to maritime IPA planning and management presents a new opportunity for expanding the marine and coastal protected area estate in

Australia, particularly at a time when proposals to establish conventional MPAs (legislated, dedicated and managed by government agencies) are often subjected to strong opposition from resource user groups and their political allies. Sea country IPAs can be more readily acceptable because they provide a non-legislated collaborative framework in which all parties share their values, resources and expertise to achieve locally, nationally and globally recognised protected areas consistent with IUCN guidelines. Multi-tenure, land/sea IPAs also represent the re-emergence of 'country' as the cultural and geographic scale at which Australia's coastal and marine environments have been managed for millennia.

The conceptual basis for these 'country-based' and 'multi-tenure' IPAs is the same as for the earlier tenure-based IPAs: they are declared by the Indigenous people traditionally associated with the area and they are managed through a package of legal and other effective means. The key innovation in these more recent IPAs is that this package includes the legal authorities and management capacities of other landowners and management agencies with rights and interests over the various tenures within the multi-tenure IPA. Of course, such IPAs can only be recognised and effectively managed with the consent and collaboration of all relevant parties with interests in the IPA. This approach provides an opportunity to 'put country back together' when it has been fragmented into multiple tenures and is providing a new, Indigenous-led pathway for negotiating co-management of existing national parks, MPAs and other government protected areas on land and sea.

Although it can be argued that country-based sea country IPAs lack the legislated 'security' of government MPAs, their collaborative nature provides unparalleled opportunities for diverse rights-holders and stakeholders to come together, exchange perspectives, share knowledge and use their collective authorities and commitments to agreed protected area objectives that can result in area-based management outcomes that are far less likely to be achieved through a legislated MPA. This is because, in many instances, the diverse interest groups would prevent the MPA being established in the first place and because government MPA management often results in a coercive rather than collaborative relationship with rights-holders and stakeholders. Furthermore, the 'security' of protected areas established under legislation can prove illusory if new governments amend legislation to reconfigure or remove protected area status, or when inadequate resources are allocated for effective management.

Future directions

While saltwater Indigenous people and their representative organisations continue to express disappointment, frustration and sometimes outrage about the failure of governments to adequately recognise and support their legitimate rights and interests in sea country governance, management and resources – as evidenced by the National Indigenous Sea Country Statement referred to above (NAILSMA 2012) – there have been substantial improvements in recognition and support in recent decades, and the trend is likely to continue.

Significant contributing factors to this trend have been the recognition of native title, and other statutory Indigenous rights regimes, that have in turn led to legislative and policy shifts within protected area management agencies that have increasingly come to see Indigenous people as rights-holders with respect to specific clan estates, rather than stakeholders with generic interests across a MPA.

At least as significant has been the increasingly proactive stance taken by Indigenous people themselves to lead the partnership-building process (e.g. through sea country IPAs) and to establish their own independent land and sea management capabilities (Indigenous

Fig. 18.4. Dhimurru Rangers with Green Turtle. (Photo: Dhimurru Aboriginal Corporation)

ranger groups – Fig. 18.4, research partnerships, resource management agreements, etc.) that in turn has resulted in policy and legislative changes. Governments are (albeit slowly) responding to Indigenous initiative and leadership, rather than to Indigenous complaints and requests for redress.

Government responses in this policy area are also part of a wider commitment to 'close the gap' between Indigenous and non-Indigenous social, health and education outcomes and a rapidly growing awareness that investment in Indigenous land and sea management is yielding multiple benefits both in terms of environmental management and Indigenous wellbeing. Indigenous caring for land and sea country is recognised not only as the continuation of an ancient and inherent cultural right, but also a propitious employment niche in 21st century Australia (Greiner 2010; Smyth 2014).

Acknowledgements

The authors wish to gratefully acknowledge assistance and advice from: Marcus Sandford, Department of the Prime Minister and Cabinet; Julie Melbourne, Nyamba Buru Yawuru Ltd; John Tapim, Great Barrier Reef Marine Park Authority; Vic McGrath and Damian Miley, Torres Strait Regional Authority.

Biographies

Irish born, **Dermot Smyth** studied zoology at the Australian National University and at James Cook University. Having encountered Indigenous people and other traditional communities during research in Papua New Guinea and West Africa, Dermot subsequently worked with many Aboriginal groups and Torres Strait Islanders to support their management of traditional estates around Australia. Dermot helped develop the concept of IPAs from the 1990s to the present time, including the establishment of multi-tenure and sea country IPAs, and has contributed to the re-emergence of Indigenous 'country' as an appropriate scale for the management of Australia's environments. He is an Adjunct Research Fellow at the Research Institute for the Environment and Livelihoods, Charles Darwin University.

Miya Isherwood has a background in law and environmental science, and has worked in the legal profession, for Indigenous organisations and government agencies in the fields of land management, conservation planning and environmental program delivery for over 15 years. Miya played a pivotal role in the establishment of the Land and Sea Management Unit within the Torres Strait Regional Authority, including successfully negotiating collaborative arrangements between all levels of government, Indigenous communities and the research sector for the delivery of nationally funded environmental programs. Miya has also worked in planning, environmental and local government law, including on the development of community legal education materials and on law reform submissions. Miya works in the Land and Sea Management Unit at the Torres Strait Regional Authority.

References

Allen D (1994) Saltwater Dreaming. In *Surviving Columbus: Indigenous Peoples, Political Reform and Environmental Management in North Australia.* (Ed. P Jull) pp 39–46. North Australia Research Unit, Australian National University, Darwin.

ATNS (2011) *Kuuku Ya'u People's Marine Park Indigenous Land Use Agreement (ILUA). Agreements, Treaties and Negotiated Settlements.* Australian Native Title Services, Melbourne, <http://www.atns.net.au/agreement.asp?EntityID=5007>.

Australian Human Rights Commission (2000) *Native Title Report 2000.* Commonwealth of Australia, Canberra.

Bauman T, Hayne C, Lauder G (2013) *Pathways to the Co-management of Protected Areas and Native Title in Australia.* AIATSIS Research Discussion Paper 32. Australian Institute of Aboriginal and Torres Strait Islander Studies, Canberra.

Brennan S (2008) Wet or dry, it's Aboriginal land: the Blue Mud Bay decision on the intertidal zone. *Indigenous Law Bulletin* **7**(7), 6–9.

Cordell J (1991) *Managing Sea Country: Tenure and Sustainability of Aboriginal and Torres Strait Islander Marine Resources.* Consultancy report for the Ecologically Sustainable Development Fisheries Working Group, Commonwealth of Australia, Canberra.

Department of Parks and Wildlife (2015) *Proposed Lalang-garram / Horizontal Falls and North Lalang-garram marine parks and Oomeday National Park.* Department of Parks and Wildlife, Perth, <http://www.dpaw.wa.gov.au/parks/management-plans/draft-plans-open-for-public-comment/359-horizontal-falls-have-your-say>.

Dhimurru Aboriginal Corporation (2006) *Yolŋuwu Monuk Gapu Wäŋa Plan of Management.* Dhimurru Aboriginal Corporation, Nhulunbuy, Northern Territory.

Dhimurru Aboriginal Corporation (2015) *Dhimurru Indigenous Protected Area Management Plan 2015–2022.* Dhimurru Aboriginal Corporation, Nhulunbuy, Northern Territory, <http://www.dhimurru.com.au/uploads/8/9/3/6/8936577/dhimurru_ipa_management_plan_2015-22.pdf>.

Greiner R (2010) The potential for a conservation economy in the tropical savannas based on 'payments for environmental services'. In *North Australian Political Economy: Issues and Agendas.* (Ed R Gerritsen) pp. 162–177. Charles Darwin University Press, Darwin.

Hiscock P (2008) *Archaeology of Ancient Australia.* Routledge, London.

Kerins S (2015) 'Kimberley conservation threatens to take a step back on Indigenous rights', *The Conversation*, 18 November 2015, <https://theconversation.com/kimberley-conservation-threatens-to-take-a-step-back-on-indigenous-rights-50778>.

Marine and Marine Industries Council (2001) *Tasmanian Marine Protected Areas Strategy.* Department of Primary Industries, Water and Environment, Hobart. <http://www.parks.tas.gov.au/file.aspx?id=6755>.

Marine Parks Authority (2010) *Aboriginal Engagement and Cultural Use of Fisheries Resources Policy.* Marine Parks Authority, Sydney, <http://www.mpa.nsw.gov.au/pdf/Cultural%20Resource%20Use%20Policy%202010.pdf>.

Marsh H, O'Shea TJ, Reynolds JE, III (2011) *Ecology and Conservation of Sirenians: Dugongs and Manatees.* Cambridge University Press, New York.

NAILSMA (2012) *The National Indigenous Sea Country Statement – A declaration from the delegates of the 2012 National Indigenous Sea Country Workshop.* North Australian Land and Sea Management Alliance, Darwin, <http://www.nailsma.org.au/sites/default/files/publications/NISCW-statement_Web.pdf>.

Parks Victoria (2003) *Victoria's System of Marine National Parks and Marine Sanctuaries Management Strategy 2003–2010.* Parks Victoria, Melbourne.

Peterson N, Rigsby B (Eds) (1998) *Customary Marine Tenure in Australia.* Oceania Monograph 48. University of Sydney, Sydney.

Rose B (2013) Indigenous Protected Areas – innovation beyond the boundaries. In *Innovation for 21st Century Conservation.* (Eds P Figgis, J Fitzsimons and J Irving) pp. 50–55. Australian Committee for IUCN, Sydney.

Ross H, Innes J, George M, Gorman K (2005) *Traditional Owner Aspirations Towards Co-operative Management of the Great Barrier Reef World Heritage Area: Community Case Studies.* CRC Reef Research Centre Technical Report No. 56. CRC Reef Research Centre, Townsville.

Sharp N (1996) *No Ordinary Judgment – Mabo, The Murray Islanders' Land Case.* Aboriginal Studies Press, Canberra.

Smyth D (1993) *A Voice in All Places – Aboriginal and Torres Strait Islander Interests in Australia's Coastal Zone.* Consultancy report to the Resource Assessment Commission, Canberra.

Smyth D (1994) *Understanding Country – The Importance of Land and Sea in Aboriginal and Torres Strait Islander Societies.* Key Issues Paper 1. Council for Aboriginal Reconciliation, Canberra.

Smyth D (1995) Caring for Sea Country – accommodating Indigenous peoples' interest in marine protected areas. In *Marine Protected Areas – Principles and Techniques for Management.* (Ed. S Gubbay) pp. 149–173. Chapman and Hall, London.

Smyth D (1997) *Saltwater Country Aboriginal and Torres Strait Islander Interest in Ocean Policy Development and Implementation Socio-cultural Considerations – Issues Paper 6.* Commonwealth of Australia, Canberra.

Smyth D (2001) Management of Sea Country – Indigenous people's use and management of marine environments. In *Working on Country – Contemporary Indigenous Management of Australia's Lands and Coastal Region.* (Ed. R Baker, J Davies and E Young), pp. 60–74, Oxford University Press, New York.

Smyth D (2008) Just add water? Taking Indigenous Protected Areas into Sea Country. In *Protecting Country: Indigenous Governance and Management of Protected Areas*. (Eds D Smyth and G Ward) pp. 95–110. Australian Institute of Aboriginal and Torres Strait Islander Studies, Canberra.

Smyth D (2011) *Guidelines for Country-based Planning*. Queensland Department of Environment and Resource Management, Cairns, <http://www.ehp.qld.gov.au/cape-york/pdf/country-based-planning-guideline.pdf>.

Smyth D (2014) Caring for Country: An Indigenous propitious niche in 21st century Australia. In *AIATSIS National Indigenous Studies Conference 2014 Papers*. Australian Institute of Aboriginal and Torres Strait Islander Studies, Canberra, <http://aiatsis.gov.au/_files/conferences/national_indigenous_studies_conference/paper_caring_for_country_an_indigenous_propitious_niche_in_21st_century_australia.pdf>.

Torres Strait NRM Ltd (2005) *Land and Sea Management Strategy for Torres Strait*. Torres Strait Regional Authority, Thursday Island, <http://www.tsra.gov.au/__data/assets/pdf_file/0017/1772/torres20strait20land_sea_mgt20strategy20final20edit20060106.pdf>.

Chapter 19

Beyond marine protected areas: marine bioregional planning and oceans policy

Chris Smyth and Geoff Wescott

Introduction

Australia's Commonwealth marine reserve network, proclaimed across five marine regions and the Coral Sea and covering 36% of the nation's Exclusive Economic Zone (EEZ), is the outcome of a marine bioregional planning process that began with the Australian Government's release of Australia's Oceans Policy in 1998 (Commonwealth of Australia 1998a,b).

The aim of the Oceans Policy was to reconfigure how Australia planned and managed its oceans by establishing integrated and ecosystem-based Regional Marine Planning. Its vision was for 'Healthy oceans: cared for, understood and used wisely for the benefit of all, now and in the future', and it had goals that included to 'understand and protect Australia's marine biological diversity, the ocean environment and its resources, and ensure ocean uses are ecologically sustainable'.

Wescott (2000) reviewed the first 2 years of the Oceans Policy's implementation, concluding that, while it was comprehensive in its coverage, it was not necessarily integrated. He further noted that the challenge ahead was to use marine bioregional planning to integrate what was up to then, at least, a sectorally driven approach to the management of the marine environment.

Seventeen years later, the Oceans Policy is largely 'dead in the water', with little action on its recommendations in recent years and no attempt to update it from its 1998 original form. It appears that successive federal governments have effectively replaced it in practice with marine bioregional planning which has primarily focused on the identification of marine protected areas while maintaining viable resource-based industries (see also Vince 2015; Vince *et al.* 2015).

Australia now has: four marine bioregional plans for the south-west, north-west, north and temperate east marine regions; a Regional Marine Plan for the south-east region; a large Commonwealth marine reserve in the Coral Sea; and an extensive marine reserves network (see Chapter 4). Even so, the way our oceans are managed – or how human use of the marine environment is managed, to be more precise – continues to be driven by the fishing, petroleum and shipping sectors, while conservation is largely subservient to those sectors' needs, or at best treated as just another competing use.

In selecting the location and zoning of the Commonwealth marine reserves referred to above, considerable adjustments were made to cater for the needs of regional stakeholders, with one of the key principles of the plans being to minimise social costs (Australian Government 2015). Several socio-economic criteria were also used in the selection or non-selection of areas for inclusion in the network. These criteria aimed to ensure the marine reserves avoided areas with potential for offshore oil and gas exploration and production, of highest value to the commercial fishing industry, and of interest to recreational, game and charter fishing. There was a preference for multiple-use zones to minimise impacts on these uses, while marine national park zones could not be located within 40 nautical miles of a port or marina (Australian Government 2011).

Australia's EEZ is the third largest, and its oceans the most diverse, in the world (Cooper 2010) but they are under pressure from commercial and recreational fishing, oil and gas exploration and production, land-based sources of pollution, coastal development, invasive marine pests and climate change (State of the Environment 2011 Committee 2011).

Oceans governance remains disintegrated – based on individual industry sectors and jurisdictions, and with a decided lack of 'vertical' integration between governments (federal and state) and 'horizontal' integration across sectors, government departments and the catchment-coastal-marine continuum (Wescott 2012).

Australia continues to fall well short of infusing ecologically sustainable development, ecosystem-based management, and effective multi-user management into its oceans planning, protection and management framework (Australian Conservation Foundation 2005).

Marine bioregional planning across Australian jurisdictions

When marine-based resource industries emerged in Australia's oceans, the response was to establish specific legislative arrangements to manage them – fisheries Acts, transport Acts, mining Acts and so on. These first appeared in state jurisdictions but, as the Australian Government became more involved in oceans planning and management after 1950, similar statutes were introduced at the federal level.

When environmental issues arose in the oceans, the legislative response – pollution Acts and wildlife Acts – were targeted at the single issue and within separate jurisdictions.

Over time, these *ad hoc* and reactive processes allowed the evolution of a complex, sector-based and disintegrated legislative and institutional framework for the planning, protection and management of Australia's oceans.

This, along with the 3-nautical-mile limit of state and territory waters, was entrenched with the signing of the Offshore Constitutional Settlement (OCS) between the Commonwealth, states and territories in 1978. The OCS was a response to a High Court battle between the states and the Commonwealth over the Great Barrier Reef, which had determined that the Commonwealth, not the states, controlled marine waters (see also Chapter 5). The OCS returned control of waters within 3 nautical miles to the states but left the waters of the Great Barrier Reef under Commonwealth control (see also Chapter 12).

But, by the late 1990s, there remained some appetite in the Australian Government to tackle difficulties arising from the sectoral and jurisdictional basis of Australia's oceans governance framework. This was stimulated by Australia's ratification of the United Nations Law of the Sea Convention, which included requirements for a signatory state to demonstrate that it was establishing mechanisms for managing the EEZ that the Convention delivered, and its separate international commitments for establishing marine protected areas under the Convention on Biological Diversity.

Commonwealth level

After a period of public consultation, the Australian Government released its Oceans Policy in 1998, which sought to establish integrated, ecosystem-based Regional Marine Plans as a prime (but not the only) mechanism of implementing the policy in a non-sector based form (Wescott 2000). Although the current official status of the policy is unclear, and in practice it is certainly moribund, it was the foundation for the marine bioregional planning process that has recently established the world's first national network of marine reserves.

Although the stakeholder focus in the Australian Government's marine bioregional planning process was the establishment of the marine reserves, those preparing them saw their purpose as being far broader. This was certainly the original intent in implementing the Oceans Policy – the zoning system was meant to apply across all of the oceans covered – not just within designated MPAs (Wescott 2000).

The *Environment Protection and Biodiversity Conservation Act 1999* (Cth) (EPBC Act) contains several provisions for the establishment of Commonwealth marine reserves, the preparation of management plans and statutory consultation periods. In contrast, the only legislative guidance to the preparation and implementation of marine bioregional plans is provided in the five short clauses of s 176 of that Act. In the mid-2000s, the Environment Minister, Senator Ian Campbell, used this section of the EPBC Act to give legislative support to marine planning – which up until then had only been supported by the Oceans Policy, a document with no statutory basis.

Under the Act, the decision to prepare marine bioregional plans is at the discretion of the Minister. While some suggestions regarding content are provided in the legislation, the scope and direction of a plan are largely up to the Minister and the bureaucrats of the day. However, once prepared, the Minister must have regard to a bioregional plan when making decisions under the EPBC Act.

The marine plan completed for the South-east Marine Region in 2004 (National Oceans Office 2004), and the marine bioregional plans (Australian Government 2012a,b,c,d) completed for the remaining four marine regions in 2012 are, excepting the marine reserves, very much strategic plans. On-the-water management in Commonwealth waters – the plans do not cover state waters – is left to existing agencies in the petroleum, shipping and fisheries sectors. The difference to the original Oceans Policy's intent was that under that Policy, certain principles and operations would apply across all uses, not just via a physical zoning map over one use area (MPAs). This was the basis for claiming that integrated planning was to apply to the whole EEZ.

Each of the plans begins by identifying the region's conservation values and threats, and is then confined to outlining largely existing marine conservation approaches and actions that can be used to conserve those values. There is no attempt to plan or manage marine-based industries or to integrate that management with the plan's marine conservation objectives.

The priority strategies within the South-west, North-west, North and Temperate East Marine Bioregional Plans include providing sound scientific data on which to underpin the decision-making processes for project assessment and approvals under the EPBC Act (Commonwealth marine waters are a declared 'Matter of National Environmental Significance' under the Act) and hopefully the work of the National Offshore Petroleum Safety and Environmental Management Authority under the *Offshore Petroleum and Greenhouse Gas Storage Act 2006*.

Other priority actions in the marine bioregional plans include collaborative scientific research to improve understanding of the region's marine environment: improved monitoring, evaluation and reporting on marine ecosystem health; and targeted collaborative programs across jurisdictions to help species recovery. By identifying marine ecosystems and key ecological features, the plans could be used to assist the development of objectives for ecosystem-based management if and when it is adopted in Australia.

The South-east Regional Marine Plan and the four marine bioregional plans are not like those envisaged under the Oceans Policy. Their remit is limited to strategic planning and protecting the conservation values of the Commonwealth waters in each of the marine regions (Christmas Island, Cocos (Keeling) Islands and the Torres Strait were excluded from the planning process, as were the existing marine reserves at Heard and McDonald Islands and Macquarie Island, and the Great Barrier Reef Marine Park) by establishing marine reserves and making better use of provisions in the EPBC Act. As such, they lack the integration that was assumed to be the basis of the Oceans Policy.

Although the plans have identified ecosystems and key ecological features and processes, they do not establish ecosystem-based management. While the individual resource sectors are influenced by the location and zonings of the marine reserves, and in some cases displaced under those zones, their main areas of use have been avoided and marine planning and management remain sector- and jurisdictionally-based across the two-thirds of the EEZ outside the network. The network covers 36% of the EEZ but marine national park zones cover only 11% and are mostly well offshore and away from areas of commercial interest (Devillers *et al.* 2015).

Ensuring that marine bioregional plans help protect a region's conservation values will require strong political will and commitment, along with adequate resourcing. It will also require stronger and more detailed legislative support than is currently provided by the EPBC Act. It could be argued that the marine reserve network management plans prepared under the EPBC Act will give this statutory support, but it remains largely unclear how conservation actions are to be implemented outside the reserves.

Even so, the EPBC Act does provide some support for the broader management of Australia's oceans. Not only is it the statute used to establish the national marine reserves network but it also includes provisions that can be used to protect whales and dolphins, marine turtles, seahorses, sharks, seabirds and migratory species. Some marine ecological communities are also listed as threatened under the EPBC Act, including Giant Kelp Marine Forests of South East Australia and Subtropical and Temperate Coastal Saltmarsh.

However, only three of the 20 threatening processes listed under the EPBC Act relate to the marine environment (Department of the Environment 2015). Two specifically cover marine issues – 'by-catch of seabirds and sea turtles', 'the ingestion of marine debris by marine invertebrate animals' – while the other, 'novel biota and their impact on biodiversity', relates to the marine environment in part. Significantly, the loss of habitat caused by greenhouse gas emissions is a listed threatening process specific to terrestrial habitats only.

Lack of resources and political commitment has limited the preparation and implementation of species and ecological community recovery plans that do exist and there is little scientific data on the threatened status of Australia's marine life more generally.

State and territory level

There is currently no marine bioregional planning occurring at the state and territory level. South Australia did begin a process of spatial marine planning in 2006, when it was developing the groundbreaking draft Spencer Gulf Marine Plan (Government of South Australia 2006). Unfortunately, the plan was never completed, abandoned by the state government in order to focus on establishing marine reserves – the tortuous path that finally came to an end in 2013 with the creation of a marine parks network (see Chapter 8).

The other states have also focused on marine reserves and not planning, with mixed results. Except for Victoria's small and highly protected marine national parks network, all states have opted to establish multi-zoned marine reserves with few highly protected areas. They continue to allow fishing in most of the reserves and, in the case of New South Wales, introduced an amnesty on shore-based recreational fishing in sanctuary zones in March 2013 that was lifted in December 2014 but only for 20 of the 30 areas affected – the other 10 lost their sanctuary status along the shorelines (see Chapter 11). Angling groups are continuing their pressure to be given access to highly protected areas in states such as Victoria (VRFish 2013) and through the Australian Government's 2015 review of the Commonwealth marine reserves network with federal ministerial encouragement: 'The Coalition Government is delivering a key election promise by scrapping Labor's plan to lock out recreational fishers from key coastal areas around the country' (Hunt and Colbeck 2013).

It could be argued that a large multi-zoned marine protected area network, such as the Commonwealth network – and the South Australian one covering 44% of state waters but with only 6% under high-level protection – is a demonstration of integrated marine planning but it covers barely one-third of marine waters and is largely 'business-as-usual' with few restrictions on extractive uses that continue to be managed by sector and jurisdiction. In fact, if some zones within MPAs are basically 'general use' and have little restriction of usage, it is difficult to argue why not apply a zoning scheme across all waters – as was the practice for General Use Zones in the Great Barrier Reef Marine Park for the first few decades after declaration (Fig. 19.1).

State government agencies, natural resource management bodies and local communities – often with federal government funding – have also been working together to improve the quality of runoff flowing into estuaries and nearshore waters, including the Great Barrier Reef, and to reduce the impacts of marine debris, such as ghost nets (Gunn *et al.* 2010).

Although most of Australia's coastal waters are in good condition, those along Australia's eastern and south-western coasts were assessed as poor by the State of the Environment 2011 Report (State of the Environment 2011 Committee 2011). The agricultural, urban and industrial development of catchments and along shorelines has led to significant and negative change.

In response to the threats that Australia's oceans face – and the severe limitations of our oceans governance framework to tackle them – we could choose to continue to use existing sector-based legislation, which has proved largely ineffective. We might also decide to tinker with existing sector-based legislation and/or the EPBC Act, or include oceans planning provisions in a new Environment Act, as recommended by the Hawke review of Australian environmental legislation (Hawke 2009).

Or, a completely new approach to oceans planning, protection and management could be taken, as below.

A new approach to Bioregional Marine Planning

In the absence of any commitment to the original objectives of Australia's Oceans Policy, Australia lacks a coordinated national vision for the planning, protection and management of our oceans. Different jurisdictions and different industry sectors have specific legislative, institutional, regulatory and management arrangements, each with their own purposes and objects.

Fig. 19.1. The waters bordering the Cape Hillsborough National Park on the Mackay coast and surrounding the islands are part of the Great Barrier Reef Marine Park and Great Barrier Reef Coast Marine Park. These MPAs consist of a range of zones from high protection no-take zones to general use zones. (Photo: James Fitzsimons)

The Australian Government should work with all jurisdictions, marine-based industry sectors, scientists, the community and other stakeholders to develop a national vision and a set of operational objectives. This document should include actions with measurable targets and timelines to guide the planning, protection and management of Australia's oceans and to assess the effectiveness of their performance.

Despite the policy of the Coalition Government elected in 2013 to delegate responsibilities for environmental approvals and actions back to the states and territories, in the case of Commonwealth marine waters the government continues to have the main assessment and approval role, with the exception of activities assessed by the National Offshore Petroleum Safety and Environmental Management Authority.

Several models have been proposed or developed that could form the basis for a new approach. For example, the rezoning of the Great Barrier Reef Marine Park in 2003 – which successfully increased the park's high-level protection to 34% – has shown the way to develop integrated, ecosystem-based spatial marine planning that has biodiversity conservation as its prime objective but also aims to ensure that marine resource use is ecologically sustainable (Great Barrier Reef Marine Park Authority 2004; see also Chapter 5). The Great Barrier Reef Marine Park Authority carried out the rezoning of the Great Barrier Reef Marine Park under a specific statute for that purpose, the *Great Barrier Reef Marine Park Act 1975* (Cth).

To achieve something similar in Australia's broader oceans would probably require specific legislative and institutional arrangements, such as a National Oceans Commission, to guide the planning, protection and management of Commonwealth, state and territory waters in Australia's oceans.

The origins of proposals for a national body with responsibility for implementation can be traced back to options discussed at consultations during the preparation of the Oceans Policy (Department of Environment, Sport and Territories 1997). It was raised again in the 2003 report, *Oceans Eleven: the Implementation of Australia's Oceans Policy and Ecosystem-Based Management* (Smyth *et al.* 2003), which reviewed the implementation of the Oceans Policy and ecosystem-based regional marine planning. *Oceans Eleven* recommended an 11-step process for integrated and ecosystem-based Regional Marine Planning, a national Oceans Act and a National Oceans Authority.

In 2005, the *Marine Legislative Review* (Australian Conservation Foundation 2005) assessed more than 200 statutes across all jurisdictions and the fishing, petroleum, shipping, tourism and conservation sectors. The review concluded that the Commonwealth and state laws in those key sectors and within separate jurisdictions were inadequate for implementing ecologically sustainable development, ecosystem-based management and multiple-user management.

The *Marine Legislative Review's* findings underpinned the oceans law reform proposals in the 2006 report, *Out of the Blue: An Act for Australia's Oceans* (Australian Conservation Foundation and National Environmental Law Association 2006). The proposal in *Out of the Blue* is being reviewed and refined consistent with the following principal components:

- an Intergovernmental Agreement on Australia's Oceans signed by the Commonwealth, states and territories that sets long-term goals for integrated and regional ecosystem-based oceans planning, protection and management
- a cross-sectoral, expertise-based National Oceans Commission
- complementary Commonwealth, state and territory legislation
- ecosystem-based, spatial regional oceans plans
- a Council of Australian Governments Ministerial Council to approve regional oceans plans, which are also tabled in jurisdictional parliaments
- a cross-sectoral, non-government National Oceans Advisory Committee and Regional Oceans Advisory Councils to advise the Ministerial Council and the Commission
- a national community-based network to inform and engage people in regional oceans planning
- incentive payments from the Commonwealth to states/territories — that is, National Partnership (Council on Federal Financial Relations 2015) facilitation and reward payments in return for implementing reform
- development of a National Coastal Policy and associated institutional and funding arrangements to implement it. This policy would link the EEZ-based recommendations above to the state and territory-based coastal zone planning and management approaches across the country.

Intergovernmental Agreement

The Intergovernmental Agreement would outline the framework for cooperation and collaboration between the Commonwealth, states and territories, as well as establishing the legislative, regulatory and policy frameworks for integrated ecosystem-based regional oceans planning across sectors and jurisdictions.

The agreement would also outline the functions and powers of the National Oceans Commission, and establish principles for marine reserves, land-based pollution, resource allocation, approvals, dispute resolution and the role of Traditional Owners and oceans research.

National Oceans Commission

The functions of the National Oceans Commission would include:

- coordinating, preparing, reviewing and monitoring integrated regional oceans plans
- overseeing resource allocation for ecologically sustainable uses in each marine region
- preparing oceans policy statements and state-of-the-oceans reports
- assessing the management performance of Commonwealth, state and territory agencies delegated tasks within the regional oceans plans.

Members of the Commission would have expertise in matters relevant to its work including in: oceans management and policy development; ecologically sustainable oceans use; oceans science; Indigenous knowledge of and relationships with the oceans environment; environmental law; conservation; and resource-based industries.

Regional oceans plans

The regional oceans plans would be ecosystem-based and adaptive, allocating resources across sectors and across state and Commonwealth waters. The plans would be developed after extensive, inclusive public and cross-sectoral consultation, and give Indigenous communities a significant role in planning and implementing the management of their sea country. Community engagement and education would be the key to a plan's success.

Each of the regional oceans plans would have operational objectives, indicators, targets and timetables, and be underpinned by comprehensive monitoring, risk analysis and review. High-level protection and other zones would be used to spatially conserve biodiversity and manage ecologically sustainable uses. Once prepared, the plans would be approved by Commonwealth, state and territory parliaments.

Monitoring and research

Reforming the oceans governance framework is just one part of the story. To ensure that marine bioregional planning is integrated and ecosystem-based, and that it protects our oceans and establishes their ecologically sustainable use, Australia needs to significantly expand its oceans scientific research effort to build a comprehensive understanding of how our oceans work. An ongoing national program that monitors ocean values and the impacts of various uses on those values should be established to support this.

As a result of the suggested reforms, oceans planning, protection and management would be integrated across sectors and jurisdictions. There would be adequate resources, legislative support and new institutional arrangements to support implementation of the plans, and ocean life, ecosystem functions and processes within ecosystems, as a whole, would be better managed and protected. Planning and management would be adaptive, reflecting advances in knowledge and understanding, based on an expansion of scientific research and monitoring.

The new governance arrangements would also be used to build community capacity – especially for Indigenous communities – to engage in planning, and be collaborative, consistent and accountable. For governments, the new framework would be more cost effective and deliver on international obligations. For industry, it would provide for investor certainty or lessen and avoid future conflicts and delays. The community would benefit from the protection of oceans assets and amenity, while oceans users could be guaranteed healthier and more productive waters. For the nation, it would optimise the economic, social and environmental wealth of oceans through more strategic decision making and resource allocation.

Without our oceans, Australia's environment, economy, society and culture would be very different. To maintain our lifestyle, we need to chart a new course in oceans planning, protection and management.

Conclusion

Australia now leads the world in the planning and protection of its marine environment through marine bioregional planning processes and the almost completed Commonwealth marine reserves network, although this is under threat from the 2015 review by the Australian Government. Even so, Australia has failed to establish integrated, ecosystem-based Regional Marine Planning across industry sectors and jurisdictions, the promise of Australia's 1998 Oceans Policy, while the Commonwealth marine reserves network has been criticised for being residual (Devillers *et al.* 2015), with protection in isolated locations well away from places of commercial interest and failing to protect the broad sweep of ocean habitats.

The challenge remains for Australia to genuinely lead the way by establishing a coordinated national vision for the planning, protection and management of its oceans.

Acknowledgements

Parts of this chapter were adapted from Smyth and Wescott (2013).

Biographies

Chris Smyth is an environmental consultant specialising in marine and coastal planning, protection and management. Chris was previously the healthy oceans campaigner for the Australian Conservation Foundation.

Geoff Wescott is Associate Professor of Environment, Deakin University. He is a past Chair of the Marine and Coastal Community Network and was a member of the National Oceans Advisory Group. He is currently President of the Australian Coastal Society.

References

Australian Conservation Foundation (2005) *Marine Legislative Review.* Australian Conservation Foundation, Melbourne.

Australian Conservation Foundation and National Environmental Law Association (2006) *Out of the Blue: An Act for Australia's Oceans.* Australian Conservation Foundation and National Environmental Law Association, Melbourne.

Australian Government (2011) *Marine Bioregional Plan for the Temperate East Marine Region: Draft for Consultation.* Department of Sustainability, Environment, Water, Population and Communities, Canberra.

Australian Government (2012a) *Marine Bioregional Plan for the North-west Marine Region.* Australian Government, Canberra.

Australian Government (2012b) *Marine Bioregional Plan for the North Marine Region.* Australian Government, Canberra.

Australian Government (2012c) *Marine Bioregional Plan for the South-west Marine Region.* Australian Government, Canberra.

Australian Government (2012d) *Marine Bioregional Plan for the Temperate East Marine Region.* Australian Government, Canberra.

Australian Government (2015) *Goals and Principles for the Establishment of the National Representative System of Marine Protected Areas in Commonwealth Waters.* Australian Government, Canberra, <http://www.environment.gov.au/resource/goals-and-principles-establishment-national-representative-system-marine-protected-areas>.

Commonwealth of Australia (1998a) *Australia's Oceans Policy Volume 1: Caring, Understanding, Using Wisely.* Commonwealth of Australia, Canberra.

Commonwealth of Australia (1998b) *Australia's Oceans Policy Volume 2: Specific Sectoral Measures.* Commonwealth of Australia, Canberra.

Cooper D (2010) Australian oceans most biologically diverse. ABC Science. 3 August 2010, ABC, Sydney, <http://www.abc.net.au/science/articles/2010/08/03/2972213.htm>.

Council on Federal Financial Relations (2015) *Agreements.* Council of Federal Financial Relations, Canberra, <http://www.federalfinancialrelations.gov.au/content/npa/ >.

Department of Environment, Sport and Territories (1997) *Australia's Oceans Policy: Report of the forum held in Canberra on 2–3 December 1997.* Department of Environment, Sport and Territories, Canberra.

Department of the Environment (2015) *Listed Key Threatening Processes.* Department of the Environment, Canberra, <http://www.environment.gov.au/biodiversity/threatened/key-threatening-processes>.

Devillers R, Pressey B, Grech A, Kittinger J, Edgar G, Ward T, Watson R (2015) Reinventing residual reserves in the sea: are we favouring ease of establishment over need for protection? *Aquatic Conservation: Marine and Freshwater Ecosystems* **25**, 480–504. doi:10.1002/aqc.2445.

Government of South Australia (2006) *Draft Spencer Gulf Marine Plan.* Government of South Australia, Adelaide.

Great Barrier Reef Marine Park Authority (2004) *Great Barrier Reef Marine Park Zoning Plan 2003.* Great Barrier Reef Marine Park Authority, Townsville.

Gunn R, Hardesty BD, Butler J (2010) Tackling 'ghost nets': local solutions to a global issue in northern Australia. *Ecological Management & Restoration* **11**, 88–98. doi:10.1111/j.1442-8903.2010.00525.x.

Hawke A (2009) *The Australian Environment Act. Report of the Independent Review of the* Environment Protection and Biodiversity Conservation Act 1999: *Final Report.* Australian Government, Canberra.

Hunt G, Colbeck R (2013) 'Supporting recreational fishing while protecting our marine parks'. Joint media release The Hon. Greg Hunt MP Minister for the Environment and Senator the Hon. Richard Colbeck Parliamentary Secretary to the Minister for Agriculture, 14 December 2013, <http://www.environment.gov.au/minister/hunt/2013/mr20131214.html>.

National Oceans Office (2004) *South-east Regional Marine Plan, Implementing Australia's Oceans Policy in the South-east Marine Region.* National Oceans Office, Hobart.

Smyth C, Wescott G (2013) A future for bioregional marine planning in Australia? *Australian Environmental Review* **28**, 761–765.

Smyth C, Prideaux M, Davey K, Grady M (2003) *Oceans Eleven: the Implementation of Australia's Oceans Policy and Ecosystem-Based Management.* Australian Conservation Foundation, Melbourne.

State of the Environment 2011 Committee (2011) *Australia: State of the Environment 2011.* Report presented to the Australian Government Minister for Sustainability, Environment, Water, Population and Communities. Australian Government, Canberra.

Vince J (2015) Integrated policy approaches and policy failure: the case of Australia's Oceans Policy. *Policy Sciences* **48**, 159–180. doi:10.1007/s11077-015-9215-z.

Vince J, Smith ADM, Sainsbury KJ, Cresswell ID, Smith DC, Haward M (2015) Australia's Oceans Policy: past, present and future. *Marine Policy* **57**, 1–8. doi:10.1016/j.marpol.2015.02.014.

VRFish (2013) *VRFish submission to the VEAC Marine Investigation Discussion Paper.* VRFish, Melbourne. <http://www.veac.vic.gov.au/submissions/published/7125-MI_Subs_Dis_ppr_0035_VRFish.pdf>.

Wescott G (2000) The development and initial implementation of Australia's 'integrated and comprehensive' oceans policy. *Ocean and Coastal Management* **43**, 853–878. doi:10.1016/S0964-5691(00)00067-3.

Wescott G (2012) Disintegration or disinterest? Coastal and marine policy in Australia. In *Environmental Policy Failures: Australian Stories.* (Eds K Crowley and K Walker) pp. 88–101. Tilde University Press, London.

Chapter 20

False polemics or real consensus? What Australians really think about marine protected areas

Paul Sheridan

Introduction

Facebook may have reinterpreted what it means to 'like' something but, thankfully, it does not accurately represent a mirror to Australian attitudes towards marine protected areas (MPAs). If it did, decision makers could make a shallow judgement that no common ground existed between people passionate about fishing and people passionate about MPAs or marine conservation.[1]

Away from the second world of social media, in the 'here and now', the Australian public is saying something quite different. While the statements and retorts posted on Facebook pages may create a perception of a chasm between the opinions of people who fish and those that do not, the reality is that significant common ground exists.

Initiatives in recent decades to implement MPAs have often led to heated debate between fishing and conservation interest groups (Burdon 2012; Fishing World 2014). This chapter explores, instead, how the majority of Australians – the 'middle ground' – view MPAs.

Recent research reveals that significant support exists among this middle ground for high levels of protection of the marine environment, including among a majority of those who identify as conservative and progressive voters (Essential Research 2012a). And, far from being bitterly opposed, a majority of people who fish appear to have similar views (Essential Research 2009b).

Too often, arguments to establish a MPA have failed to engage a broad constituency of Australians (Essential Research 2008a) because the messages they receive are 'framed' in a way that does not fit with their world view (FrameWorks Institute 2005). However, far from being a contest between lifestyles and conservation, most Australians believe MPAs are a logical, sensible and practical step to take (Essential Research 2008a).

Understanding the middle ground

During the Australian Government's assessment and consultation on proposed increases in marine protection in Commonwealth waters between 2009 and 2011, qualitative research in metropolitan and regional areas on the east and west coasts identified a range of attitudes and values that the general public and people who fish share (Essential Research 2009a; Patterson Market Research 2011) (Box 20.1 provides details of polls undertaken by Essential Research from 2008 to 2012).

Levels of public awareness of what lies beneath the waves remains low, despite being a nation of coastal dwellers. However, Australians share an appetite to learn about what is unique about the marine life and marine environment, particularly near to where they live (Essential Research 2008a).

Qualitative research has also found that widespread concern exists that so little of Australia's oceans are protected from the impacts of oil and gas drilling, and over-fishing (Essential Research 2008a).

In follow-up quantitative research, people randomly selected from among the middle ground of Australian opinion and also people who identified as recreational fishers were asked how much of the oceans around Australia's coast should be protected from such key threats. Few nominated low levels of protection (20% of less) and few opposed very high levels of protection (50%+) (Table 20.1).

In fact, across the spectrum from those who never fish to those who fish regularly, more than half of respondents believed at least 50% and as high as 60% or more of our oceans should be protected (Essential Research 2009b).

While there is broad agreement that fishing and oil and gas resources continue to be important for economic and social reasons, there is almost no disagreement that an imbalance exists between resource extraction and conservation of marine biodiversity (Patterson Market Research 2011; Essential Research 2008b).

Quantitative research in 2009 (Essential Research 2009b; national online poll, sample size $n = 4074$) and then later in 2011 (Patterson Market Research 2011; Western Australian telephone poll, sample size $n = 604$) identified significant common ground between people who fish and the rest of the community on the question of how much protection should be put in place to protect the marine environment.

Irrespective of whether people fished or not, very low numbers of people believed only small amounts – 30% or less – of our marine environment should be protected.

The survey results reflected the findings of qualitative research conducted both in metropolitan areas (Sydney, Perth and Brisbane) and in regional areas (Albany and Esperance) in 2008 (Essential Research 2008b).

Participants in the qualitative research program were recruited on the basis that they did not self-identify as environmentalists or supporters of MPAs, but most expressed shock when informed of the small percentage of Australia's ocean territory that was protected from extractive impacts.

When asked how much they thought should be protected, participant responses reflected the survey results in 2009, that at least 50% of Australia's oceans should be protected.

Attitudes shift through personal experiences

What social media can offer, however, is a reminder of the entrenched debates in the public arena that have raged between interest groups over MPAs in state and Commonwealth waters in recent decades (Gladstone 2014).

Although there is often initial concern and also opposition when MPAs are first established, awareness and appreciation of the benefits on offer evolves over time (e.g. Hoisington 2013; see also Chapters 7 and 21, although a contrary view is seen in Chapter 9).

Box 20.1. Details of polls undertaken by Essential Research from 2008–2012 relating to the marine environment

Essential Research (2008a).
Six focus groups were conducted with a total of 41 participants. Two groups were conducted in Perth, two in Darwin and two in Sydney's Central Business District. Groups were split between males and females and participants were drawn from an age bracket of 20–55 years, a mix of tertiary and no-tertiary level education, middle incomes and were recruited on the basis they had no connection to environment organisations.

Essential Research (2008b).
An online poll of 1008 people, aged 18 years plus, were selected randomly from the Your Source national consumer panel of 109 500 Australians. Results were weighted to match the Australian population distribution in terms of age, gender and state/territory. Margin of error was less than 2%.

Essential Research (2009a).
Four focus groups – two each in Perth and Esperance – totalling 37 participants were conducted in December, 2008, with participants drawn from among male and females aged 20–50 years, low to middle incomes, trade or TAFE-level education with no connection to environment organisations.

Essential Research (2009b).
Two questions related to protection of Australia's marine environment and people's fishing behaviour were asked of a sample total of 4074 participants nationally over 4 consecutive weeks in July and August, 2009. The questions were: (1) How often do you go fishing? and (2) How much of the marine environment do you think should be protected?

Essential Research (2012a).
Three online focus groups were conducted in November, 2012, with a total of 39 participants recruited from east and west coast metropolitan centres and regional areas of Western Australia, Northern Territory, South Australia, Queensland, Victoria and New South Wales. Participants were aged 20–60 years, a 50:50 split of males/females, and were recruited on the basis that they had voted for the Liberal or National Parties in the previous two elections. The online groups were conducted over 3 days each, with participants committing to logging on for 2 hours each day.

Essential Research (2012b).
Essential Research conducted four face-to-face focus groups in Brisbane and Gladstone with 'soft' Labor and LNP voters in February. Two online, over time focus groups were conducted with respondents from regional Queensland and Brisbane suburbs. All respondents were aged 25–65 years.

The New South Wales experience

In two key New South Wales marine parks – Solitary Islands and Jervis Bay – where local opinion had been mixed particularly during the establishment period, government-commissioned surveys have tracked the evolution of community attitudes. Support for the

Table 20.1. Results of a national online survey, conducted over 4 weeks, of people aged 18+ years (sample size $n = 4074$), in July and August 2009, which asked: 'How much of Australia's marine environment do you think should be protected?'

(Source: Essential Research 2009b)

How much of Australia's marine environment should be protected?	Response total	Total that fish at least once a year	Fish at least once a month	Fish about once every 3 months	Fish about once every 6 months	Fish about once a year	Fish less often	Never go fishing
Less than 10%	2%	3%	4%	5%	2%	2%	2%	2%
At least 10%	5%	6%	8%	6%	5%	5%	4%	5%
At least 20%	6%	8%	11%	9%	12%	5%	7%	5%
At least 30%	8%	11%	9%	13%	13%	10%	9%	7%
At least 40%	5%	6%	7%	6%	5%	6%	5%	4%
At least 50%	24%	24%	18%	24%	28%	24%	22%	25%
More than 60%	40%	34%	30%	32%	31%	39%	41%	42%
Don't know	10%	8%	13%	5%	4%	9%	10%	10%

marine parks is now high across local communities, with support for no-take sanctuary areas at 80% or higher (McGregor Tan Research 2008a; see also Chapter 11).

In the region surrounding Solitary Islands Marine Park, surveys have found the marine park is widely accepted and supported by the general community. In January 2008, an attitudinal survey of 407 local residents found 87% support for the conservation of the marine environment (McGregor Tan Research 2008a). This level of support was demonstrated among both non-fishers (88% in favour of the marine park) and fishers (82% in favour of the marine park) (McGregor Tan Research 2008a).

Sanctuary areas in the marine park were also highly supported – by 80% of all respondents. When asked to rate a list of reasons for protecting the marine park, half (50%) of those surveyed indicated that it was to support a diversity of marine and bird life. The next two most supported reasons were: allowing future generations to enjoy the marine park (44%) and protecting natural habitats (42%) (McGregor Tan Research 2008a).

In January 2008, an attitudinal survey of 402 local residents surrounding Jervis Bay Marine Park found that 84% supported the marine park, with only 4% opposed. Sanctuary zones in the marine park were supported by 82% of the respondents, with only 5% opposed. Among fishers, 76% supported the sanctuary zones, while support among non-fishers was higher at 84% (McGregor Tan Research 2008b).

When asked to nominate a benefit of marine parks, an overwhelming majority (89%) of those surveyed were able to name at least one benefit. The four main benefits identified were: 'preservation of the environment for future generations' (51%); 'protection of marine habitats' (47%); 'a better environment' (21%); and 'increase in fish stocks' (20%) (McGregor Tan Research 2008b).

The Queensland experience

Three years after the Great Barrier Reef Marine Park was rezoned in 2004 and the no-take sanctuary area increased from 4.5% to 33% of the Park (see Chapter 5), a survey of recreational fishers was conducted. A majority of fishers believed the rezoning was necessary and had little impact on their fishing activity (Sutton and Tobin 2009). The survey results revealed the high levels of support for conservation measures among recreational fishers, particularly if they are consistent with their values.

The findings of qualitative research by Essential Research in Queensland in February 2012 both reinforced the findings above and provided further insights into the common ground that exists between fishers and non-fishers. Participants for four focus groups were recruited in Brisbane and in Gladstone to test attitudes to established MPAs at Moreton Bay (see also Chapter 12) and on the Great Barrier Reef, as well as to explore opinions of proposals for new protected areas off the Queensland coast (Essential Research 2012b). A mix of male and female participants were selected on the basis that they were not members of an environmental organisation or were supporters of MPAs. Instead, recruitment was narrowly focused on people who identified as being a fisher or were interested in issues affecting the marine environment.

Most believed there was no single main threat, but several threats to the marine environment. Unprompted, the reasons, in order of priority, were 'pollution from the land', 'overfishing' and 'damage from oil and other industrial spills'.

Among the Gladstone participants and, more specifically, the recreational fisher participants, there was agreement that fish stocks were nowhere near as good as they used to be and, as one said, 'you can't even reach a bag limit for many species any more'.

The recreational fishers considered their pastime to be a 'right', a natural and important thing to be able to do. Many maintain, however, that they are 'conservation-minded' and misunderstood by 'extreme greenies'. While many acknowledged that over-fishing is a threat, they tended to blame 'the professionals', specifically commercial operators who used destructive techniques such as gill-netting and bottom trawling.

Recreational fishers expressed a preference for seasonal closures as an approach to improve the health of fish stocks and marine life, but there was almost universal agreement that sanctuary zones protecting feeding and breeding areas are necessary and beneficial. The majority of participants agreed that MPAs would help to sustain fishing and the fishing industry. Regional participants also nominated a further benefit of MPAs was that they would enhancing tourism opportunities.

Moving on from conflict: reflecting the middle ground

Since 2009, when national attention turned to an election commitment by the then Labor Australian Government to increase protection of Australia's Commonwealth waters, public campaigns in support of MPAs focused on developing a positive public narrative about the value of safeguarding the oceans and marine life for the benefit of all Australians now and in the future.[2]

Doing so demonstrated a deeper understanding of the 'middle ground', or majority of Australians' opinion about MPAs and the values most people share and reflect when making decisions about propositions to establish MPAs.

Equally important was an avoidance of narrowly communicating to fishers in an attempt to head off conflict. Much of the opposition from fishing groups to MPAs has centred on the location of 'no-take' sanctuary zones and communication efforts in the past have centred on efforts to avoid conflict by highlighting benefits of sanctuary zones to fisheries (Dreher *et al.* 2013).

Sanctuary zones are, however, not designed to be fisheries management tools. Further, the communications appeal to fishers about fisheries benefits has failed to satisfy (Voyer *et al.* 2013). Instead, successful efforts to tap into the views of the majority of Australians – and local residents in close proximity to MPAs – have involved a process of raising awareness about the uniqueness and diversity of marine life found in our waters and providing evidence of the full range of benefits that sanctuaries are designed to, and have been proven to, deliver.

Opponents of improved or increased marine protection have in the past successfully established a negative public messaging frame by arguing that proposals to tighten regulations, establish marine parks or sanctuary areas means a choice has to be made between pursuing 'our lifestyle' and protecting the environment (Essential Research 2008a; Banks and Skilleter 2010).

Recent campaigns to improve protection of the marine environment have been able to shift public debate away from a focus on an 'either/or' scenario – that to create a MPA requires a choice to be made between maintaining lifestyle or increasing protection – and instead reinforce an inter-dependent scenario – that to maintain lifestyle depends on maintaining a healthy marine environment.

Conservatives love nature too

Traditional environmental campaigns have focused on mobilising an environmentally concerned audience to take actions that demonstrate a constituency of support. This has been done by targeting core environmental values, such as a 'life on Earth' value that describes a belief that the health of the natural environment is fundamental to all life and should be protected.

This core environmental value is most commonly shared among supporters or members of environmental groups and also progressive voters. However, the expression of this value also occurs widely among people who identify as economic rationalists and vote conservatively as well (Essential Research 2012a). Qualitative research of people identifying themselves as Liberal/National Party supporters in November 2012 found that despite their conservative voting intentions, this 'life on Earth' value was clearly evident (Essential Research 2012a). When asked directly about the challenges facing Australia's natural environment, participants spoke passionately about the need to protect the environment. Specifically, the main environmental challenges identified were the 'need to protect what makes Australia unique and special' (Essential Research 2012a).

The power of positivity

Recent successful campaigns to establish MPAs have been able to demonstrate that a majority of people support high levels of marine protection.

The 'either/or' message frame pursued by opponents to MPAs has also been shown to resonate among only a very limited section of Australians. For example, surveys conducted in 2008 found only 14% of 1008 people surveyed believed 'people and jobs come before the environment and animals' whereas 73% believed 'a strong economy depends on maintaining a healthy environment' (Table 20.2).

During the Australian Government's assessment of areas of Commonwealth waters for further protection between 2009 and 2012, the Save Our Marine Life alliance of conservation groups adopted this positive, inter-dependent message frame to great effect (e.g. Kiernan 2011).

Recreational fishing lobby groups have consistently claimed to represent an estimated 5 million Australians who fish (Fishing World 2013a,b). There is, however, no evidence that concerns or opposition to MPAs determines how five million recreational fishers in Australia vote at a state or federal election. If indeed there are 5 million people who identify as recreational fishers in Australia, further research of their views of marine conservation issues in general, MPAs specifically, and also their voting intentions is necessary. Surveys of recreational fishers have been limited to specific regions where MPAs are located and where fishers visit (e.g. McGregor Tan Research 2008a) and follow up national research is now necessary to further explore attitudes and priorities.

Far from controversial: the middle ground view

Propositions to create MPAs have either genuinely generated localised controversy or proponents have often failed to prevent a perception becoming entrenched that their creation is controversial (Meder 2013).

Most of Australians – the middle ground – do not consider MPAs to be controversial and, instead, support high level of protection for the marine environment. A not dissimilar majority of people who fish recreationally also support MPAs, including no-take sanctuary zones.

In places where MPAs have been established, initial concerns or opposition have dissipated and first-hand experiences among the general public and fishers who frequent the protected areas are overwhelmingly positive (Fig. 20.1) (see Chapter 21). The middle

Table 20.2. Results of an online national survey of 1008 people aged 18 to 65 on their attitudes towards environmental protection and economic development conducted in September 2008 and then repeated in 2009 asked: 'Which of the following statements is closest to your view?'

(Source: Essential Research 2008b, 2009b)

		Age						Gender	
	Total	**18–25**	**26–35**	**36–45**	**46–55**	**56–65**	**65+**	**Male**	**Female**
People and jobs come before the environment and animals	14%	20%	13%	15%	8%	11%	19%	18%	9%
A strong economy depends on maintaining a healthy environment	73%	61%	69%	67%	82%	85%	73%	69%	76%
Don't know	13%	19%	18%	18%	10%	4%	8%	13%	15%

		Australian state					Region	
	Total	**NSW**	**VIC/TAS**	**QLD**	**SA**	**WA**	**Metro**	**Regional**
People and jobs come before the environment and animals	14%	15%	16%	7%	11%	20%	13%	15%
A strong economy depends on maintaining a healthy environment	73%	68%	69%	82%	82%	73%	73%	71%
Don't know	13%	17%	15%	11%	7%	7%	14%	14%

Fig. 20.1. Recreational fishing within Batemans Marine Park, New South Wales. (Photo: James Fitzsimons)

ground of Australian opinion of MPAs reflects an attitude that they are a logical and sensible step to take to improve the health of the marine environment and prevent a decline of species. Successfully connecting with and reflecting back to decision makers the views of the middle ground towards marine protection and MPAs enables proponents to also demonstrate that opponent interest groups' claims about representing a majority view are inflated at best.

Biography

Paul Sheridan manages communications for the Pew Charitable Trusts in Australia. He has more than 20 years experience as a media and strategic communications adviser focused on environmental issues. Paul started his career as a reporter on suburban and metropolitan daily newspapers in Sydney before specialising in environmental communications for the NSW National Parks and Wildlife Service, NSW Environmental Protection Authority, as Press Secretary for a NSW Environment Minister, at Clean Up Australia and at a strategic communications agency Essential Media Communications.

Endnotes

1. See for example https://www.facebook.com/KeepAustraliaFishing?fref=ts and https://www.facebook.com/SaveourSAMarineParks
2. For example, the Save Our Marine Life campaign – www.saveourmarinelife.org.au

References

Banks SA, Skilleter GA (2010) Implementing marine reserve networks: a comparison of approaches in New South Wales and New Zealand. *Marine Policy* **34**, 197–207. doi:10.1016/j.marpol.2009.06.006.

Burdon D (2012) 'Fishing lobby slams marine parks'. *Tweed Daily News* 15 June 2012, <http://www.tweeddailynews.com.au/news/fishing-lobby-slams-marine-parks/1419348/>.

Dreher T, Gladstone W, Goodall H, Voyer M (2013) Dodgy science or global necessity: local media reporting of marine parks. In *Environmental Conflict and the Media* (Eds B Hutchins and L Lester) pp. 153–168. Peter Lang Publishing, New York.

Essential Research (2008a) *Report on Qualitative Research of Public Perceptions and Attitudes to Unspoiled Areas of Australia, June 2008*. Report to Pew Environment Group and The Nature Conservancy, Sydney.

Essential Research (2008b) *National Poll of Perceptions and Attitudes to Environment Issues, September 2008*. Report to The Nature Conservancy and Pew Environment Group, Sydney.

Essential Research (2009a) *Report on Qualitative Research Testing of Attitudes to the Marine Environment of Australia's Southwest and Issues Affecting It, January 2009*. Report to Pew Environment Group, Perth.

Essential Research (2009b) *Results of National Polling, Conducted Online Over Four Consecutive Weeks, Testing Support Among the General Public and Fishers on Marine Protection. Sample Size: 4072, August 2009*. Report to Save Our Marine Life alliance, Perth.

Essential Research (2012a) *Report on Qualitative Research of Conservative Voters' Attitudes Toward Australia's Natural Environment, October 2012*. Report to Pew Charitable Trusts, Sydney.

Essential Research (2012b) *Report on Qualitative Research of Public Attitudes Towards Queensland's Marine Environment and Fisheries, February 2012*. Report to Graeme Wood Foundation, Brisbane.

Fishing World (2013a) 'Keep Australia Fishing launches election campaign'. *Fishing World*, 22 August 2013, <http://www.fishingworld.com.au/news/keep-australia-fishing-launches-election-campaign>.

Fishing World (2013b) 'Aussie rec fishers call for "balance" on marine parks'. *Fishing World*, 23 February 2013, <http://www.fishingworld.com.au/news/aussie-rec-fishers-call-for-balance-on-marine-parks>.

Fishing World (2014) 'Anglers slam Labor over Sydney lock-outs'. *Fishing World*, 2 October 2014, <http://www.fishingworld.com.au/news/anglers-slam-labor-over-sydney-lock-outs>.

FrameWorks Institute (2005) *Framing Public Issues, April 2005*. FrameWorks Institute, Washington DC, <http://www.frameworksinstitute.org/assets/files/PDF/FramingPublicIssuesfinal.pdf>.

Gladstone W (2014) Criticisms of science, social impacts, opinion leaders, and targets for no-take zones led to cuts in New South Wales' (Australia) system of marine protected areas. *Aquatic Conservation: Marine and Freshwater Ecosystems* **24**, 287–296. doi:10.1002/aqc.2469.

Hoisington C (2013) *The Marine Protection Dividend, November 2013*. Centre for Policy Development Occasional Paper. Centre for Policy Development, Sydney, <http://cpd.org.au/wp-content/uploads/2014/01/CPD-OP-34-Marine-Protection-Dividend-v3.pdf>.

Kiernan I (2011) 'Sea parks are good for people and fish'. *The Age*, 20 January 2011, <http://www.theage.com.au/it-pro/sea-parks-are-good-for-people-and-fish-20110119–19wj2.html>.

McGregor Tan Research (2008a) *Final report, Solitary Islands Marine Park Community Survey, for the NSW Government, February 2008.* McGregor Tan Research, Sydney, <http://www.mpa.nsw.gov.au/pdf/SIMP-Community-survey-2008.pdf>.

McGregor Tan Research (2008b) *Final report, Jervis Bay Marine Park Community Survey, for the NSW Government, February 2008.* McGregor Tan Research, Sydney, <http://www.mpa.nsw.gov.au/pdf/JBMP-Community-Survey-2008.pdf>.

Meder A (2013) Why have Australian environment groups, the marine science community and the recreational fishing community been oceans apart over marine parks? In *Australia's Marine Environment: Progress, Challenges and Opportunities.* (Eds K Zischka, A Leverington and P Figgis). Australian Committee for IUCN, Sydney, <http://aciucn.org.au/wp-content/uploads/2013/10/16-Meder.pdf>.

Patterson Market Research (2011) *Western Australian Community Attitudes Towards Marine Sanctuaries, April 2011.* Save Our Marine Life alliance, Perth, <http://www.saveourmarinelife.org.au/wp-content/uploads/2014/07/pmr_attitudes_to_marine_sanctuaries_final_report.pdf>.

Sutton SG, Tobin RC (2009) Recreational fishers' attitudes towards the 2004 rezoning of the Great Barrier Reef Marine Park. *Environmental Conservation* **36**, 245–252. doi:10.1017/S0376892909990270.

Voyer M, Goodall H, Gladstone W (2013) 'Go fish: why fishers don't care for marine parks'. *The Conversation*, 4 June 2013, <http://theconversation.com/go-fish-why-fishers-dont-care-for-marine-parks-14558>.

Chapter 21

'The issue was that big, I swear!': evidence for the real impacts of marine protected areas on Australian recreational fishing

Adrian Meder

Introduction

Australia already has an extensive marine protected area (MPA) network, but has yet to secure the development of a spatial conservation network that effectively represents the full range of marine habitats and ecological features (Devillers *et al.* 2015). Although the ecological evidence for the value of MPAs in Australia, even with our relatively world-class fisheries management, is abundant and rapidly growing (Ling *et al.* 2009; Babcock *et al.* 2010; McCook *et al.* 2010; Harrison *et al.* 2012; Olds *et al.* 2014), we have fallen behind in documenting what happens to the human communities along the coastlines of the extensive MPAs we already have.

Recreational fishers are key stakeholders in Australian marine conservation and fisheries management, given that recreational fishing is one of our most popular pastimes, with an estimated 3.4 million participants each year (Stephan and Hobsbawn 2014). Fishing is strongly woven into the fabric of the ocean-loving culture that has in large part driven Australia's progress towards becoming a leader in fisheries and conservation management. An estimated 20% of Australians are recreational fishers (Griffiths *et al.* 2014), who hold a strong connection with and understanding of their local marine environment in ways that can surpass that of ecological and fisheries scientists. Critically they are asked to surrender spatial access (no-take areas/zones are at the centre of recreational fishers' concerns in regard to MPA planning) in the interests of conservation management, through processes often driven by expertise and public support that originates outside their local communities, yet they are not as explicitly entitled to financial (or otherwise) compensation for any reductions in spatial access as other extractive users. For these reasons, they have generally been regarded as being opposed to the establishment of highly protected MPAs and public statements from lobby groups with regard to no-take zones support this.

This chapter examines the growing body of research that has documented Australian recreational fisher's attitudes, perceptions and experiences around MPAs, and the nature of recreational fishing politics in relation to MPA planning. There is a clear need for a shift in Australian MPA planning towards a model that explicitly emphasises empirical evidence around how recreational fishers have already been impacted by the extensive suite of existing MPAs, in order to compare the more rhetorical claims at either end of the spectrum of the debate around their establishment with the 'lived experience' of the large proportion of the recreational fishing community that already fishes in MPAs. The current approach has been responsive and reactionary decision making based on the relative levels of 'noise' from supportive versus oppositional voices within the community in the latter stages of planning. A new approach can allow more realistic predictions of the level of impacts on recreational fishing communities, both initially and how they might be expected to change over time. Furthermore, and perhaps most importantly, it can enable less polarising discussions and debates within affected communities during the planning stage.

'Political perception of Australian fishing is flawed'

Recent political decisions around MPAs and recreational fishing have apparently assumed that recreational fishers broadly oppose no-take MPAs, and are largely in line with the views of peak recreational fishing groups and many opinion-leading individuals that are active in recreational fishing 'politics'. Such decisions include: the 'reopening' of no-take MPA zones to recreational fishing in NSW in 2012, later largely revoked as a result of widespread community opposition and an apparent lack of significant political 'reward' or demonstrable community benefit (ABC 2014); similar proposals in Queensland in 2014 (Newman *et al.* 2014) that were not put into action due to an apparent lack of support from recreational fishers (Crawley 2014) and strong opposition from the wider community and tourism industries (and defeat at the by-election the announcement was associated with; Dickson 2014); and the decision to locate only 4% of no-take MPA protection in the Federal Labor Government's 2012 Commonwealth Marine Reserve Network zoning plan within 100 km of the Australian coastline (Bax and Creswell 2012).

Politicians, decision makers and conservation advocates have been guilty of treating recreational fishers as a 'subculture'. In reality, recreational fishers are more accurately viewed as a broad cross-section of the wider general public, because such a large proportion of the general public enjoys fishing (particularly in the regional areas away from urban centres where MPAs are mostly located), and because their spectrum of values, personal attitudes towards conservation and concerns from a political perspective are likely to reflect the full spectrum in the broader community (McPhee 2008). If there are any defining characteristics, they would include only a slight tendency towards living outside metropolitan areas, being predominantly male and slightly older than the median age of the Australian population (Henry and Lyle 2003; Lyle *et al.* 2009; Taylor *et al.* 2012).

What Australian recreational fishers really feel about MPAs – the empirical evidence case

Australian recreational fisher's attitudes and perceptions about no-take MPAs – in principle and in practice (in the years after their implementation in local waters) – have been investigated many times in academic and government studies, and market research surveys commissioned by stakeholder groups. The finding that a significant majority of recreational fishers support no-take MPA protection in their waters, once their establishment is embedded and in principle, is common to every published example of such a study this author is aware of. These findings are summarised in Table 21.1.

Market research polling commissioned by stakeholder groups has also demonstrated clear majority support for no-take MPAs among recreational fishers. For example, a survey using a representative sample of New South Wales recreational fishers in 2014 found that:

Fig. 21.1. Luderick fishers at Port Stephens Marine Park, NSW, 2014. It is clear that there are abundant and rewarding family fishing opportunities in our existing suite of marine parks, despite continuing claims that marine parks 'lock out' recreational fishers. (Photo: Adrian Meder)

91% broadly support no-take MPAs; 74% said existing no-take zones in NSW have had no impact on their participation in and enjoyment of recreational fishing (13% said there was reduced enjoyment/participation); 79% said the existing no-take zones have had no impact on their choices for fishing trip holiday destination (9% said there was a negative impact); and 86% said their spend on fishing and related boating and camping equipment, fishing tackle and bait, has been similarly unaffected by existing no-take zones (8% said their spend had declined) (Galaxy Research 2014) (Fig. 21.1). It is notable that this strong support followed an unprecedented national campaign aimed at building opposition for no-take MPAs, centred around a simplistic 'Don't lock us out' message. These findings also indicate that no-take MPAs have little perceived negative impact among this recreational fishing community on tourism and the recreational fishing-related economy.

A survey of ~300 Australian recreational fishers by Australia's Fisheries Research & Development Corporation during the height of debate over the potential impacts of expansion of the Commonwealth marine reserve network (see also Chapter 4) found broad positivity for the future of fishing (57% positive versus 14% negative) (FRDC 2011). While a part of reported negative sentiment related to concern over spatial access, one of the main themes in responses to a question around the 'most successful advancement' for the fishing industry that has come from research and development was:

> *'Success in and through the creation of marine parks; this was mentioned by a number of recreational fishers as a significant long term benefit for the industry'.* (FRDC 2011, p. 17)

The only exception recorded (that the author is aware of at the time of writing) in the scientific literature to this overwhelmingly positive sentiment in favour of protection was

Table 21.1. Summary of published academic or government surveys of recreational fishers' attitudes/perceptions to no-take MPA zones or effects on their fishing activity

Measure of support	Location	Number surveyed	Proportion of MPA zoned in no-take at the time	Years after zoning change/introduction that survey was undertaken
73% of local fishers said Great Barrier Reef Marine Park had either no effect, or a positive effect on their recreational fishing activity (Sutton and Li 2008)	Great Barrier Reef local fishers (northern Queensland)	> 800	33.6%	3
Most fishers reported believing that there was a positive effect on fisheries and sustainability associated with the zoning plan (76% agree, 9% disagree) and the no-take zones in general (67% agree, 10% disagree) (Arias and Sutton 2013)	Great Barrier Reef local fishers/ Townsville	144	33.6%	5
64.7% of boaters in Moreton Bay Marine Park said the no-take zones would be positive for the biology of Moreton Bay (Department of Environment and Resource Management 2012)	Moreton Bay (Brisbane) boaters[a]	200	16%	up to 24 months
88% of local fishers are neutral or support no-take zones in Solitary Islands Marine Park. 68% 'strongly support' or 'support' no-take zones in general (McGregor Tan Research 2008a)	Local region fishers	97	12%	6
76% of local fishers 'strongly support' or 'support' no-take zones in Jervis Bay Marine Park (McGregor Tan Research 2008b)	Local region fishers	125	20%	6
90% of fishers supported Lord Howe Island Marine Park no-take zones (NSW Marine Parks Authority 2010)	Lord Howe Island local/visitor fishers[a]	< 500	30%	5
88% of fishers gave in principle support to the establishment of no-take 'biodiversity conservation sanctuaries' in Western Australia (Prior and Beckley 2007)	South-western Australia, local/visitor fishers	512	N/A	N/A

In year after rezoning, 80% of Ningaloo Marine Park fishers claimed to be 'affected' by no-take zones – 99% said they would return again. Reported visitor satisfaction along Ningaloo coast was 95–98% (Northcote and Macbeth 2008a)	Ningaloo Reef Marine Park visiting fishers[a], Western Australia	< 358	34%	1
Three years after rezoning, 75% of repeat visitor fishers surveyed said Ningaloo Marine Park no-take zones did not affect their fishing (Smallwood and Beckley 2012)	Ningaloo Reef Marine Park visiting fishers, Western Australia	328	34%	3
54.5% of local residents were generally unhappy with no-take zone changes (Northcote and Macbeth 2008a)	Ningaloo Reef Marine Park local residents[a], Western Australia	135	34%	< 1
79.6% and 45.9% of 'wilderness campers' at two locations opposed sanctuary zones (Northcote and Macbeth 2008b)	Ningaloo Reef Marine Park visiting campers[a], Western Australia	347	34%	1.5

[a]Note: sample is of a population that contains, but is not exclusively, recreational fishers

at Ningaloo Marine Park (at the time of writing, one of Australia's largest nearshore highly protected MPAs, with 34% no-take zoning), where 54.5% of local residents and at one location 79.6% of visiting campers (not explicitly recreational fishers) were unhappy with the zoning changes (Northcote and Macbeth 2008a,b). However, these findings came very soon after that marine park's expansion, and those studies note that, among those dissatisfied with the marine park zoning, opposition was likely directed at the perceived process by which the no-take zone management plan was devised and implemented, more so than the principle of no-take zone protection itself (see also Chapter 7). Furthermore, very high overall satisfaction rates accompanied this perception for visiting campers, with 98.2% of the most aggrieved survey respondents also rating their visit as either good or excellent (Northcote and Macbeth 2008b). A more recent management audit, found that 8 years after the marine park expansion, visitor numbers had almost doubled and increasing recreational fishing effort had become a major management priority (Marine Parks and Reserves Authority 2013), strongly indicating that initial perceptions of opposition and inconvenience had scant lasting negative impacts.

Studies that have explored how Australian recreational fishing has been affected (and the perception of effects) by MPAs reveal some interesting results. Most available data comes from Australia's (currently) largest MPAs in inshore waters: at Ningaloo in the north-west and the Great Barrier Reef in the north-east. In the first year after the expansion of Ningaloo Marine Park, ~80% of recreational fishers reported that their activity had been affected by the no-take zones (in that they had to change fishing locations), but 98% also reported very high satisfaction rates, and 99% said they would return again (Northcote and Macbeth 2008a). Three years later, 75% of recreational fishers surveyed said their fishing activity was not affected (Smallwood and Beckley 2012). It should be noted that fishing activity being 'affected' in this study does not discriminate between simply changing fishing locations and more consequential negative impacts such as reduced ability to participate in fishing or catch fish. The latter would likely be a smaller proportion; again, a government audit undertaken 9 years after expansion found that visitor numbers had almost doubled and recreational fishing effort was increasing (Marine Parks and Reserves Authority 2013). In the Great Barrier Reef Marine Park, 73% of local fishers surveyed reported that their fishing had not changed, or was positively affected, 3 years after the zoning was put in place (Sutton and Li 2008). A study of Townsville recreational fishers (the largest population centre nearest to the Great Barrier Reef Marine Park) 2 years later found that most fishers reported believing that there was a positive effect on fisheries and sustainability associated with no-take zones (76% agree) and the zoning plan in general (67% agree) (Arias and Sutton 2013).

In an extensive survey of on-water users of the Moreton Bay Marine Park, only 6% of recreational fishing activity was displaced (in a MPA with 16% no-take protection) documented in a field study based on direct observations of fishing (Fig. 21.2). However, greater than 40% perceived they were displaced in opinion surveys (Pascoe *et al.* 2014). Furthermore, studies of catch rates and distance travelled to fish in that marine park indicate that, contrary to predictions before establishment that the MPA would lead to tens of millions of dollars in negative economic impacts on local recreational fishers (McPhee *et al.* 2007), empirical data following implementation suggested instead a positive economic benefit (and certainly not a negative impact) to recreational fishing, arising after only 3 years of protection (Pascoe *et al.* 2014; see also Chapter 12).

A study of long-term changes in recreational fishing effort in a temperate marine park (Jervis Bay) showed that fishing effort was not redistributed significantly on a spatial scale

Fig. 21.2. 'Mums and Dads' fishers at Moreton Bay Marine Park. On any given weekend hundreds, if not thousands, of families can be found enjoying recreational fishing in Moreton Bay, a marine park in the backyard of a major capital city. Predictions about the impacts of MPAs on recreational fishing can, and must be, informed by information available from the extensive suite of existing MPAs in Australian waters. (Photo: Adrian Meder)

within the MPA (beyond the no-take zones), but a sharp decline in fishing effort over several years followed MPA establishment (Lynch 2006, 2008, 2014; see also Chapter 11). This is the only published Australian example of a significant decline in local fishing participation following local MPA zoning changes, but also likely reflects other social and fishery management externalities, such as the introduction of large areas known as 'recreational fishing havens' (a resource reallocation mechanism allowing exclusive access for recreational fishing) throughout NSW waters, including a basin directly adjacent to Jervis Bay, in the year following the marine park's establishment.

A study of the redistribution of effort following the implementation of the Great Barrier Reef Marine Park (De Freitas *et al.* 2013) showed that 56% of surveyed fishers in two major coastal recreational fishing communities reported spatial displacement as a result of the 2004 rezoning (to 33.6% no-take; see also Chapter 5). Although the study suggested that such displacement could potentially result in increased pressure on fish stocks in non-restricted areas, this requires further monitoring to determine if negative impacts on fish stocks will arise. Subsequent empirical studies point to potential direct fishery benefits to key recreational target species (Harrison *et al.* 2012), and there is little objective evidence of reported recreational fishery declines since the MPA's rezoning in fishery management reporting (Queensland Government 2012) or ecological surveys (Russ *et al.* 2008), although stock status information is generally lacking. Interestingly, fishing effort redistribution resulted in fishers travelling 27% less distance overall, a result reflecting findings in the Moreton Bay Marine Park (and attributed as a benefit to fishers, see Pascoe *et al.* 2014).

This suggests that redistribution of recreational fishers as a result of MPAs does not, as often claimed in the public debate, necessarily force fishers to travel further (which is potentially less safe). Further, attitude surveys of one of those communities showed that most recreational fishers believed that, despite any displacement, there were positive fisheries and sustainability benefits as a result of the no-take zones (Arias and Sutton 2013).

The nature of recreational fishing opposition to MPAs in Australia

Recreational fishing interest groups have formed oppositional 'campaigns' of some sort against almost every significant MPA proposal in Australian history. These have been based on broadly consistent themes:

- MPAs represent an attack on the 'right to fish', by a movement (conservation groups) with an anti-fishing agenda.
- MPAs affect 'mums and dads' by reducing their participation in fishing, values of enjoyment around fishing, and ability to catch their own seafood. They also force fishers further offshore, threatening their safety.
- MPAs negatively affect regional economies by reducing tourism income generation, and possibly by depletion of fish stocks.

These arguments have been repeated in very similar fashion since the community debate over MPAs began. There is evidence of such arguments going back to the 1970s, with one headline about the declaration of a MPA in a small New Zealand coastal town reading 'Nothing to do at Goat Island Bay anymore' (Warne 2007). It is notable that this particular MPA now features in New Zealand Government-backed international tourism marketing,[1] and attracts in excess of 300 000 tourist visits annually (Hunt 2008). The town remains a regionally important commercial and recreational fishing port.

These arguments have been subject to little objective critical analysis of scientific evidence – around how Australian recreational fishers feel about MPAs, or what the effects of actual MPAs on actual Australian communities have been. Neither, it would appear, has the historical and anecdotal evidence – such as the regular and continued featuring of historically renowned fishing destinations in fishing magazines (after their becoming MPAs). For example, Ningaloo Reef's excellent recreational fishing opportunities are featured in at least five online articles since 2011 in a widely circulated Australian fishing magazine. Ningaloo Reef and its lagoon (all of which were placed inside a highly protected MPA network around 7 years prior) are promoted in one article (Fishing World 2012) as '...a veritable wonderland of fishing opportunities'. Elsewhere, the main town in the region is described:

> *'Exmouth is the new frontier of Aussie sport and gamefishing. I can't think of anywhere else that's got the same buzz about it as Exmouth, or anywhere that offers such a diversity of fishing'.*

Clearly, such sentiment contrasts with much recreational fishing media opinion predicting how communities are to be affected by new MPAs. A decade earlier, when the Ningaloo Marine Park was significantly expanded, recreational fishing interest groups claimed that 'local businesses will suffer because anglers will no longer visit the area' (ABC 2004).

A 'common sense' conclusion available to the casual observer is that Australia has maintained strong recreational fishing communities and tourism economies following the

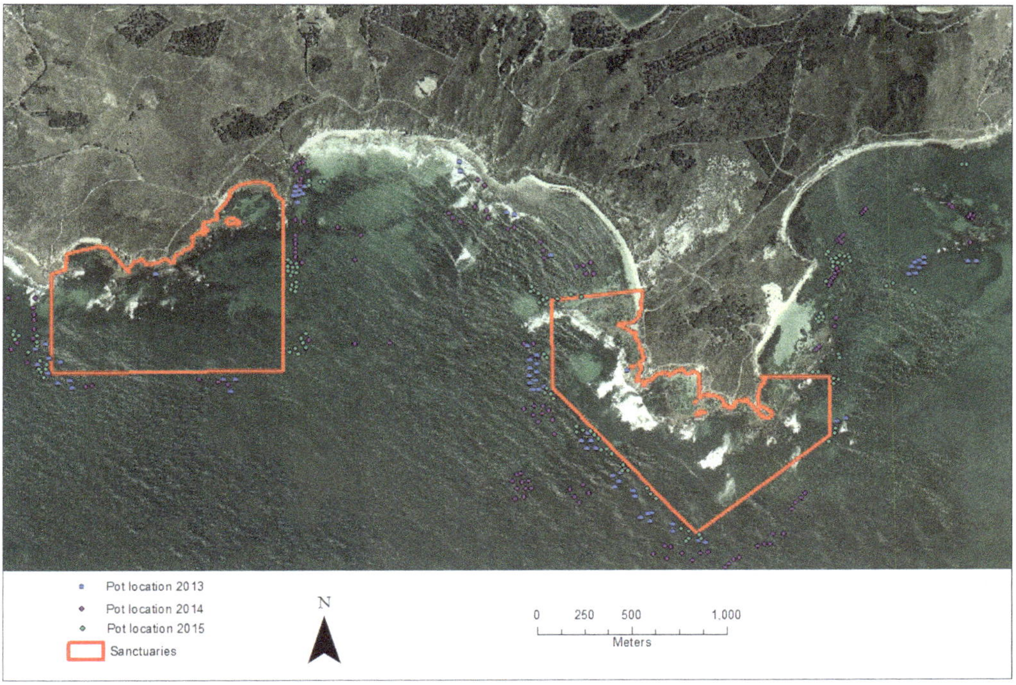

Fig. 21.3. Location of recreational cray pots deployed around no-take zones ('sanctuaries') in Rottnest Island Marine Park, Western Australia, in annual surveys from 2013 to 2015. This documents a pattern often anecdotally described of recreational fishing behaviour – that fishing effort can be concentrated around the boundaries of no-take zones – indicating from fishers actions that there is at least a perceived benefit for fishing associated with these zones. A crayfisher is more likely to deploy a cray pot in an area perceived, based on their knowledge and experience, to deliver better odds of a successful catch. (Source: Pia Bessell-Browne *et al.*, pers. comm.)

implementation of marine parks. For example, throughout Western Australia, many iconic recreational fishing destinations – such as the Rowley Shoals, the Montebello Islands, Ningaloo Reef, Shark Bay, the Abrolhos Islands, as well as important 'family fishing' destinations such as Rottnest Island (Fig. 21.3) and the Busselton Jetty – have remained just that in the years after becoming MPAs and apparently with little clear correlation to levels of spatial access restriction.

There has been little objective debate or investigation in Australia of the relationship between spatial access to water in which to fish, and access to the actual values that define the perception of good sustainable fishing (which can be seen as the essence of the 'right' or 'privilege' to fish): the ability to participate in fishing, derive enjoyment from fishing, and successfully catch fish. These perceptions have been dynamic over time quite independently of MPA effects, exemplified by the increasing importance of catch and release fishing, increasing willingness to travel further offshore as fishing technology and safety improves and affluence increases, and the changing importance of particular target species over time.

The summary of Australian recreational fisher's perceptions provided in Table 21.1 provides key information in this regard – that existing Australian MPAs have apparently not reached a level of spatial access restriction that caused widely held and long-lasting perceived negative impacts to these values, even when levels of spatial access restriction through no-take zones are in the order of one-third of accessible waters. Regardless, perceived benefits of no-take MPA protection (such as a perception that there will be increased access to 'fish to catch' in the long run) have apparently outweighed the impacts caused by this loss of 'water to fish in' (Arias and Sutton 2013). If there were an anti-fishing agenda behind the advocacy for MPA network expansion, it appears to have overwhelmingly failed to curtail Australian recreational fishing catches, effort, values of enjoyment or associated economies.

This raises questions around why recreational fishing interest groups continue to campaign against MPAs when their core constituency appears to hold considerable – even overwhelming – support for the mechanism after a decade or more of 'lived experience'. For example, surveys in 2014–2015 of supporter bases for campaigns by environmental non-government organisations calling for increased MPA protection in Australia have counted 30–40% of supporters describing themselves as recreational fishers: a higher proportion than in the general populace (author's unpublished data). This author cannot speak for those recreational fishing interest groups, but speculates that:

- The symbolic value of portraying reduced spatial access as equivalent to a reduction in 'rights' to conduct an activity of cultural value is easy to communicate and easy to understand at face value.
- There is a natural, and perhaps not entirely unjustified, scepticism and suspicion of the 'messenger': environmentalists, ecological scientists and government employees among a portion of the wider Australian community on any issue that undoubtedly includes some recreational fishers. There has been no MPA planning in Australia that has not heavily involved these 'messengers'. Recent anti-MPA campaigns and political opportunism including that during the Commonwealth Marine Reserve planning process were aimed as much (indeed, more) at criticising the motivations of the various groups supporting conservation protection as any predicted real impacts on recreational fishing caused by actual MPA zoning (e.g. Anon 2010)
- There has been an extension of a latent, deep-seated, 'city versus country' bias in Australian communities, as many of the supporters of MPA protection – most often established near country communities – can be described as 'experts from the city' for which there is an innate suspicion (Pritchard and McManus 2000).
- Because recreational fishing in Australia is strongly associated with values of escapism, relief from everyday and political pressures, interest groups associated and active in the 'politics of recreational fishing' may be likely to attract a concentration of individuals with hold stronger, more homogenous political viewpoints than their grassroots constituency. Furthermore, because of the free-time demand required, those active involved in the representation of recreational fishers in political and other management issues may tend to a narrower demographic of 'grey-haired men in their 50s, 60s and 70s', leading to 'a growing generation gap which is resulting in the rec-fishing hierarchy becoming somewhat out-of-touch with what younger fishos want or expect' (Harnwell 2015). This may partially explain the apparent disconnect between the opposition of recreational fishing interest groups and the wide support for no-take MPAs in scientific and market-research documentation of attitudes and perceptions.
- As more and more Australian recreational fishers' perceptions and attitudes towards MPAs are shaped by direct experience, rather than information/communications from interest groups, it is likely that fishers with more 'moderate' views fall away

from the core support base of oppositional campaigns. In order to satisfy the remaining core audience, it is possible that these campaigns have been pressured into a less moderate public position than they would prefer.

Governments have been typically disinclined, for reasons of cost and convenience, to genuinely test the mandate of stakeholder representative groups during MPA planning or political positioning. This comes because governments have seemingly approached the socio-economic aspects of each new MPA planning process in isolation from the information from the extensive suite of existing MPAs already affecting Australian communities.

Conclusions and future directions

The findings presented here are strong evidence that Australian recreational fishers on the whole support the establishment no-take MPAs and are for the most part not negatively impacted (in perception or in behaviour) in the way in which their activity is shifted spatially by no-take MPAs. Support for no-take MPAs appears far stronger than is acknowledged in the oppositional campaigns that purport to represent this key stakeholder group, or by the actions of decision makers concerned at the political implications of recreational fishing sentiment (Sutton and Tobin 2009).

However, there is also clear and consistent evidence that many recreational fishers are left dissatisfied by the planning and consultation processes behind MPA designation. For example, 44% of recreational fishers disagreed that they were adequately consulted about the 2004 Great Barrier Reef Marine Park zoning plan, and did not trust the Great Barrier Reef Marine Park Authority to consider the concerns of recreational fishers (Sutton and Li 2008). Similar concerns were held by local residents near Ningaloo Marine Park about the quality of decision making and consultation, as discussed earlier (Northcote and Macbeth 2008a). This remains the major source of discontent among recreational fishers with MPA implementation.

The period of a few years after the gazettal of a MPA is the period where there is the most 'down side' for recreational fishers, because memories of an often polarising community debate are fresh, and adjustment time to zoning impacts has been low. It would also be too soon for any direct or indirect fishery benefits to manifest, such as research that benefits fishery/conservation management (providing a tangible 'return' for lost access). Given these factors, this is the time period where we would expect recreational fishers to be least satisfied with MPAs in their waters. But critically, this is also the time period (< 5 years after implementation) where most socio-economic research has been conducted. Thus, the bulk of socio-economic data, including that presented here, captures recreational fishing attitudes near their most negative towards MPAs.

There appears to be little clear link between the 'quantity' or 'quality' of consultation of a MPA planning process and the levels of satisfaction among recreational fishers. For example, the Great Barrier Reef Marine Park rezoning in 2004 was based on 3 years of consultation and restricted recreational fishing over 34% of its area, whereas the consultation process for South Australia's state waters marine park planning occurred over a decade, and resulted in a much lower relative spatial restriction of recreational fishing activity (6%). Both generated significant protest from recreational fishing interest groups in the latter stages of planning. Consultation, planning and process will always be divisive when the groups that seek to gain influence from increasing the politicisation of MPAs have not had their advocacy held accountable by planners and decision makers to the information – particularly the socio-economic information – that is available from our extensive existing suite of MPAs. This highlights the importance of better documenting the socio-economic benefits and impacts of MPAs – and continuing to do so beyond the first few years following implementation.

Biography

Adrian Meder is a marine conservation campaigner for the Australian Marine Conservation Society, working on issues of mutual concern for recreational fishing and conservation. He has worked extensively in an advocacy role during the later stages of Australia's Commonwealth Marine Reserve Network planning process. He has a background in marine science and is a passionate recreational fisher.

Endnote

1 http://www.newzealand.com/au/scenic-highlights+marine-reserves/

References

ABC (2004) 'Ningaloo Marine Park to be extended'. Australian Broadcasting Corporation, 27 November 2004, <http://www.abc.net.au/news/2004-11-27/ningaloo-marine-park-to-be-extended/593008>.

ABC (2014) 'The New South Wales Government announces a decision on fishing in marine sanctuary zones'. Australian Broadcasting Corporation, 23 December 2014, <http://www.abc.net.au/news/2014-12-23/fishing-amnesty/5985344>.

Anon (2010) Boswell: 'electoral disaster over marine parks'. *Ausmarine* **33.2**, 14.

Arias A, Sutton SG (2013) Understanding recreational fishers' compliance with no-take zones in the Great Barrier Reef Marine Park. *Ecology and Society* **18**(4), 18. doi:10.5751/ES-05872-180418.

Babcock RC, Shears NT, Alcala AC, Barrett NS, Edgar GJ, Lafferty KD, *et al.* (2010) Decadal trends in marine reserves reveal differential rates of change in direct and indirect effects. *Proceedings of the National Academy of Sciences of the United States of America* **107**, 18256–18261. doi:10.1073/pnas.0908012107.

Bax N, Creswell I (2012) 'Marine reserves not about closing fisheries, but about preserving ocean health'. *The Conversation* 27 August 2012, <http://theconversation.com/marine-reserves-not-about-closing-fisheries-but-about-preserving-ocean-health-8936>.

Crawley M (2014) 'No changes to Great Sandy Marine Park recreational fishing'. *Fraser Coast Chronicle* 5 March 2014, <http://www.frasercoastchronicle.com.au/news/the-state-government-will-not-make-changes-to-allo/2188465/>.

De Freitas DM, Sutton SG, Moloney JM, Lédée EJI, Tobin RC (2013) Spatial substitution strategies of recreational fishers in response to zoning changes in the Great Barrier Reef Marine Park. *Marine Policy* **40**, 145–153. doi:10.1016/j.marpol.2012.12.038.

Department of Environment and Resource Management (2012) *Moreton Bay Marine Park Monitoring Program February 2012.* Department of Environment and Resource Management, Brisbane.

Devillers R, Pressey RL, Grech A, Kittinger JN, Edgar GJ, Ward T, *et al.* (2015) Reinventing residuals reserves in the sea: are we favouring ease of establishment over need for protection? *Aquatic Conservation: Marine and Freshwater Ecosystems* **25**, 480–504. doi:10.1002/aqc.2445.

Dickson S (2014) 'Balance for the Bay'. Media statement by the Minister for National Parks, Recreation, Sport and Racing, The Honourable Steve Dickson, 9 May 2014, <http://statements.qld.gov.au/Statement/2014/5/9/balance-for-the-bay/>.

Fishing World (2012) The Wild West: Exmouth, WA. *Fishing World,* 1 August 2012, <http://www.fishingworld.com.au/news/the-wild-west-exmouth-wa>.

FRDC (2011) *2011 Recreational Fishers Survey.* Report for Fisheries Research and Development Corporation. Intuitive Solutions, Melbourne.

Galaxy Research (2014) *Community Attitude Survey.* January 2014. Prepared for the Dive Industry Association of Australia, Sydney.

Griffiths S, Sahlqvis P, Lyle J, Venables W, Pollock K, Sawynok W (2014) *A Coordinated National Data Collection for Recreational Fishing in Australia.* FRDC Final Report 2011/036. CSIRO, Brisbane.

Harnwell J (2015) Young blood needed now! *Fishing World* April 2015, p. 15.

Harrison HB, Williamson DH, Evans RD, Almany GR, Thorrold SR, Russ GR, *et al.* (2012) Larval export from marine reserves and the recruitment benefit for fish and fisheries. *Current Biology* **22**, 1023–1028. doi:10.1016/j.cub.2012.04.008.

Henry G, Lyle J (2003) *The National Recreational and Indigenous Fishing Survey.* Project No. 1999/158. Australian Government Department of Agriculture, Fisheries and Forestry, Canberra.

Hunt L (2008) *Economic Impact Analysis of the Cape Rodney Okariri Point (Leigh) Marine Reserve on the Rodney District.* Department of Conservation, Wellington, New Zealand.

Ling SD, Johnson CR, Frusher SD, Ridgway KR (2009) Overfishing reduces resilience of kelp beds to climate-driven catastrophic phase shift. *Proceedings of the National Academy of Sciences of the United States of America* **106**, 22341–22345. doi:10.1073/pnas.0907529106.

Lyle JM, Tracey SR, Stark KE, Wotherspoon S (2009) *2007/08 Survey of Recreational Fishing in Tasmania.* Tasmanian Aquaculture and Fisheries Institute, Hobart.

Lynch TP (2006) Incorporation of recreational fishing effort into design of Marine Protected Areas. *Conservation Biology* **20**, 1466–1476. doi:10.1111/j.1523-1739.2006.00509.x.

Lynch TP (2008) The difference between spatial and temporal variation in recreational fisheries for planning of Marine Protected Areas: response to Steffe. *Conservation Biology* **22**, 486–491. doi:10.1111/j.1523-1739.2008.00888.x.

Lynch TP (2014) A decadal time-series of recreational fishing effort collected during and after implementation of a multiple use marine park shows high inter-annual but low spatial variability *Fisheries Research* **151**, 85–90. doi:10.1016/j.fishres.2013.09.014.

Marine Parks and Reserves Authority (2013) *Ningaloo Marine Park/Muiron Islands Marine Management Area Management Plan Periodic Audit Report.* Marine Parks and Reserves Authority, Perth.

Marine Parks Authority NSW (2010) *Lord Howe Marine Park Summary of Research and Monitoring.* Marine Parks Authority of New South Wales, Sydney.

McCook LJ, Ayling AM, Cappo MC, Choat JH, Evans RD, de Freitas DM, *et al.* (2010) Adaptive management of the Great Barrier Reef: a globally significant demonstration of the benefits of networks of marine reserves. *Proceedings of the National Academy of Sciences of the United States of America* **107**, 18278–18285. doi:10.1073/pnas.0909335107.

McGregor Tan Research (2008a) *Solitary Islands Marine Park Community Survey Final Report.* Project No: 8353. Prepared for NSW Marine Parks Authority, Sydney.

McGregor Tan Research (2008b) *Jervis Bay Marine Park Community Survey Final Report.* Project No: 8353. Prepared for NSW Marine Parks Authority, Sydney.

McPhee DP (2008) *Fisheries Management in Australia.* Federation Press, Sydney.

McPhee DP, Buxton C, Knuckey I, Hundloe TJA, Stone S, Williams KA (2007) *Part 2: Final Submission: A Participatory and Coordinated Fishing Industry Response to the Proposed Rezoning of the Moreton Bay Marine Park*. Project 2007/053, Fisheries Research and Development Corporation (FRDC), Canberra.

Newman C, McVeigh J, Dickon S (2014) 'Proposal to allow recreational fishing at Scotts Point'. Joint media statement from the Premier The Honourable Campbell Newman, The Honourable John McVeigh, Minister for Agriculture, Fisheries and Forestry, and the Minister for National Parks, Recreation, Sport and Racing The Honourable Steve Dickson. 16 February 2014. <http://statements.qld.gov.au/Statement/2014/2/16/proposal-to-allow-recreational-fishing-at-scotts-point>.

Northcote J, Macbeth J (2008a) *Socio-economic Impacts of Sanctuary Zone Changes in Ningaloo Marine Park: a Preliminary Investigation of Effects on Visitation Patterns and Human Usage*. Cooperative Research Centre for Sustainable Tourism, Gold Coast.

Northcote J, Macbeth J (2008b) Threshold of tolerability: the impact of management changes to recreational fishing in Ningaloo Marine Park. *Tourism Review* **63**, 28–35. doi:10.1108/16605370810861026.

Olds AD, Pitt KA, Maxwell PS, Babcock RC, Rissik D, Connolly RM (2014) Marine reserves help coastal ecosystems cope with extreme weather. *Global Change Biology* **20**, 3050–3058. doi:10.1111/gcb.12606.

Pascoe S, Doshi A, Dell Q, Tonks M, Kenyon R (2014) Economic value of recreational fishing in Moreton Bay and the potential impact of the marine park rezoning. *Tourism Management* **41**, 53–63. doi:10.1016/j.tourman.2013.08.015.

Prior SP, Beckley LE (2007) Characteristics of recreational anglers in the Blackwood Estuary, a popular tourist destination in southwestern Australia. *Tourism in Marine Environments* **4**, 15–28. doi:10.3727/154427307784835679.

Pritchard B, McManus P (Eds) (2000) *Land of Discontent: The Dynamics of Change in Rural and Regional Australia*. UNSW Press, Sydney.

Queensland Government (2012) *Annual Status Report 2011: Coral Reef Fin Fishery*. Queensland Government, Brisbane.

Russ GR, Cheal AJ, Dolman A, Emslie MJ, Evans RD, Miller IR, *et al.* (2008) Rapid increase in fish numbers follows creation of world's largest marine reserve network. *Current Biology* **18**, 514–515. doi:10.1016/j.cub.2008.04.016.

Smallwood CB, Beckley LE (2012) Spatial distribution and zoning compliance of recreational fishing in Ningaloo Marine Park, north-western Australia. *Fisheries Research* **125–126**, 40–50. doi:10.1016/j.fishres.2012.01.019.

Stephan M, Hobsbawn P (2014) *Australian Fisheries and Aquaculture Statistics 2013*. Fisheries Research and Development Corporation project 2010/208. ABARES, Canberra.

Sutton S, Li O (2008) Attitudes of recreational fishers to the rezoning of the Great Barrier Marine Park. *Great Barrier Reef Research News – Special Edition* **5**, 1–4.

Sutton SG, Tobin RC (2009) Recreational fishers' attitudes towards the 2004 rezoning of the Great Barrier Reef Marine Park. *Environmental Conservation* **36**, 245–252. doi:10.1017/S0376892909990270.

Taylor S, Webley J, McInnes K (2012) *2010 Statewide Recreational Fishing Survey*. Queensland Department of Agriculture, Fisheries and Forestry, Brisbane.

Warne K (2007) Blue haven: New Zealand marine reserves are a model for the world. *National Geographic* April 2007, <http://ngm.nationalgeographic.com/2007/04/new-zealand-coast/warne-text>.

Chapter 22

Has south-eastern Australia's marine reserve network reduced risk and allowed the commercial fishing industry to create value? An industry perspective

Simon Boag

Introduction

The South-east Commonwealth Marine Reserves Network has been in existence since 2007. A review of the costs and benefits of this particular marine reserve network gives insight as to how the remainder of the Australian Commonwealth marine reserve networks might be implemented at greater gain with less cost. Given this author's experience in the Commonwealth Trawl Fishery, this chapter will draw heavily, but not exclusively, on lessons from this fishery. The Commonwealth Trawl Fishery, often called the South-east Trawl Fishery, is Australia's second largest Commonwealth fishery and runs from north of Sydney, southward through New South Wales, Victoria and Tasmania to Cape Jervis in South Australia. It is the largest supplier of fresh fish to the Melbourne and Sydney fish markets.

What are fishing businesses trying to achieve?

Before considering the cost–benefit of marine reserves, it is worthwhile understanding what the commercial fishing industry is trying to achieve and this is informed by understanding what motivates investment. All businesses, including fishing businesses, are funded by both debt and/or equity investors and they are motivated by different aims.

Debt investors are guaranteed payment through security over assets, meaning their risk is lower and therefore so are their expectations of return. With Reserve Bank rates in Australia at all-time lows, debt investors' expectations are currently low but following the Global Financial Crisis they are increasingly careful about securing their loan. Calculating the expectations of debt investors is easy: you go to banks and get the best rate you can. Debt investors get paid first, meaning that there may be a little or a lot remaining for equity investors.

Equity investors are not guaranteed returns, and therefore have higher risk and understandably hold expectations of higher returns. Equity investors consider the risk and return of other businesses when deciding whether to invest.

The 'capital asset pricing model' can be used to assess the expectations of both types of investors. The model assesses how a listed subject business' share price reacts to broader movements in the wider share market. Businesses that have greater volatility than share market indexes are high risk and vice versa. This volatility can be summarised into something called a β value. A β lower than 1.0 means the business is lower risk and investors should have a lower expectation of return, whereas a β larger than 1.0 indicates the opposite.

By weighting the cost of debt from the bank and by applying the business' β value on the equity component of the investment, the weighted average cost of capital (WACC) can be calculated. If a business' profits are higher than their WACC they are creating value, if lower they are destroying value.

At first instinct, most observers would think that commercial fishing is a risky business: biological vagaries, weather, mechanical breakdown, crew injury, volatile commodity markets, stock assessments, complex management arrangements, harvest strategies and environmental pressure surely make it so. However, when you calculate the β value of large listed fishing business in New Zealand (where they exist), these companies have betas of less than 1.0. It is difficult to find information on the β values of fishing businesses because they are often grouped with the agriculture sector, but the contention that fishing businesses have sub 1.0 betas is confirmed by international rankings of fishing, hunting and trapping businesses (Damodran 1997). My calculations are that a reasonable β for a (listed) fishing business in New Zealand is something in the order of 0.75. Somehow, fishing equity investors consider fishing to be less risky than one would think and, although returns are often only moderate, it appears that more often than not they are steadily mediocre. Fishing is a food business and these generally have lower betas. To some extent, customers have to buy seafood, or another protein. Economists call this relationship between demand and supply price inelasticity.

To create value, the ultimate goal of for-profit businesses is to deliver profits that are greater than those of businesses that hold the same level of risk. Marine reserves hold the potential to reduce investment risk, to reduce management costs and are widely reported as increasing fish stocks, all of which should allow businesses to create value.

What value have marine reserves helped fishing businesses create?

In the Commonwealth waters of south-eastern Australia (herein 'the South-east'), 14 marine reserves cover 388 464 km^2 or 9% of the fishery[1] (Pitcher 2015) (Fig. 22.1). Bottom trawling is not allowed in any South-east Commonwealth Marine Reserve Network (Director of National Parks 2013). Thus there is no risk of losing more fishing grounds to reserves so the investment landscape has one less variable. However, the rest of the Australian commercial fishing industry lives in risky times. Existing fishing industry equity investors are considering whether they should get out or stay in and potentially be compensated by the Australian Government when marine reserves are implemented. A third option is to make a speculative purchase and buy into a fishery threatened with the implementation of marine reserves in the hope of receiving compensation. In my experience, this third speculative investment only occurs very rarely and only when a fire sale is on offer. Uncertainty is uncomfortable but very occasionally it provides opportunity.

The Australian Bureau of Agricultural and Resource Economics and Science (ABARES) calculates that net economic returns (NER),[2] a measure of profitability that includes the opportunity cost of capital, in the Commonwealth Trawl Sector over the last few years have been: A$1.6 million in 2005/06 to A$7.1 million in 2007/08, A$2.5 million in 2009/10, A$6.82 million in 2010/11 and preliminary results of A$5.51 million in 2011/12 (Vieira *et al.* 2010; George and New 2013). Why then in a relatively profitable fishery has there not been a new build vessel enter the Commonwealth Trawl Fishery since the stern trawler *Western Alliance* was commissioned over a quarter of a century ago in 1989? (G Richie and M Miriklis *pers. comm.* 2015).

The answer is because, even with the likelihood of profit, that the initial implementation in the south-east of 14 marine reserves in 2007, followed by actions including (i) management arrangements such as fisheries closures, (ii) the more recent temporary banning of large vessels in the Commonwealth Small Pelagic Fishery and (iii) the rise of the recreational fishing movement which is overtly anti-commercial fishing, have all weakened the statutory fishing right and quota right. This means that, depending on the whim of the Australian Government and the Australian community, that equity investors could have their assets stripped away and be left with nothing. In other words, the Australian fishing industry's β value has increased and investors want the potential of more reward than historical NERs suggest is available.

As a member of the Temperate East Bioregional Advisory Panel for the Commonwealth Marine Reserves Review (Department of the Environment 2015), I have been moved by fishermen who have pleaded with the panel for the process to end. Fishermen are equity investors so want certainty to be able to re-establish their businesses, or walk away, to know if their sons will be fishermen, but mostly they just want their sanity back. The most recent roll-out of a network of Commonwealth marine reserves in Australia outside the South-east has taken more than 8 years if the start date was when Goals and Principles were agreed. A review of the zoning was called for but the process has taken a significant human toll that will not be measured in dollars or kilos of fish. Recent work by King *et al.* (2014) into the mental health of fishermen noted that a particular concern for fishers is the insecurity around fishing concessions. The research found that this ultimately renders fisher livelihoods insecure, and insecure livelihoods are a big contributor to poor mental health. Fishermen struggling with mental health issues are unlikely to be confident investors in the fishery.

The concept of cost-effective management of uncertainty in fisheries can be examined using the catch, cost, risk (CCR) trade-off framework. This framework sets down that where there is a low commercial catch relative to the stock, there is lowered risk and there should also be lowered management cost (and vice versa) (Sainsbury 2005) (Fig. 22.3). Theoretically, therefore, the existence of marine reserves should reduce risk in the fishery by reducing environmental risk and there should be a corresponding reduction in management cost. It is worthwhile exploring whether this happened in the South-east.

The cost of running Commonwealth fisheries is almost fully levied on fishermen and quota owners. In 1996, the South East Trawl Fishery had 109 licences and cost A$2.9 million per annum to manage. In 2014, after the implementation of the South-east Commonwealth Marine Reserve Network, the remaining 52 statutory fishing licences were levied A$3.4 million. Adding a little for inflation, this means that the fishery's management costs remain roughly the same, even though both the number of vessels (risk) and catches (risk) have both more than halved. The average trawl vessel in the fishery is 50 feet long, employs 2–3 crew members and pays A$1300 per week in management levies (Fig. 22.2). These

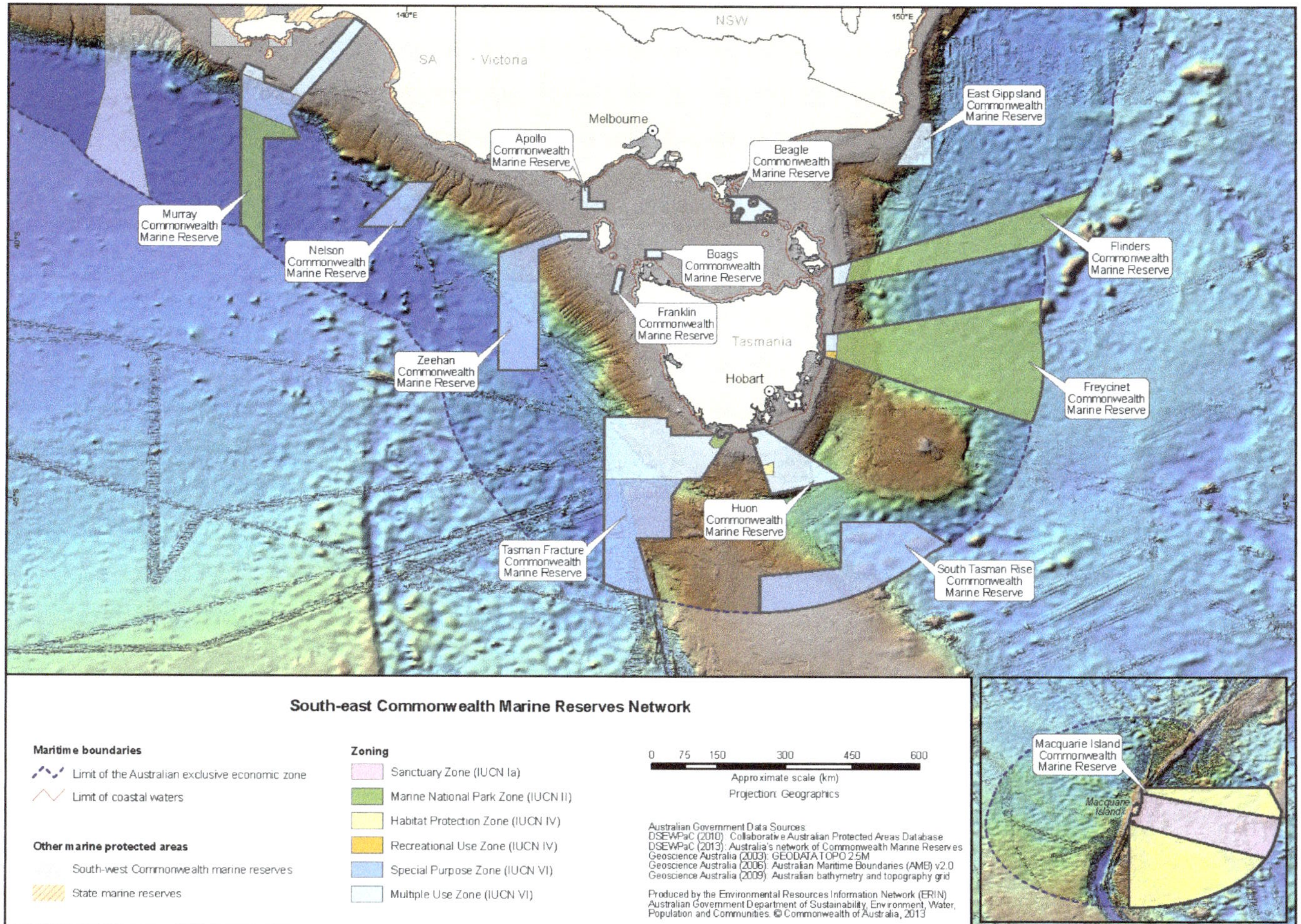

Fig. 22.1. South-east Commonwealth Marine Reserve Network. (Source: Department of the Environment http://www.environment.gov.au/topics/marine/marine-reserves/south-east)

Fig. 22.2. The trawler *Rockfish I* hauls a bag. (Photo: Simon Boag)

figures are not insignificant and comparing the fishery's annual levies (~$3.0 million) with the fishery's annual NERs (A$2.5–$7.1 million) shows that if levies were reduced by 50% that there would be a corresponding and very significant 21–60% increase in NER. Reducing risk and lowering recovered management costs and a subsequent increase in profits by 21–60% would add significant value to fishing businesses in the South-east. However, for some reason, setting aside 9% of the trawl fishery in no-take marine reserves appears to have had no effect on risk, management cost or profits.

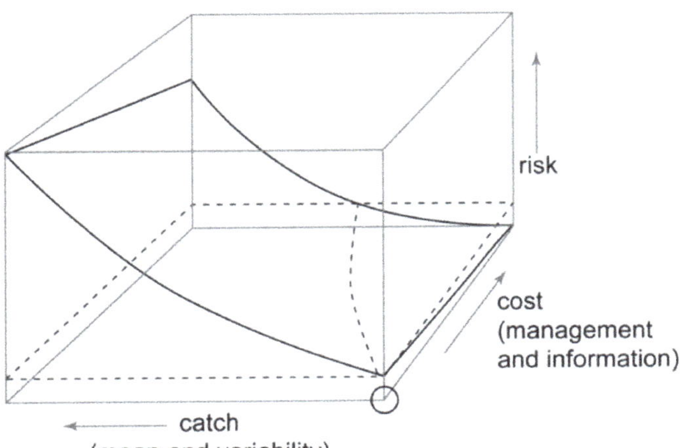

Fig. 22.3. Schematic for the interaction between risk, management cost and catch in a fishery. (Source: Sainsbury 2005)

In 2005, the Australian Government spent A$220 million on a structural adjustment of the Commonwealth fishery including a A$149 million fishing concession buyback removing licences in four Commonwealth fisheries including the South-east. This process was bound up with the implementation of marine reserves in the South-east and it was clear to fishermen who stayed that they would return to a fishery where 9% of the grounds were closed by marine reserves. The generous structural adjustment compensation removed commercial fishing over-capacity, which allowed a dignified move to sustainable catches, improved the status of targeted fish stocks and vessel level net economic returns improved substantially (Vieira *et al.* 2010). The removal of fishing over-capitalisation did position the industry to be able to create value. However, the process of removing over-capacity is a separate action to the introduction of the South-east Commonwealth Marine Reserve Network. It would therefore be wrong to suggest that marine reserves increased profits in the Commonwealth Trawl Fishery.

Fisheries closures are spatial closures that manage risks or emerging issues within the fishery. Interestingly, unlike marine reserves for which fishermen are generally compensated, fishermen receive no compensation for fisheries closures because they are a consequence of commercial fishing. Between 2003 and 2015, 39% of the South-east's trawl fishing grounds were then closed by various fisheries closures (Pitcher 2015), to protect target fish stocks such as Orange Roughy, deepwater sharks, upper-slope dogfish and Pink Ling, to separate incompatible fishing methods and to protect fish breeding areas. In the South-east fishery, closures and marine reserves overlap in places so the total portion of the fishery closed by marine reserves and fishery closures is 44% (Pitcher 2015). The fishing industry in the South-east is frustrated that, even with a representative network of marine reserves in place, approximately another 10 new fisheries closures and extensions to existing fisheries closures were required following the declaration of the South-east marine reserve network in 2007. The marine reserve network did not lay a pathway for value creation by materially mitigating some of these fishery management issues.

In 2004, 28% of the Great Barrier Reef was closed to commercial fishing (Fletcher *et al.* 2015). Advice to the Australian and Queensland governments at the time was that the closures would generate minimal initial reductions in catch and landed value and that both would recover in 3 years (Fletcher *et al.* 2015). However, both catch and landed value declined by about a third (roughly the same portion of the fishery that was closed) and 9 years later there are no signs of this recovering. In fact, there is growing evidence that dredging, agricultural runoff, climate change and its extreme weather events, (all things other than fishing), are causing the reef to deteriorate (GBRMPA 2014; see also Chapter 5). Since 1985, the reef has lost more than half of its corals with two-thirds of the loss occurring from 1998. UNESCO recognised the Great Barrier Reef as a World Heritage site in 1981 and in 2015 Australia narrowly avoided UNESCO classing the reef as 'in danger'.

Presumably, if commercial catches in the Great Barrier Reef had recovered, this would have been explained by the so-called 'spill-over effect' and there is overwhelming evidence of this occurring in over-exploited fisheries (Stobart *et al.* 2009). However, there is also evidence that in well-managed fisheries the net effect of the marine reserve is negative (Buxton *et al.* 2014). As a general rule, Australian fisheries are well managed. A comprehensive review of 49 fish stocks that contribute more than 80% of the value of Australian wild capture fisheries production and ~70% of the annual catch found that only two were over-fished (Flood *et al.* 2014). This would explain why commercial catches in the (not-overfished) Great Barrier Reef fishery did not recover as expected to pre-reserve levels. If marine reserves created an operating environment that allowed value creation then equity investors in the fishery would welcome them because they would increase commercial catches. During my 6 years working in the South-east, I have never met a fishermen, manager or assessment scientist who stated that the South-east Commonwealth marine reserve network had increased biomasses or allowable catches (quotas). The reality in the South-east is that if fish stock biomasses had increased past targeted levels (known as the target reference point) then the Commonwealth Fisheries Harvest Strategy would have increased commercial quotas in order to reduce fish stocks back to target levels. Using marine reserves as a tool to reduce commercial catch is only valid where catches are too high and there are no other tools available.

Marine reserves should have mitigated the risk of commercial fishing especially potentially damaging trawl fishing. Pitcher (2015), in his study *Predicting benthic impacts and recovery to support biodiversity management in the South-east Marine Region*, overlaid three datasets or maps to assess trawl fishing's impact on benthic habitat-forming invertebrates:

1 The distribution of 10 different benthic habitat-forming invertebrate groups, which were derived from surveys of invertebrates and computer modelling of depth, currents, temperature and latitude.
2 Maps of 15 different seafloor habitat types, or assemblages, determined using records of fish captures as an indicator of different habitat type and by again using computer modelling.
3 A map of the area of seafloor touched by trawls, known as the swept area. This was possible because deepwater trawl fishing vessels log their position each time they shoot and haul their net and record the width of their fishing gear.

The work was able to quantify the extent that fishing grounds were lost to closures including fisheries closures and marine reserves.

In my view, the four most relevant findings of Pitcher's (2015) work were that:

1 Trawling in the South-east only currently contacts the seafloor on the same relatively constant 6% of the fishery.[3] Some areas are trawled more than once a year. Trawling's current 6% footprint compares favourably with the 26% of the Australian landmass that has been converted to farming (Jayasuriya 2005). The relatively low trawl footprint is probably because not all areas are productive for fishing – many areas are too far from port or the seafloor is too rough to allow trawling. Seven of the largest marine reserves in the South-east Commonwealth marine reserve network funnel outwards, being thinnest on the continental shelf where fishing occurs and largest and widest in the deep sea abyss where fishing has never occurred and probably never will (see Fig. 22.1).
2 At the peak of the fishery in 2005, after almost 100 years of trawling the abundance of benthic habitat-forming invertebrates was 80–93% of their unfished abundance (biomass). In my contention that 80–93% is a very healthy number and Australia would be a very different place if its terrestrial fauna and flora enjoyed this level of abundance. The reader can decide how the South-east's current 80–93%, and likely still recovering, invertebrate abundance sits in the context of the loss of 40% of Australia's forest (Bradshaw 2012) and 50% of its wetlands (Finlayson 2000) since the arrival of Europeans.
3 Some habitats are not heavily fished and are highly protected while others are heavily fished and have no protection.
4 Marine reserves and fishery closures totalling 44% of the fishery had improved the invertebrate abundance by only 2% to 82–95% of pre-trawled levels. Further this 2% increase in bottom invertebrate abundance is a consequence of both types of closures and, because fishery closures are larger, most of the gain has come from fisheries closures not marine reserves.

Why is public support for marine reserves so strong?

So if the South-east's marine reserve network has not materially increased the abundance of bottom-dwelling invertebrates and Australia's fisheries are already well managed, then why does the Australian community remain largely supportive of marine parks?

At face value, the argument is compelling: 'do you (the Australian community) want to protect the marine environment?' 'Yes'.

'Do you think we should set aside an area of the marine environment and limit the activities, such as fishing, that can occur there, perhaps at the expense of a small amount of seafood (that we can import anyway)?' 'Yes'.

Many in the fishing industry will find it hard to ever forgive former Environment Minister Burke for his messaging leading up to the implementation of the national network of marine reserves. At times Minister Burke was publically supportive of Australia's fisheries management and the state of its fisheries. However, in an infamous press conference that was a turning point in Joe Ludwig's (the then Minister for Agriculture and indirectly fishing) public support for the operation of large vessels in the small pelagic fishery, Minister Burke subtly linked overfishing and marine reserves stating: 'This Government has a strong commitment to protection of the oceans, we've shown it in marine national parks, we have shown it in our treatment of the Great Barrier Reef, today we also show it in our concerns about the potential operation of the Super-trawler…' (Burke 2012).

In May 2013, Minister Burke announced the national rollout of marine reserves at the Sydney Aquarium (Burke 2013). While explaining that the reserves would protect against acidification, chemical changes in sea water and plastic pollution, hundreds of colourful fishes finned around happily behind his podium but in shot. During this press conference, he made his strongest linkage between overfishing and marine reserves to date by offering a choice between marine reserves or our children only ever seeing fish in tanks.

The media also blurred the lines between the need to mitigate risks to the Great Barrier Reef, the need to protect biodiversity across Australia, the small pelagic fishery and the state of our fisheries. Most media articles on the subject at the time would contain commentary from conservation groups about the existence of large vessels in the small pelagic fishery, note concern for the Great Barrier Reef and its UNESCO listing and the ongoing process of implementing a national network of marine reserves and its effect on commercial fishing.

I would argue that Minister Burke understood the community better than most and for political reasons gave the community reasoning based on logic they already held: this being that without marine reserves Australian fish and marine life were at risk from extractive activities such as commercial fishing.

At the launch, Minister Burke then presented Sir Richard Branson with an award for ocean protection. Sir Richard made a more direct linkage between overfishing and marine reserves, explaining that Manta Rays and sharks were being slaughtered for their fins in other parts of the world and this was why the Australia should be thanked for implementing marine reserves (even though finning is banned in Australia).

Sir Richard is not the only celebrity to endorse marine reserves and, in the same press conference, the Australian Government unveiled a video with cameos by a plethora of celebrities including Ian Kiernan AO (founder of Clean Up Australia), Brooke Hanson (an Olympic swimming medallist), Jackie Chan (Kung Fu genre actor and singer), Jean-Michel Cousteau (ocean adventurer and son of Jacques), all of whom espoused the degraded state of Australia's oceans and the need to establish no-take zones.

Conservation non-government organisations (NGOs) understand the power of celebrity endorsement and simple, but compelling, messaging (see also Chapter 23). In 2012,

Kelly Slater (a world surfing champion) signed a petition against the operation of a particular type of vessel in Australia's Small Pelagic Fishery and this was another straw in the loss of Australian Government support.

As a result of political statements, celebrity endorsements and conservation NGO messaging, the Australian community has linked marine reserves with commercial over-fishing, much to the fishing industry's horror. Perhaps the strongest evidence of this is research that shows that only 37% of surveyed respondents stated that they believed the Australian commercial fishing industry was sustainable, 37% were unsure and 26% stated that their perception was that the Australian commercial fishing industry was not sustainable (FRDC 2013).

Selling the fishing industry is tough. Unlike terrestrial farming, where there are fewer native species because the land has been cleared and populations have been killed off, commercial fishing operates in a natural environment full of dolphins, seals, seabirds, large sharks and invertebrates. Some, such as invertebrates and large sharks with no commercial value are an unfortunate by-catch. Seals, dolphins and seabirds are aggressive pursuers of fishing vessels who, through natural instincts and acclimatisation to fishing vessels as a food source (Fig. 22.4), put themselves at risk and are sometimes caught during fishing. An image of dead dolphin, seabird or baby seal in a fishing net is a powerful image and can be streamed around the world at very low cost.

Public tolerance for interactions seems to depend on the species. An internet search with high results or an animal's appearance on a tourist brochure is a sure sign that community tolerance is very low. An internet search on sea snakes returns low results and they do not feature on the Northern Territory's tourism websites, so tens of thousands of interactions in Australian fisheries is not considered a 'problem', but eight dolphin interactions is front page news (Darby 2015). The resultant public rush to outrage is channelled to support for marine reserves and donations.

In the pro-fishing documentary *Drawing the Line*,[4] Dr Caleb Gardener explains that the situation is more complicated. The marine environment is special and is threatened by

Fig. 22.4. 'At length did cross an Albatross, Thorough the fog it came; As if it had been a Christian soul, We hailed it in God's name' (*The Rime of the Ancient Mariner*, Samuel Taylor Coleridge, 1798). (Photo: Simon Boag)

fishing, siltation and nutrients in runoff, oil spills, climate change, pollution (including plastics) and invasive species. However, the zoning of marine parks only manages fishing and to a lesser extent oil and gas exploration and mining. The other threats do not respond to geography and marine parks are not the solution to mitigate them.

I contend that the risk that fishing poses to marine mammals is not mitigated by marine reserves because marine mammals are mobile and will not usually restrict their movements to marine reserves. The same amount of fishing effort in smaller fishing grounds outside marine reserves does not reduce the risk to marine mammals. The solution lies in changed fishing behaviours, strategic fishery closures (perhaps around breeding or feeding areas), modifications to fishing gear and finally with the acceptance that fishing, like all human activities, will never have zero environmental impact.

The South-east marine reserve network has not mitigated the risk to marine mammals such as seals. The Commonwealth Trawl Sector's seal interaction rate was quantified pre-reserves in 2002 (Knuckey *et al.* 2002) and more recently in 2014 (Koopman *et al.* 2014) and remains unchanged. Likewise, there is no reason to believe that the presence of reserves reduced the significant threat that trawl fishing poses to seabirds. Seabird interactions with trawlers are very hard to gauge because they generally go unseen. Long after marine reserves were implemented in the South-east, seabird interactions were reduced by 75% by protecting seabirds from the cables used to tow trawl nets (Pierre *et al.* 2014). Moving the fishing industry, and the seals and seabirds that follow it, outside of marine reserves does not solve the problem: it just moves it.

What have we learned so far from marine reserves in Australia?

Marine reserves do not promote investment in commercial fisheries because they remind equity investors that their property right is tenuous. There are serious mental health issues developing in the fishing industry because fishermen are concerned about their livelihoods, and this also discourages investment.

The catch–risk–cost framework states that if risk is removed then management costs should fall. Although marine reserves reduce risk by protecting areas from potentially damaging fishing methods, they have not reduced levied management costs in the South-east's Commonwealth Trawl Fishery.

The South-east's share of the A$149 million buyback money improved profits for the fishing industry, reduced the volume of Australian seafood, thus reducing Australia's GVP and increasing imports, allowed a move to sustainable catches (Vieira *et al.* 2010) but only increased the abundance of benthic biodiversity by less than 1% (Pitcher 2015).

The Australian tax payer spent A$213 million buying out commercial reef fishing licences on the Great Barrier Reef as part of a range of mechanisms to protect the reef. The reality is that the Australian economy has then lost A$58 million each subsequent year in fishing turnover (Fletcher *et al.* 2015) but the status of the reef has declined. None of this is a compelling value proposition for the tax payer. As per Dr Gardner's argument, spatial management of fishing only mitigates some of the risks to the marine environment.

However, my contention in the South-east is that marine reserves have not protected benthic biodiversity because it did not really need protection and because marine parks were poorly placed. This poor placement is evidenced by the failure of reserves to materially rebuild invertebrate abundance.

Reserves have reduced South-east catches forever, perhaps not in line with the Great Barrier Reef's 'grounds lost = lost revenue forever' rule because in the South-east marine reserves were placed where fishing did not occur and made to look bigger than they effectively are by extending them into the commercially worthless abyss. Their implementation consumed tax payer dollars. These funds could have been directed to other parts of the environment so perhaps the South-east's marine reserves were implemented at the expense of other and larger non-marine environmental gains. Against the law of CCR, although marine reserves have reduced risk and catch, somehow they did not reduce management cost.

Marine reserves hold public support because they are a simple and compelling social argument supported by scientists, politicians, conservation NGOs and celebrities, but somehow marine reserves in the South-east failed to build desperately needed social licence for the commercial fishing industry.

If marine reserves in the South-east had been better placed...

If the South-east Commonwealth marine reserve network had been better placed, outcomes could have been better. The process should have considered the Australian Fisheries Management Authority's ecological risk evaluation. Ideally it should have used Pitcher's (2015) work on invertebrates, habitats and the fishing footprint but unfortunately, this work did not exist at the time. It should have considered all threats, including those that cannot be managed with marine reserves. It should have been part of a larger marine threat planning process that would have considered oil and gas exploration more strongly. The current zoning in the South-east prohibits bottom trawling in all parts of all 14 reserves but allows mining and oil and gas extraction in parts of 12 of these 14 reserves.

Trying to extract sustainable levels of different fish species is challenging because the volume of sustainable take is based on the life history of a species and does usually not match the catch mix. A network of reserves in the South-east that was representative of habitat would have mitigated more of these fisheries issues. There would likely have been gains in benthic biodiversity. This might have brought about a fishery with less risk running at lowered management cost. In short, a properly placed network of marine reserves could have added value to fishing businesses, but it did not.

The fishing industry has no option but to embrace marine reserves and use them to their advantage

In the process for determining marine reserves in the other Commonwealth Marine Planning Regions (see Chapter 4), the discussion to date has been mostly about where to allow and disallow fishing. There has been some consideration of 'key ecological features' but these are not habitats (see also Chapters 3, 16). Rather they are interesting features that can be sold to the public such as canyons, seamounts and whale breeding grounds. Unless marine reserves are carefully placed and protect habitats, they will not enhance biodiversity.

It is clear that there is overwhelming community support for marine reserves. This is augmented by celebrity endorsement. Politicians have realised this and are generally pro-marine reserves. As at 2015, the current Federal Coalition Government does not publicly support marine reserves but did not roll back the national network, possibly due to this community support. Rather, in 2014, the Federal Government instigated a review of the zoning within the existing boundaries (see Chapter 4). But, considering a Coalition Government implemented the marine reserve network in the South-east, a national network of marine reserves generally has political support from both sides of politics and is therefore inevitable.

There has in the past been a separation of fisheries management and management of the environment and this is effectively now over. The merging of these disciplines creates interesting questions for the structure of relevant government departments.

Fishermen in Commonwealth fisheries are managed under full cost recovery and it is likely that state fisheries will follow in the near future. Fishermen should use the catch–risk–cost model to make the compelling argument that reserves must reduce levied management costs.

Habitat mapping work is already underway in many Commonwealth fisheries. This research can be used to protect a portion of all habitats from potentially damaging fishing gear. It is inevitable that, into the future, as environmental issues caused by commercial fishing are discovered, that only a truly representative network has a chance of mitigating these problems. The earlier the intervention in fisheries management issues the better.

The fishing industry could learn from the actions of the two main Australian supermarkets who, rather than attempting to compete with fishmongers on freshness or product knowledge, will create a new product attribute and now compete on sustainability (Coles 2015; Woolworths Ltd 2015).

The fishing industry must mobilise and follow the lead of conservation oriented NGOs in seeking celebrity endorsement for the sustainability benefits that marine reserves bring to fisheries. But the fishing industry must also acknowledge by-catch issues and work hard to mitigate them. Ultimately, the industry must accept and embrace the concept of social licence and encourage business to create value by using the new reserves to sell its products (because it cannot compete with imported seafood on price), to gain social licence and to reduce management cost. All of these actions reduce risk, increase profits and position the industry to create value.

Biography

Simon Boag is Chief Executive Officer with the South East Trawl Fishing Industry Association (SETFIA, www.setfia.org.au), a body representing the interests of deepwater trawl fishermen, quota owners and sellers in South-east Australia. He has previously worked in operational and general manager roles in commercial fisheries operations in New Zealand and Argentina. He holds an MBA, a B.Ed (Environmental Science) and a Postgraduate Certificate in Applied Science (fisheries). He is a panel member on the Temperate East Bioregional Advisory Panel for the Commonwealth Marine Reserves Review, a member of the Victorian Fisheries Advisory Council, a member of the Australian Fish Names Committee, holds board positions on South-east fishing and quota investment companies, is a Commonwealth Fishing Association board member, a South-east Management Advisory Committee member and a member of all South-east Resource Assessment Groups.

Endnotes

1 The fishery is defined as the beginning of Commonwealth waters at 3 nautical miles to 1500 m deep.

2 Net economic returns consider revenue from fish sales, operating costs, the opportunity cost of capital and an owner's wage, even if one is not drawn.

3 3 nautical miles from land (generally the end of 'state' waters) to 1500 m.

4 *Drawing the Line* http://drawingthelinemovie.com/

References

Bradshaw C (2012) Little left to lose: deforestation and forest degradation in Australia since European colonization. *Journal of Plant Ecology* **5**, 109–120. doi:10.1093/jpe/rtr038.

Burke T (2012) *Press Conference: Supertrawler – Murray Darling Basin Reform - Tasmanian Forestry - Wheat Export Marketing Bill – Live Animal Exports.* 11 September 2012, High Beam Research, Chicago, IL, http://www.highbeam.com/doc/1G1–302237105.html>.

Burke T (2013) Commonwealth Marine Reserve Networks launch by Minister for Sustainability, Environment, Water, Population and Communities, the Hon Tony Burke, MP, <www.youtube.com/watch?v=vLwbtsKme8E>.

Buxton C, Hartmann K, Kearney R, Gardner C (2014) When is spillover from marine reserves likely to benefit fisheries? *PLoS ONE* **9**, e107032. doi:10.1371/journal.pone.0107032.

Coles (2015) *Responsibly Sourced Seafood.* Coles Ltd, Melbourne, <https://www.coles.com.au/corporate-responsibility/responsible-sourcing-and-sustainability/responsibly-sourced-meat-and-seafood/responsibly-sourced-seafood>.

Damodran A (1997) *Corporate Finance: Theory and Practice.* Stern School of Business, New York University, New York.

Darby A (2015) Super-trawler Geelong Star kills another dolphin. *Sydney Morning Herald.* 22 June 2015, <http://www.smh.com.au/environment/supertrawler-geelong-star-kills-another-dolphin-20150622-ghuihi.html>.

Department of the Environment (2015) *Commonwealth Marine Reserves Review.* Department of the Environment, Canberra, <http://www.environment.gov.au/marinereservesreview/home>.

Director of National Parks (2013) *South-east Commonwealth Marine Reserves Network Management Plan 2013–23.* Director of National Parks, Canberra.

Finlayson CM (2000) Loss and degradation of Australian wetlands. Paper presented at *LAWASIA Conference: Environmental Law Issues in the Asia-Pacific Region 16–19 May 2000.* LAWASIA, Darwin. <http://www.environment.gov.au/science/supervising-scientist/publications/internal-reports/loss-and-degradation-australian-wetlands>.

Fletcher W, Kearney R, Wise B, Nash W (2015) Large-scale expansion of no-take closures within the Great Barrier Reef has not enhanced fishery production. *Ecological Applications* **25**, 1187–1196. doi:10.1890/14-1427.1.

Flood M, Stobutzki I, Andrews J, Ashby C, Begg G, Fletcher R, *et al.* (Eds) (2014) *Status of Key Australian Fish Stocks Reports 2014.* Fisheries Research and Development Corporation, Canberra.

FRDC (2013) *Community perceptions of the sustainability of the fishing industry in Australia.* Intuitive Solutions, Melbourne, <http://frdc.com.au/research/market_research/Documents/2013%20Community%20Perceptions%20Australian%20Fishing%20Industry.pdf>.

GBRMPA (2014) *Great Barrier Reef Outlook Report 2014.* Great Barrier Reef Marine Park Authority, Townsville.

George D, New R (2013) *Australian Fisheries Surveys Report 2012: Financial and Economic Performance of the Eastern Tuna and Billfish Fishery, the Commonwealth Trawl Sector and the Gillnet, Hook and Trap Sector.* Australian Bureau of Agricultural and Resource Economics and Sciences, Canberra.

Jayasuriya RT (2005) *Farming systems in Australia.* AGRIPPA. Food and Agriculture Organisation of the United Nations, Rome, <http://www.fao.org/docrep/article/agrippa/686_en.pdf>.

King T, Kilpatrick S, Willis K (2014) *Staying Healthy: Industry Organisations' Influence On Behaviours and Services Used by Fishers.* Report to the Fisheries Research and Development Corporation Project No. 2012/402. Deakin University, University of Tasmania and University of Sydney, Canberra.

Knuckey I, Eayrs S, Bosschietter B (2002) *Options for Reducing Incidental Catch of Seals on Wet-Boats in the SETF: a Preliminary Assessment.* ARF Project R01/0887. Final Report to the Australian Fisheries Management Authority. Marine and Freshwater Resources Institute, Department of Natural Resources and Environment, Queenscliff, Vic., http://www.fishwell.com.au/app_cmslib/media/lib/1012/m4522_v1_sealbycatchfinalreport.pdf>.

Koopman M, Boag S, Knuckey I (2014) *Assessment of the Use of Shortened Codends to Mitigate Seal Interactions in the SESSF.* Report for Australian Fisheries Management Authority, Canberra.

Pierre J, Gerner M, Penrose L (2014) *Assessing the Effectiveness of Seabird Mitigation Devices in the Trawl Sectors of the Southern and Eastern Scalefish and Shark Fishery in Australia.* Australian Fisheries Management Authority, Canberra, <http://www.afma.gov.au/wp-content/uploads/2014/12/Seabird-Mitigation-Assessment-Report.pdf

Pitcher R (2015) Predicting benthic impacts and recovery to support biodiversity management in the South-east Marine Region. In *Marine Biodiversity Hub, National Environmental Research Program, Final Report 2011–2015* (Eds NJ Bax and P Hedge) pp. 24–25. Report to Department of the Environment, Canberra.

Sainsbury K (2005) *Cost Effective Management of Uncertainty in Fisheries.* Outlook 2005 Conference, ABARES, Canberra, <http://pandora.nla.gov.au/pan/45562/20050622-0000/PC13024.pdf>.

Stobart B, Warwick R, César González C, Mallol S, Díaz D, Reñones O, *et al.* (2009) *Long-term and Spillover Effects of a Marine Protected Area on an Exploited Fish Community.* Instituto Español de Oceanografía, Centro Oceanográfico de Baleares, Palma de Mallorca, Spain and Plymouth Marine Laboratory, Plymouth, UK, <http://www.int-res.com/abstracts/meps/v384/p47-60/>.

Vieira S, Perks C, Mazur K, Curtotti R, Li M (2010) *Impact of the Structural Adjustment Package on the Profitability of Commonwealth fisheries.* Research report 10.01 ABARE, Canberra.

Woolworths (2015) *Sustainable Fish and Seafood.* Woolworths Ltd, Sydney, <http://www.woolworthslimited.com.au/page/A_Trusted_Company/Responsibile_Sourcing/Sustainable_Fish_and_Seafood/>.

Chapter 23

Commonwealth marine reserves: big and bold campaigns but business-as-usual outcomes?

Chris Smyth

Introduction

Poet Judith Wright once described the Great Barrier Reef as 'still the closest most people will come to Eden' during a 10-year community campaign to save it from limestone mining and oil production that she spearheaded with artist John Büsst and scientist Leonard Webb (Foster 2014). Together they enlisted the support of environment non-government organisations (eNGOs), trade unionists, artists, politicians, scientists and the wider community in a campaign that resulted in what was then the world's largest marine park.

The iconic status of the Great Barrier Reef enabled a regional campaign to become national, then international, and evolve into Australia's largest community-based marine campaign, providing the foundation and inspiration for the progressive establishment of marine protected areas around Australia.

This chapter reviews just two of the campaigns associated with the recent Commonwealth marine reserve planning process in Australia's Exclusive Economic Zone (EEZ). Although campaigns by eNGOs were also run for the North West, North and Temperate East marine regions, this chapter focuses on the two prominent ones: those for the South-west Marine Region and the Coral Sea. In the South-west Marine Region the campaign aimed for the establishment of a comprehensive, adequate and representative network of marine reserves zoned marine national park (no-take). For the campaign in the Coral Sea, the objective was the creation of the world's largest marine reserve in one marine national park zone.

The chapter begins with a background to the campaigns and their highlights, followed by an analysis of campaign objectives and achievements, the factors that undermined the level of those achievements, and some lessons for future marine conservation campaigns.

Build up to the campaigns

In 2007, the Australian Coalition Government established the South-east Commonwealth Marine Reserve Network, the first under the Commonwealth marine bioregional planning process (see also Chapter 19). That process was continued for the remaining marine regions by the Australian Labor Government elected later that year.

Although eNGOs had engaged in the South-east Marine Region process, the campaign was muted by a lack of resources, the difficulty of engaging the community on a little known area of Australia's oceans, and the groups' limited influence on the Australian Coalition Government. The resulting network gave insufficient protection to the region's marine ecosystems (Wescott and Fitzsimons 2011), especially the continental shelf. Campaign capacity was also constrained by the relatively low priority given to marine issues by Australian eNGOs that had for many years campaigned on forest, threatened species (admittedly some marine) and, more recently, climate change issues.

The continuing lack of resources was set to undermine efforts to build the community's awareness and appreciation of Australia's ocean values and support for their protection. This was overcome when Pew Charitable Trusts, in partnership with The Nature Conservancy, established the Wild Australia Program in 2007, which, among other aims, sought to increase protection for the South-west Marine Region (Pew Charitable Trusts 2007). At the same time, Pew's Global Oceans Legacy project established its Coral Sea campaign, which was articulated in the report *An Australian Coral Sea Heritage Park* (Zethoven 2008).

Pew viewed Australia as a stable and affluent democracy with a well-deserved international reputation in marine conservation, providing the opportunity to significantly expand the nation's marine reserves, increase the international coverage of marine reserves and act as a model for other nations. Its marine campaigning experience was considerable, having secured protection for the North-western Hawaiian Islands Marine National Monument in 2006 (Pew Charitable Trusts 2006), and later reserves for the Marianas Trench Marine National Monument and the Chagos Marine Reserve that together covered ~1 500 000 km^2.

To direct its Australian marine campaigns, Pew recruited two highly experienced lobbyists with considerable expertise in coordinating successful environmental campaigns. Imogen Zethoven, director of the Coral Sea campaign, had also directed the successful WWF-Australia campaign for rezoning the Great Barrier Reef Marine Park. Michelle Grady, director of the South West campaign, was the former CEO of the Conservation Council of South Australia and had been a marine campaigner with the Australian Conservation Foundation and a member of the National Reference Group of the Marine and Coastal Community Network.

The 'Save our Marine Life' campaign

Pew and The Nature Conservancy initially brought together the Conservation Council of Western Australia, the Australian Marine Conservation Society, the Australian Conservation Foundation, The Wilderness Society and WWF-Australia to form the Perth-based Save our Marine Life (SOML) alliance, the focus of which was the Commonwealth waters of the South-west Marine Region between Kalbarri in Western Australia and Kangaroo Island in South Australia. The alliance gradually grew to more than a dozen organisations.

The campaign was publicly launched in May 2009 with the release of *Protecting Western Australia's Big Blue Backyard* (Save our Marine Life 2009), a coffee table-styled report with full-page photographs that outlined the region's biodiversity, social and economic values and the case for marine protection.

Key campaign strategies and tactics included:

- the commissioning of *Systematic Conservation Planning: A Network of Marine Sanctuaries*, the University of Queensland Ecology Centre's Marxan analysis of options for marine protected areas (Ecology Centre 2010). Marxan is conservation planning software that can be used to assess multiple options for marine reserve design, with each option achieving a set of biodiversity conservation goals. It can also be used to identify reserve options that minimise socio-economic effects. Never before available to eNGOs, Marxan analysis enabled the campaign to match and understand similar work by the Australian Government's environment department and provide more substance to submissions (see also Chapter 16)

- activities such as sustainable seafood barbeques in coastal towns such as Bunbury, Busselton and Geraldton from 2009 to 2011, and a roadshow in August 2009 that had 1400 people attend 15 screenings of *The End of the Line* (Murray 2009), a film about international over-fishing. These activities helped promote the campaign in coastal towns, gave an identity to the nearby offshore places proposed as reserves, and helped counter local opposition
- recruitment of a recreational fisher with a marine science background who helped allay fears and build support for the reserves within the Western Australian angling community
- communications materials, including *Atlantis Found – Underwater Icons of Australia's Unique South West* (Save our Marine Life 2010), which used colourful imagery to highlight 10 marine hotspots and the case for their protection
- an interactive website that encouraged people to support the campaign writing submissions, signing online petitions and joining what was then the 40000-strong 'blue army', a clever campaign device to recruit and retain supporters by giving them a collective, 'war-footing' identity and allowing them to show and share their commitment (in June 2015 it numbered 128 269 supporters, with the Save our Marine Life campaign having become national and incorporating the Coral Sea campaign)
- commissioned reports *The Economics of Marine Protected Areas* (Allen Consulting 2010) and *Stocking Up: Securing our Marine Economy* (Eadie and Hoisington 2011) to balance the debate and show decision makers that the campaign understood economics and that marine reserves benefited the Australian economy
- focus group research and polling (Essential Research 2008, 2009) that helped develop targeted campaign messages and demonstrate strong support for marine reserves, even among anglers: 70% support across the community, with angler support almost as high
- development of constructive relationships with the Western Australian business community and anglers that helped constrain the local influence of fishing bodies.

The 'Protect our Coral Sea' campaign

Although the Coral Sea was already an icon to older Australians who remembered the pivotal naval battles there during World War II, the challenge was to turn it into a natural icon. To meet that challenge and inspire community support required a big vision – the world's largest highly protected marine reserve covering almost 1 million km^2 of diverse marine habitats that were home to many threatened species.

The Sydney-based 'Protect our Coral Sea' campaign was launched in early 2008 and quickly established contracts with the Australian Marine Conservation Society in Brisbane, and the Cairns and Far North Environment Centre in Cairns, to provide an effective and visible campaign at the regional level. It also began to build an alliance with a variety of partner eNGOs including the Australian Conservation Foundation and others involved in the SOML alliance.

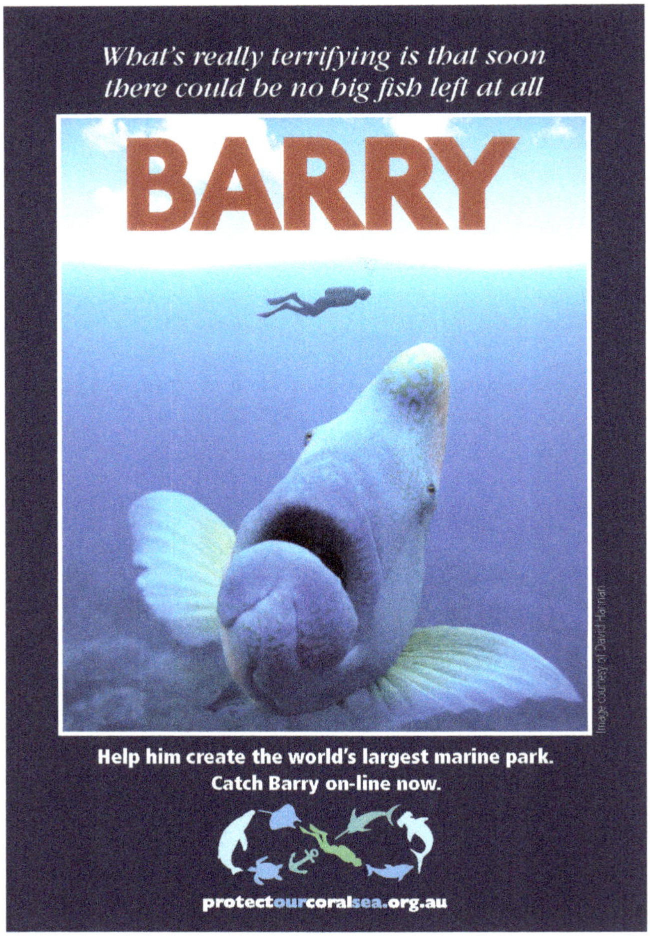

Fig. 23.1. Barry the Wrasse in the 'Protect our Coral Sea' campaign.

Key campaign strategies and tactics included:

- securing the Australian Government's separation of the Coral Sea from the East Marine Region under the interim conservation zoning of the area, enabling a singular campaign focus on the Coral Sea
- innovative communications materials, including TV advertising with a talking Barry the Wrasse (Protect our Coral Sea 2010) to brand the campaign as very friendly and non-threatening (Fig. 23.1)
- community outreach in towns along the coast, especially in Cairns where opposition from anglers and charter boat operators was loudest, helping to 'localise' the campaign, respond to negative commentary and build regional support
- commissioned expert reports that provided a strong scientific basis to the campaign and showed that the proposed reserve was the most cost-effective management approach and of economic benefit (Ban *et al.* 2009; Ceccarelli 2011)
- the national and international mobilisation of support, which generated the bulk of the more than 700 000 submissions (Protect our Coral Sea 2013) to the Commonwealth

marine reserve planning process. One of the campaign supporters, Avaaz, an international web-based organisation that enables people to engage in decision-making processes around the world announced in February 2012 that 235 000 of its supporters, including 28 000 from Australia, had sent a submission on the Coral Sea Marine Reserve (Avaaz 2012)

- collaborative campaigning with Coral Sea commercial Yellowfin Tuna fishers who publicly backed the proposed reserve (Cleary 2011; The Courier Mail 2011), serving to undermine opposition from commercial fishing bodies
- promoting the iconic naval history of the Coral Sea and the inspired recruitment of two retired vice admirals of the Australian Navy as champions in the media and with the Department of Defence and members of parliament.

Achieving the campaign objectives?

The two campaigns substantially expanded Australia's marine campaigning capacity. At the height of the two campaigns, there were between 15 and 20 full-time and part-time campaigners working on them, as well as staff contributions from other alliance partners.

Together they helped secure what at the June 2012 proclamation was the world's largest marine reserve network covering 36% of Australia's EEZ, with 11% in marine national park zones (see also Chapter 4). But how did the outcome compare with their initial campaign objectives?

Place-based targets

The SOML campaign initially sought high-level protection for places such as the Abrolhos Islands, Perth Canyon, Geographe Bay, Naturaliste Plateau, the Diamantina Fracture Zone, Albany Canyons and the Recherche Archipelago in a network that covered more than 30% of the South-west Marine Region. Percentage targets were never used in the campaign's public commentary to avoid a 'percentage' debate that would have taken the focus away from places and the need for their protection. The network proclaimed in June 2012 covered 39% of the marine region's waters, with 14% under marine national park zoning (down from the 20% in the government's draft proposal after changes to the South-west Corner reserve), although that was largely well offshore (Director of National Parks 2013a).

The Protect our Coral Sea campaign aimed for all of the Coral Sea's Australian waters to be included in one marine national park zone but, like the South-west campaign, never used percentage targets in campaign communications. Although the proclaimed reserve covered those Coral Sea waters, the marine national park zones covered only 49%, almost wholly comprising the eastern half of the reserve that lay furthest from the coast (Director of National Parks 2013b), with many of the sea's coral reefs along the western boundary missing out on high-level protection due to resistance from anglers and charter boat operators.

Comprehensive, adequate and representative networks

A comprehensive, adequate and representative (CAR) network, one of the principles of the National Representative System of Marine Protected Areas (ANZECC TFMPA 1998), was an objective of the SOML campaign. In contrast, the Protect our Coral Sea campaign was advocating one large highly protected reserve, not a network in itself, but part of a larger national network.

Leading marine scientists have expressed concerns that the regional networks are 'residual', located in areas remote and/or of little extractive/commercial interest and failing to satisfy CAR principles (e.g. Pressey 2013; Devillers *et al.* 2015). This criticism has largely come after the network proclamations and rarely articulated during the planning process. Marine campaigners are well aware of the shortcomings of the networks, including that proclaimed earlier in the South-east Marine Region. An eNGO-commissioned report (Centre for Conservation Geography 2011) analysed the proposed network in the South-west Marine Region against the CAR principles, found it inadequate and recommended areas in need of high-level protection. Campaigners used these recommendations when seeking increased protection on the continental shelf where they made small but important gains: Perth Canyon, Geographe Bay and the Eastern Recherche, Twilight and Murat Marine Reserves. The Coral Sea campaign also made small gains on the reefs along the marine reserve's western boundary.

Review of the marine reserve network management plans

Soon after its election in 2013, the new Australian Coalition Government announced a review of the Commonwealth marine reserve management plans and that it was 'scrapping Labor's plan to lock out recreational fishers from key coastal areas around the country' (Hunt and Colbeck 2013).

This review (Department of the Environment 2015) threatens to significantly weaken protection levels in the proclaimed network in the South-west Marine Region and the Coral Sea Marine Reserve: membership of the review's bioregional advisory panels is largely representative of commercial and recreational fishing interests; eNGOs were not invited to join. The focus of the review is the management zones within the reserves – the external reserve boundaries were left untouched but all levels of protection within them are open to change. This means that the external boundaries are simply lines on maps that could eventually become 'paper parks' where protection levels are so downgraded that the area's use is largely what existed before proclamation.

Campaign adjustments

The campaigns for the South-west and the Coral Sea adjusted their campaign objectives when it became clear that the Australian Labor Government and the Coalition Opposition were not going to support what they had advocated for the spatial extent and protection levels for the reserves. Eventually the focus became more one of defending the proclaimed reserves and seeking relatively small amendments to increase zone protection to marine national park status, especially on the continental shelf in the South-west and several reefs in the Coral Sea.

Campaign objectives versus campaign outcomes

There were several external factors that undermined the achievement of the campaign objectives in the South-west and Coral Sea campaigns.

Underestimating the opposition

Australia was considered an ideal place for a big marine conservation campaign, but this underestimated the level of political and stakeholder discontent that would be generated. For example, previous big international successes were in geographically and political isolated parts of the world's oceans with few stakeholder issues. Factors that assisted the rezoning and increased protection of the Great Barrier Reef Marine Park in 2003, such as the reef's internationally iconic status, a statutory body charged with its protection, a clear set of scientific objectives and largely accepted protection target, widely recognised threats and bipartisan support, largely did not apply to the rest of Australia's oceans.

Government conservatism

Although Commonwealth, state and territory governments agreed to establish a national network of marine reserves in the 1990s, progress has been slow and the outcomes less than ambitious due to the influence of commercial and recreational fishers (see Chapters 8, 9). With the exception of the Great Barrier Reef Marine Park, most of the earlier Commonwealth marine reserves had been established in isolated regions well offshore. The Federal Labor Government's proclaimed marine reserve networks were largely consistent with this conservativism, as was the Australian Coalition Government's South-east Commonwealth Marine Reserves network in 2007. This conservatism now appears to have been replaced by political intransigence in Canberra and New South Wales (Friedlander 2014).

Electoral fortunes and misfortunes

At the time the two campaigns were being established, few could have predicted how the federal political landscape would change. Rather than two terms of political stability, the Federal Labor Government elected in 2007 was beset by leadership tensions that resulted in a period of minority government (2010–2013) and eventually its demise, while the Coalition Opposition, which had an impressive legacy of marine conservation from its times in government, saw stakeholder resistance to the marine reserves as a means by which it could possibly gain marginal seats, many of which were coastal, at the 2013 federal election.

The collapse of bipartisanship was dramatic and coincided with the ascension of Tony Abbott to the position of Opposition Leader in 2009, who said in the 2013 federal election campaign that: 'We do not want to lock up our oceans' (Fishing World 2013). The loudest Opposition voices were Coalition members from Queensland (Thomas and Harnwall 2009), such as Senator Ron Boswell, who targeted the Coral Sea Marine Reserve. He also attacked the Ecology Centre's marine conservation planning report on the South-west Marine Region (Baird Maritime 2010), which recommended extensive areas of high-level protection with minimal impact on commercial fishing.

Try as they did, the two campaigns were unable to change the Opposition's negative stance and the Coalition parties went to both the 2010 (Business Chamber NSW 2010) and 2013 (Heath and Rollason 2013) federal elections committing to winding back the marine protection proclaimed by the Federal Labor Government. At one point the Labor Government was seriously considering abandonment of the marine reserve process. The campaign's response was to intensify their lobbying – regular visits to Canberra by teams of campaigners were a feature of both campaigns and included presentations by leading scientists (e.g. Dr Sylvia Earle), seminars, films and well-designed information packs – and to use polling to demonstrate community support for marine reserves. It changed the government's mind.

This lack of bipartisanship and minority government made the Labor Government nervous and cautious, concerned about the perceived electoral influence of anglers in marginal coastal electorates and the potential for seat losses, not wishing to be seen as causing job losses in the fishing industry during and after the global financial crisis, and wanting to limit the budget implications of transitional arrangements for affected commercial fishers (eventually A$100 million was committed).

The US connection

Claims that Pew was a US-based, oil industry-funded conspiracy to end fishing struck a chord with some in regional communities, especially in northern Queensland, as demonstrated by the heated public meetings held in Cairns (Fishing World 2009) at the beginning of the Protect our Coral Sea campaign. The role of Pew as an 'outsider' organisation continued to be raised by Queensland Coalition members in their opposition to the Coral Sea

Marine Reserve, with Senator Ron Boswell saying that: '...the American green group has come over here and is leading the government by the nose' (Boswell 2012). Warren Entsch, the federal member for the seat of Leichardt, referred to Pew as 'putrid' and 'gangrenous' during a speech to Parliament (Entsch 2013). Pew's response was that it was initially established in the 1940s from trusts created by the children of Sun Oil Co. founder Joseph Pew, and that it was an independent not-for-profit. Pew's involvement in the Save our Marine Life campaign does not appear to have had the same impact, with Coalition MPs in Western Australia making few public statements about the issue.

The size and timing of the process

The Commonwealth marine reserve planning process involved the preparation of draft and final Regional Marine Plans, proposed marine reserve networks and six periods of public consultation (see Chapter 4). It was a complex task for departmental bureaucrats to manage, tested the stamina of marine campaign staff, supporters and budgets, and gave those opposing the reserves more time to get organised. The lengthy process also meant that by the time of the 2013 federal election, the management plans for the reserves were stalled in the Senate, providing the new Federal Coalition Government with the opportunity to scrap them.

The strength of stakeholder opposition

To counter the alliances formed by eNGOs, oppositional stakeholders formed their own including the Australian Recreational Fishing Foundation (formed in 2012), the Boating Industries Alliance Australia (formed in February 2011) and the Keep Australia Fishing campaign (established in 2012), joining with the Australian Fishing Trade Association and commercial fishing organisations to fight the marine reserves.

Their campaign had two key targets: anglers and politicians. To anglers, their messages portrayed the marine reserves as 'the biggest lockout of Aussie anglers in history' (ARFA 2013) that would prevent 'Mum, Dad and the kids fishing in a tinnie' (Warwick Daily News 2012) and deny access to favourite fishing spots. In visits to Canberra, they met with politicians and held events to argue that the socio-economic impacts would be large, there was no scientific evidence to support the reserves, there had been inadequate consultation and, they claimed, the protection was unnecessary because marine life was not threatened and fishing was sustainable.

Their campaigns may have been passionate, emotive and often misleading, such as publication of maps (Fishwrecked 2013) that added eNGO proposals to the proclaimed marine reserves to suggest a much bigger 'no-take' impact on recreational fishers than was in fact the case – but they were also effective. In response, the Labor minority government sought to minimise socio-economic – and electoral – impacts by avoiding areas of highest value to the commercial fishing industry, and of interest to recreational, game and charter fishing.

Out of sight, out of mind

The two campaigns worked hard to increase the community's appreciation of the values of Australia's oceans, because, other than their awareness of the Great Barrier Reef, Australians have scant knowledge of the oceans surrounding them. For example, Victorians polled in 2007 felt they knew little about the marine environment (Ipsos Consultants 2007), and this is borne out by various Australian and international polls. This meant that the campaigns could not capture the public imagination like that for the Great Barrier Reef and so were unable to garner the same broad sweep of community support of that 10-year campaign. In effect, there was no 'across the spectrum' constituency for establishing marine reserves despite the general community's professed support in polling (see Chapter 20 for detailed statistics).

Lessons for other campaigns

There are several lessons to be taken from the campaigns to secure the Commonwealth marine reserve networks in Australia:

- identify early the key decision makers and influential stakeholders on whom to focus your attention
- try to develop a dialogue with those who may be resisting your campaign. Even if that fails, you may at least have a better idea of their arguments
- avoid spreading the campaign too thin. If the planning process extends over a large area, which was the case with the Commonwealth marine reserves, then any hard decisions about campaign priorities should be made early and be maintained. Limited resources were allocated for much smaller campaigns in the remaining marine regions but with very limited success in the face of advocacy from the oil and gas industry in the North West, and recreational and commercial fishers in the North and Temperate East marine regions
- establish one campaign. Having separate campaigns running concurrently at opposite sides of the nation can create issues for their integration, including competition for media attention, gathering campaign partners, building community support, messaging and lobbying time for federal members of parliament
- limit the size of alliances. Bringing together partners for a campaign is useful for resource pooling, integration of strategies, consistency of messaging and a demonstration of broad support. However, bigger is not necessarily better, and each partner needs a clear role; otherwise the larger the alliance becomes, the more complex the administrative tasks to manage and maintain it
- set realistic objectives on spatial protection targets after carefully assessing the socio-political landscape and be prepared to adjust them if required
- ensure all key campaign decisions are local ones. In the case of campaigns funded by international organisations, ensure that decisions on such matters as budget expenditure, media statements and campaign strategies are made in the region where the campaign is conducted to ensure timely decision making and campaign spontaneity
- do not underestimate your opposition.

Conclusion

When Judith Wright and her colleagues secured the protection of the Great Barrier Reef, they all but swept the opposition aside. The reef was iconic – a household name in Australia – and the threat was very real.

Since then, commercial and recreational fishing have become entrenched in many parts of Australia's oceans. Any campaign that would displace fishing was always going to resisted but was made more difficult by the reluctance of Australia's governments to establish marine protected areas in places of commercial and recreational interest, the fracturing of bipartisanship, minority government politics, and the community's general lack of oceans awareness.

Even so, Australia currently has the world's largest network of marine reserves – thanks in large part to eNGO campaigns – but this may change after the current government's review. It is not perfect, but with time the Great Barrier Reef Marine Park was also shown to be inadequate. Thirty years later, the reef's protection levels were strengthened and the park's zoning plan is now recognised as world's best practice (e.g. Fernandes *et al.* 2005).

Australia is not yet ready for world's-best-practice in the rest of our oceans but the Commonwealth marine reserve networks are a major step towards it. We will just have to wait a little longer to achieve it.

Biography

Chris Smyth was for 9 years the healthy oceans campaigner at the Australian Conservation Foundation where he was involved in both campaigns described in this chapter. He is now an environmental consultant working with eNGOs.

References

Allen Consulting (2010) *The Economics of Marine Protected Areas*. Allen Consulting, Perth.

ANZECC TFMPA (1998) *Guidelines for Establishing the National Representative System of Marine Protected Areas*. ANZECC Task Force on Marine Protected Areas, Canberra.

ARFA (2013) 'Labor government confirms the biggest lock out of Aussie anglers in history'. Media release by Australian Recreational Fishing Association, 15 March 2013, <http://www.recreationalfishing.com.au/index.php/rss/26-labor-government-confirms-the-biggest-lock-out-of-aussie-anglers-in-history>.

Avaaz (2012) *Save our Oceans*. Avaaz, <http://www.avaaz.org/en/save_the_coral_sea_4/>.

Baird Maritime (2010) *Australia: "Electoral Disaster Over Marine Parks"*. Baird Maritime, Melbourne, <http://bairdmaritime.com/index.php?option=com_content&view=article&id=8494:australia-electoral-disaster-over-marine-parks&catid=75:fisheries&Itemid=68>.

Ban N, Adams V, Pressey R (2009) *Marine Protected Area Management Costs: An Analysis of Options for the Coral Sea*. Report prepared for the Protect Our Oceans Campaign. Centre of Excellence for Coral Reef Studies, James Cook University, Townsville.

Boswell R (2012) 'Matters of public importance: marine sanctuaries'. Speech by Senator Ron Boswell, 20 June 2012, Australian Parliament, Canberra, <http://parlinfo.aph.gov.au/parlInfo/genpdf/chamber/hansards/ 3aedd8f7-b2ee-4aa5-a160-a9d5c110b724/0113/hansard_frag.pdf;fileType=application%2Fpdf>.

Business Chamber NSW (2010) *Federal Election 2010 Policy Comparison*. NSW Business Chamber, Sydney.

Ceccarelli D (2011) *Australia's Coral Sea: A Biophysical Profile*. Protect our Coral Sea, Sydney.

Centre for Conservation Geography (2011) *Critical Gaps in the Marine Sanctuary Network Proposed by the Federal Government for Australia's South-west*. Centre for Conservation Geography, Sydney. <http://conservationgeography.org/sites/default/files/CCG_South-west_sanctuaries_critical_gaps_June_2011.pdf

Cleary P (2011) Marine conservation, with a catch. *The Australian*, 6 January 2011, <http://www.theaustralian.com.au/news/inquirer/marine-conservation-with-a-catch/story-e6frg6z6-1225982587200>.

Department of the Environment (2015) *About the Commonwealth Marine Reserves Review*. Department of the Environment, Canberra, <http://www.environment.gov.au/marinereservesreview/about>.

Devillers R, Pressey RL, Grech A, Kittinger JN, Edgar GJ, Ward T, *et al.* (2015) Reinventing residuals reserves in the sea: are we favouring ease of establishment over need for protection? *Aquatic Conservation: Marine and Freshwater Ecosystems* **25**, 480–504. doi:10.1002/aqc.2445.

Director of National Parks (2013a) *South-west Commonwealth Marine Reserves Network Management Plan 2014–24*. Director of National Parks, Canberra.

Director of National Parks (2013b) *Coral Sea Commonwealth Marine Reserve Management Plan 2014–24*. Director of National Parks, Canberra.

Eadie L, Hoisington C (2011) *Stocking Up: Securing our Marine Economy.* Centre for Policy Development, Sydney.

Ecology Centre (2010) *Systematic Conservation Planning: A Network of Marine Sanctuaries.* Ecology Centre, University of Queensland, Brisbane.

Entsch W (2013) 'Coral Sea Commonwealth Marine Reserves Network Management Plan Disallowance'. House of Representatives Private Members' Business. Speech by Warren Entsch, 4 June 2013, <http://parlinfo.aph.gov.au/parlInfo/genpdf/chamber/hansardr/e40cc95f-8da2-4485-8b5c-f83df5fee682/0135/hansard_frag.pdf;fileType=application%2Fpdf>.

Essential Research (2008) *Report by Essential Research on Focus Groups.* Essential Research, Melbourne.

Essential Research (2009) Pew Environment Group/Save Our Marine Life. Two poll questions related to protection of marine areas and people's fishing behavior. Essential Research, Melbourne.

Fernandes L, Day J, Lewis A, Slegers S, Kerrigan B, Breen D, *et al.* (2005) Establishing representative no-take areas in the Great Barrier Reef: large-scale implementation of theory on marine protected areas. *Conservation Biology* **19**, 1733–1744. doi:10.1111/j.1523-1739.2005.00302.x.

Fishing World (2009) Garrett: Government committed to MBPP, no favours for Pew. *Fishing World*, 14 May 2009, <http://www.fishingworld.com.au/news/garrett-govt-committed-to-mbpp-no-favours-for-pew>.

Fishing World (2013) Coalition promise scientific review of marine parks. *Fishing World* 27 August 2013, <http://www.fishingworld.com.au/news/coalition-promise-scientific-review-of-marine-parks>.

Fishwrecked (2013) New map reveals extent of angler lockout proposals. Fishwrecked.com, Perth, <http://fishwrecked.com/forum/new-map-reveals-extent-angler-lockout-proposals>.

Foster T (2014) Change and damage beyond belief: Judith Wright's The Coral Battleground. *Verity La Reviews* 21 June 2014, <http://verityla.com/change-and-damage-beyond-belief-judith-wrights-the-coral-battleground/>.

Friedlander J (2014) Marine reserves ditched despite tide of research. *Sydney Morning Herald* 13 September 2014, <http://www.smh.com.au/environment/marine-reserves-ditched-despite-tide-of-research-20140825-1083js.html>.

Heath J, Rollason A (2013) Coalition to suspend Labor's marine parks: Abbott. *Australian Financial Review* 26 August 2013, <http://www.afr.com/news/politics/national/coalition-to-suspend-labors-marine-parksabbott-20130825-j0gud>.

Hunt G, Colbeck R (2013) 'Supporting recreational fishing while protecting our marine parks'. Joint Media Release by the Minister for the Environment and the Parliamentary Secretary to the Minister for Agriculture. 14 December 2013, <http://www.environment.gov.au/minister/hunt/2013/mr20131214.html>.

Ipsos Consultants (2007) *Victorian Coastal and Marine Environment Community Attitudes and Behaviour Research: Wave 3, Final Report*. Prepared for Victorian Coastal Council, Ipsos Consulting, Melbourne.

Murray R (2009) *The End of the Line: Imagine a World Without Fish*. Film website, <http://endoftheline.com/>.

Pew Charitable Trusts (2006) *Global Ocean Legacy – Hawaii*. The Pew Charitable Trusts, Philadelphia.

Pew Charitable Trusts (2007) 'Global conservation groups announce major effort to protect Australia's bush, deserts and oceans'. Media release, 23 July 2007, The Pew Charitable Trusts.

Pressey R (2013) Australia's new marine protected areas: why they won't work. *The Conversation*, 17 January 2013. <http://theconversation.com/australias-new-marine-protected-areas-why-they-wont-work-11469>.

Protect our Coral Sea (2010) *Barry the Wrasse*. Protect our Coral Sea, Sydney, <https://www.youtube.com/watch?v=DTkYcCvkEW8>.

Protect our Coral Sea (2013) *Coral Sea Marine Park: Fact Sheet and Mythbuster*. Protect our Coral Sea, Sydney, <http://saveourmarinelife.org.au/coral_sea_resources/fact-sheet-coral-sea-mythbuster/>.

Save our Marine Life (2009) *Protecting Western Australia's Big Blue Backyard*. Save our Marine Life, Perth.

Save our Marine Life (2010) *Atlantis Found – Underwater Icons of Australia's Unique South West*. Save our Marine Life, Perth.

The Courier Mail (2011) 'It's time to protect the Coral Sea'. [Advertisement cosigned by Imogen Zethoven, Protect our Coral Sea, and Gary Heilman, De Brett Seafoods]. *The Courier Mail* September 2011.

Thomas S, Harnwell J (2009) 'Nats to block Garrett on Coral Sea'. *Fishing World*, 25 August 2009, <http://www.fishingworld.com.au/news/nats-to-block-garrett-on-coral-sea>.

Warwick Daily News (2012) Fishers say marine park will 'lock out mums and dads'. *Warwick Daily News*, 16 November 2012, <http://www.warwickdailynews.com.au/news/fishers-say-marine-park-will-lock-out-mums-and-dad/1624859/>.

Wescott G, Fitzsimons J (2011) Stakeholder involvement and interplay in coastal zone management and marine protected area planning. In *Marine Resources Management* (Eds W Gullett, C Schofield and J Vince). pp. 225–238. LexisNexis Butterworths, Sydney.

Zethoven I (2008) *An Australian Coral Sea Heritage Park*. Pew Environment Group, Sydney.

PART 5
Synthesis

Previous page: Estuary at Cape Tribulation, Great Barrier Reef Coast Marine Park. (Photo: James Fitzsimons)

Chapter 24

Lessons from the expansion of Australia's marine protected area networks: a synthesis

James Fitzsimons and Geoff Wescott

Introduction

Although Australia was one of the first nations in the world to apply the modern concept of a protected area over a marine environment, the implementation of networks of dedicated marine protected areas (MPAs) has only occurred in more recent decades. This rapid expansion, across all Australian jurisdictions, has resulted from the recognised need to better manage marine ecosystems for a range of threats (where possible) and to fulfil international and national commitments to establish such networks.

This expansion has also provided a substantial opportunity to begin to acquire real and substantial lessons that can improve the Australian MPA system and may well be of significant value to other jurisdictions around the world as they seek to expand their MPA systems.

In this synthesis chapter, we have attempted to extract the 'big messages' – the repeated recurring themes as well as highlighting points of difference – that the authors of the preceding chapters have outlined from their extensive experience. These are summarised below.

Marine protected areas are always contested ... at first

In almost all Australian jurisdictions, and particularly in the last two decades, the establishment of MPAs has been contested, especially where this has placed restrictions on fishing or other extractive industries (see Chapters 7, 8, 9, 11, 13 and 21). This is reflected in other countries as well (Chapter 2).

In the most recent example, some 740 000 submissions were received in the public consultation periods process for the establishment of the Commonwealth Marine Reserve networks and related management plans, the majority of which were campaign submissions for and against the marine reserve proposals (Chapters 4 and 23). This in non-coastal waters most Australian's have never seen or are likely to ever visit. The subsequent review of the reserve network (and further rounds of public consultation) was prompted by the considerable pressure from user groups, with the appointment of two 'expert panels [to] help restore confidence in Commonwealth marine reserves' by providing 'advice to the Government, based on the best available science and after genuine consultation with stakeholders' (Hunt and Colbeck 2014).

This resistance has been driven by fishing groups in particular, despite apparent strong in principle public support for MPAs (Chapter 21). Strong fishing lobbies are considered the reason for lack of concerted establishment of networks of MPAs in Tasmania (Chapter 10) and the Northern Territory (Chapter 13).

However, experience in other states suggests that this early discontent and negative reactions from fishers and some locals, mostly changes to neutral or positive perceptions over 5–10 years after a MPA is established (e.g. Western Australia, New South Wales, Great Barrier Reef; Chapters 5, 7, 11 and 21). Despite this, there have been campaigns by the fishing lobby in some states for an opening up of strict MPAs (Chapters 9 and 11). This pressure to actually reverse reservation decisions although relatively unknown in terrestrial protected area systems seems to be more common in the marine environment.

Jurisdiction-wide investigations versus single MPA declarations

MPAs in Australian jurisdictions have varied in the way they have been created. These differences in establishment are reflected in legislation, which falls into three main categories: specific MPA legislation, fisheries legislation, and general national parks legislation (Chapter 14).

In most jurisdictions, early MPA declarations were a 'one off', typically protecting a small area of particular interest. The advent of the Great Barrier Reef Marine Park heralded the beginning of considering the entire extent of a jurisdiction for the designation of a MPA network at the same time. The Victorian process, completed in 2002, was the first Australian jurisdiction to systematically consider its entire marine waters for the establishment of a 'comprehensive, adequate and representative' reserve system followed soon after by the 2003 Great Barrier Reef Marine Park rezoning. South Australia (Chapter 8), the Commonwealth Regional Marine Planning processes (Chapter 4) and the Antarctic MPA proposals being considered by the Commission for the Conservation of Antarctic Marine Living Resources (Chapter 6) are other examples of MPA networks created from jurisdiction-wide processes. Some states such as Western Australia and Tasmania have done this for parts of their jurisdiction (the Kimberley in Western Australia – Chapter 7; and South-east Tasmania – Chapter 10), whereas New South Wales and Queensland have implemented relatively large zoned MPAs in various parts of their jurisdiction but have not assessed all of their waters for this purpose (Chapters 7 and 11). The Northern Territory, which has only two legislated marine parks, seems to be the furthest from developing a comprehensive, adequate and representative MPA network.

Establishing a MPA network based on considering the entire jurisdiction at once has several advantages, including (i) the consideration of the comprehensiveness, adequacy and representativeness of their MPA estate with consistent ecological data and ability to make adjustments to boundaries to accommodate other uses, (ii) a dedicated process with public consultation and public submissions (see also Wescott 2006).

Disadvantages include a potential inability to campaign and make the case of worthwhile MPAs in the entire jurisdiction by MPA advocates due to lack of resources or limited public attention (particularly if the region is large) (Chapter 23), or claims that resource use stakeholders have not been properly consulted or drained by the process (Chapter 22).

The challenge of federalism

As with many federal nations, Australia divides the control over its marine environment between a federal (national) government and governments of states and territories. The result in Australia is that coastal waters come under the state/territory government's control while waters beyond the three nautical mile limit, in general, come under national government control.

Juxtaposed to this delineation are the practical realities of human use of these two water masses. The nearshore coastal waters are available and accessible to many more uses and many more people – they are accessible to both people standing onshore, using waters that they can wade, swim or surf in and further off shore by small boats and yachts. Hence coastal waters are an extension in human use terms, of onshore economic and recreational uses (see Chapters 7–13).

Far offshore, the uses will tend to be more 'industrial' – commercial fishers, oil and gas extraction, large shipping, and so on (Chapters 4, 19 and 22). The outcome of this human use distribution is that any restrictions that occur near to shore will impact a much greater number of people.

In realpolitik terms this means that state and territory governments will bear the brunt of any anger and disputes over the loss of access to a previously available resource (e.g. fish) while the national government will see far less pressure from voting members of the electorates.

This in reality results in far more disputation directed at the more poorly resourced state/territory level and hence greater difficulty for state and territory governments in implementing MPAs, in Australia at least. It also means it can take considerably longer time to establish a system of MPAs in coastal waters (see e.g. Chapters 8, 9 and 13), where their need is higher because of greater human use impacts, than in national waters (Chapter 17).

This would appear to be an argument for the national government to directly (i.e. financially) assist the states in establishing their MPA systems if Australia is to establish a genuine comprehensive, adequate and representative MPA system covering all the nation's waters.

Fishery reserves as the first MPAs 'muddy the waters' for the objectives of modern MPAs

In most Australian jurisdictions (see Chapters 7–14) either the first reservations covering marine areas were specifically declared to protect fish resources (e.g. nursery grounds, or temporary closures of fisheries to prevent over-fishing); or fishery regulations were used to establish MPAs for biodiversity conservation reasons (perhaps because other legislative or regulatory processes were not available for reservation in the sea).

This historical legacy may well be the reason for the continued perception by some, such as those in fishery circles, that MPAs are fisheries management tools rather than biodiversity conservation tools (see e.g. Chapters 6, 7 and 22).

This confusion has led to significant heat in the debate about placement of MPAs, with those from a fishery perspective often debating the need for a MPA based on its value for fisheries management (Chapters 6 and 7). This in turn creates 'heat' in politicians and senior decision makers minds as well – particularly if they see conservation in terms of the sustainable use of natural resources rather than including nature conservation for the intrinsic value of the nature itself (e.g. Taylor 2013).

Some jurisdictions do have some element of their MPA networks focused on fishery management and this may complicate the purpose of MPAs more generally. For example, Queensland's network of Fish Habitat Areas consists of 70 declared areas covering more than 1 million ha and is spread across the state (Chapter 12). It should also be noted that conservation groups do promote MPAs as a means to prevent overfishing (Chapters 20 and 23; see also Chapter 22) and it may be that the public do perceive MPAs for that purpose, at least in part, as well (Chapter 20).

This may in fact be a communication problem rather than an institutional one and if the nomenclature system and the sea use classification system more clearly separated the reasons for a specific reservation (e.g. biodiversity conservation, temporary fishery grounds closure, habitat protection for breeding stock, etc.), an entire area of conflict might potentially be eliminated.

Multitude of categories, zones and uses has also confused the MPA concept

There are more than 30 different named categories given to MPAs in Australia (e.g. marine parks, marine national parks, marine sanctuaries, etc.). Even where a named category is the same in multiple jurisdictions (e.g. 'marine park') they might have quite different allowable activities (see Chapters 5, 8 and 11). This potentially adds to the confusion among the public about what MPAs are and what they are trying to achieve. It is likely the concept of a 'national park' or 'nature reserve' on land has much greater community understanding of purpose and allowable activities.

There are two general models followed by jurisdictions in Australia for MPAs. One is the zoned MPA model where generally larger MPAs have a variety of legally defined zones within their outer boundaries, which typically include a mix of high protection (no-take) zones, semi-restrictive zones and zones that are only lightly restrictive of extractive uses such as recreation and commercial fishing (Fig. 24.1). The benefit of such an approach (i.e. multiple use in MPAs) is that a greater variety of users can access and appreciate a MPA, including those that might be in many cases (initially) opposed to MPAs. Conversely, having such a diversity of activities within a single protected area can make it hard for the public, who are not regular users, to determine a clear and distinctive identity for such a sea use. This zoning approach is adopted in Queensland, New South Wales, South Australia, the Great Barrier Reef Marine Park and some of the Commonwealth Marine Reserves.

The historical use of the term 'marine park' for the Great Barrier Reef to describe what was in reality a spatial zoning system with a majority of areas initially designated for 'general use' (i.e. containing few restrictions) certainly confused the public in other jurisdictions in Australia, and perhaps globally, and may well have made it harder to convince the community of the value of MPAs (see Wescott 2006).

The other model is a non-zoned approach such as Victoria's no-take system of marine national parks and marine sanctuaries (Chapter 9) (Fig. 24.2), or the multi-use but not-zoned MPAs in the Northern Territory and some Tasmanian reserves (Chapters 10 and 13). The benefit of this approach is a clear articulation of the purpose of the reserve or the network (see Wescott 2006), although these areas are typically smaller and potentially engage fewer users.

The Commonwealth Government, Western Australia and Tasmania have MPAs that are both zoned and not zoned (Chapters 4, 7 and 14). The proposed East Antarctic Representative System of Marine Protected Areas is based on multiple-use management, with conservation measures to be applied to ensure activities are consistent with the

Fig. 24.1. Great Barrier Reef Coast Marine Park near Cairns – a marine protected area with multiple zones. (Photo: James Fitzsimons)

Fig. 24.2. Twelve Apostles Marine National Park, south-west Victoria – a spectacular no-take marine protected area. (Photo: Dean Ingwersen)

maintenance of the MPA's conservation objectives (Chapter 6), although the difficulties with convincing some parties was noted.

In arguing the need for integrated ocean planning, Chapter 19 suggests that if some zones within MPAs are basically 'general use' and have little restriction of usage (see also Chapter 17), then applying a MPA zoning scheme across all waters could be a means to achieve this integration. The development of a single management body, such as the Marine Estate Management Authority in New South Wales could be another means to achieve better integration (Chapters 11 and 14).

Multi-use versus no-take: proving the benefits and the need for clear targets for each

Much emphasis in the development of MPA networks by scientists and conservationists focuses on the establishment of strict protection MPAs, particularly no-take zones (Chapters 3, 16 and 23). This in part may be due to the historic prevalence of multi-use MPAs and the clearer ecological benefits of MPAs where there is less human exploitation.

For example, Edgar *et al.* (2014) found that complete protection from human exploitation was one of five key factors for an effective MPA. Another global study found that only a quarter of coastal countries have no-take MPAs and that 94% of all MPAs allow fishing (Costello and Ballantine 2015). There is a current lack of international criteria or guidance on the mix of no-take and multi-use MPAs in reservation targets such as the Convention on Biological Diversity's Aichi Target 11 of 10% by 2020. The final declaration of the 2014 World Parks Congress called for bolder targets – 'at least 30 per cent of each habitat type to be afforded strict protection' – although this is not binding.

In Australia, the *Guidelines for Establishing the National Representative System of Marine Protected Areas* are silent on percentage targets, but state 'NRSMPA will aim to include some highly protected areas (IUCN Categories I and II) in each bioregion' (ANZECC TFMPA 1998). Although this has not yet been achieved (Chapter 17), approximately one-third of the total area of the MPA estate in Australia falls within these high protection categories (Chapter 3; although, somewhat confusingly IUCN I and II do not always equate to 'no take', see Fitzsimons 2011).

Placement of MPAs: pure science versus realpolitik?

The principles of comprehensiveness, adequacy and representativeness guide the establishment of MPAs in Australian jurisdictions and it is the stated wish that the full range of habitats are included in Australia's MPA network. More recently, Aichi Target 11 of the Convention on Biological Diversity has included the need for connectivity and consideration of ecosystem services in addition to representation of habitats. Globally, MPA coverage appears to show some level of inverse correlation with coastal population densities, despite many ecosystem services being tightly linked to local human populations who receive direct benefits from coastal protection, food provision, tourism income, and so on (Chapter 2). Nonetheless, ecosystem services and other socio-economic benefits have been shown for several Australian MPAs (Chapter 15; see also Marshall 2015).

Different groups of people have quite different perspectives on (i) the reasons for declaring MPAs, (ii) the methodology and decision-making processes for deciding where to place these MPAs and (iii) the various restrictions on use of different categories of MPAs. This is highlighted in this book in Parts 3 and 4 where the different perspectives of different groups are clear. In particular, there seems to be one most overt dichotomy: the difference between the belief among some that the scientific data should solely determine, or at least be the primary determiner (e.g. Devillers *et al.* 2015), of MPA location and extent, and nearly all other perspectives that in various forms recognise the political reality of the difficulty of achieving a change in sea use.

In the end it is a political decision to decide how, where and when a system of MPAs is to be established for biodiversity conservation purposes, and how this is balanced with other legitimate uses of the sea. Although such decisions are difficult at the best of times (there are going to be clear winners and losers), bipartisan support is usually required to see the declaration of MPAs through a parliament of the relevant jurisdiction. This usually involves some compromises in locations and uses (see e.g. Chapters 4, 5, 9 and 11).

If there is not bipartisan or multi-partisan political support, then a second level of politics occurs – alternatives are offered up to the electorate complete with electoral tactics of marketing and misleading information. This has certainly occurred in Australia in recent years and, while science is predominantly used to justify the establishment of MPA networks, it is naïve to think science alone will overcome party electoral rhetoric.

Indigenous involvement in MPAs and marine Indigenous Protected Areas

Much like the slower development of protected areas in the marine realm compared with the terrestrial realm, recognition of Indigenous rights, ownership and management of sea country has been slower than on land. But this is changing. Legal recognition in the Torres Strait, the intertidal zones of the Northern Territory and various other jurisdictions is seeing greater Indigenous involvement in formal management arrangements, including MPAs (Chapter 18).

The current lack of recognition of Indigenous Protected Areas over marine environments towards national conservation targets (such as the National Representative System of Marine Protected Areas: ANZECC TFMPA 1998; Rose 2012) is inconsistent with the treatment of IPAs on land. With the number of sea country IPAs likely to increase, including over existing MPAs such as the Great Barrier Reef Marine Park and Commonwealth Marine Reserves (Chapters 5 and 13), this policy position seems antiquated and in need of updating.

Future directions

One of the striking contradictions in the attempt to increase MPAs is the outcome of public polling and surveys that suggest strong general community support for declaring protected areas in the marine environment (e.g. Chapters 20 and 21) juxtaposed to the strident public opposition from other interests. Apparently, from the decision makers point of view, the negativity towards declaring MPAs evidenced in the popular media (e.g. Compas *et al.* 2007) is not compensated by the polling suggesting a majority of constituents want MPAs declared (Chapter 20). Exploring the sociological element of the 'MPA debate' is a potentially fruitful and useful area for further study and research.

Clearer guidelines on reservation targets are needed at the national and international level. As suggested in Chapter 2, the benefits provided by strict protection might be used to set a sub-target for no-take MPAs, to ensure the development of a network of sites delivering highest possible value in terms of biological controls, fisheries enhancement and recreational value.

Considering international obligations under the Convention on Biological Diversity to meet spatial targets for protected areas on land and sea, the need for accurate geospatial data is critical if we are to assess progress. In developing summary statistics for Chapter 4, it was discovered several significant errors occur within Australia's national protected area dataset relating to MPAs (i.e. the 2014 Collaborative Australian Protected Areas Database; Department of the Environment 2014). These included counting the entire area of several mostly terrestrial protected areas in the marine figures and including most of the Victorian MPA network in the terrestrial geospatial dataset. Although we have made these corrections in this book, use of the publicly available data should be done with care and future iterations of this data should rectify these fundamental problems.

Although primarily terrestrial protected areas on the coast may include intertidal areas down to low-water mark or estuaries and some MPAs may extend to high-water mark, there are few examples of single protected areas that incorporate significant areas of both land and sea (Kent Group National Park in Tasmania, Nooramunga Marine and Coastal Park in Victoria and Dhimurru Indigenous Protected Area in the Northern Territory are examples of exceptions to this). This has the impact of fragmenting management, particularly on the coastal fringe where use of a MPA is likely to be most exposed to public use and potential misunderstanding. Here, lessons for government institutions may be drawn from the use and management of land and sea country by Indigenous Australians. Indigenous Australians do not see a rigid boundary between the sea and the land but a continuum of 'country', which reflects the ecology of the transition far better than rigid legalistic and cultural views of western based approaches. At the very least, stronger management of intertidal zones in primary terrestrial coastal MPAs (which occur in all Australian states and the Northern Territory) for their marine values should be considered.

Finally, by way of conclusion, two observations as editors: first we were somewhat surprised by a range of matters that were not discussed in detail across the chapters despite the breadth of views we deliberately set out to include. These included enforcement, adequacy of management and the need for considering ecological connectivity and climate change.

The final observation is that a possible explanation for why it appears to have been so difficult to establish networks of MPAs around the globe may be that the development, understanding and support for marine protected areas lags a century behind terrestrial protected areas. Nonetheless, this 'gap' appears to be rapidly closing.

If this book contributes in small part to further closing the gap, we will have achieved the purpose with which we set out.

Biographies

James Fitzsimons is the Director of Conservation for The Nature Conservancy's Australia Program where he oversees the organisation's conservation planning, science and policy functions. He is an Adjunct Professor at the School of Life and Environmental Sciences, Deakin University, with particular research interests in the fields of protected area policy, practical conservation planning and mechanisms to integrate conservation outcomes on public and private lands. He has previously worked in the fields of conservation policy and planning for government environment departments and agencies, and for non-government environment organisations.

Geoff Wescott is Associate Professor of Environment at Deakin University's Melbourne Campus. He is a previous Chair of the Marine and Coastal Community Network which he also represented on the National Oceans Advisory Group. He was a Deputy Chair of Parks Victoria and has served on the steering committee of the past three Nature Conservation Reviews of the Victorian National Parks Association. He is current President of the Australian Coastal Society, a Director of Zoos Victoria and in mid-2015 was appointed to the Victorian Environmental Assessment Council. He has published over 100 articles on marine and coastal matters including a popular field guide to rocky shores in south-eastern Australia.

References

ANZECC TFMPA (1998) *Guidelines for Establishing the National Representative System of Marine Protected Areas.* Australian and New Zealand Environment and Conservation Council, Task Force on Marine Protected Areas. Environment Australia, Canberra.

Compas E, Clarke B, Cutler C, Daish K (2007) Murky waters: media reporting of marine protected areas in South Australia. *Marine Policy* **31**, 691–697. doi:10.1016/j.marpol.2007.03.001.

Costello MJ, Ballantine B (2015) Biodiversity conservation should focus on no-take Marine Reserves: 94% of Marine Protected Areas allow fishing. *Trends in Ecology & Evolution* **30**, 507–509. doi:10.1016/j.tree.2015.06.011.

Department of the Environment (2014) Collaborative Australian Protected Area Database Marine 2014. Department of the Environment, Canberra, <http://www.environment.gov.au/land/nrs/science/capad/2014>.

Devillers R, Pressey RL, Grech A, Kittinger JN, Edgar GJ, Ward T, *et al.* (2015) Reinventing residuals reserves in the sea: are we favouring ease of establishment over need for protection? *Aquatic Conservation: Marine and Freshwater Ecosystems* **25**, 480–504. doi:10.1002/aqc.2445.

Edgar GJ, Stuart-Smith RD, Willis TJ, Kininmonth S, Baker SC, Banks S, *et al.* (2014) Global conservation outcomes depend on marine protected areas with five key features. *Nature* **506**, 216–220. doi:10.1038/nature13022.

Fitzsimons JA (2011) Mislabelling marine protected areas and why it matters – a case study of Australia. *Conservation Letters* **4**, 340–345. doi:10.1111/j.1755-263X.2011.00186.x.

Hunt G, Colbeck R (2014) 'Review of Commonwealth marine reserves begins'. Joint Media Release from The Hon. Greg Hunt MP Minister for the Environment and Senator the Hon. Richard Colbeck Parliamentary Secretary to the Minister for Agriculture. 11 September 2014, <http://www.environment.gov.au/minister/hunt/2014/pubs/mr20140911a.pdf>.

Marshall PA (2015) Value of ecosystem service benefits from the marine environment and the importance of MPAs. In *Valuing Nature: Protected Areas and Ecosystem Services.* (Eds P Figgis, B Mackey, J Fitzsimons, J Irving and P Clarke) pp 84–89. Australian Committee for IUCN, Sydney.

Rose B (2012) Indigenous Protected Areas – innovation beyond the boundaries. In *Innovation for 21st Century Conservation* (Eds P Figgis, J Fitzsimons and J Irving). pp. 50–55. Australian Committee for IUCN, Sydney.

Taylor L (2013) Coalition to set up scientific panel to review marine park boundaries. *The Guardian,* 26 August 2013, <http://www.theguardian.com/world/2013/aug/26/marine-life-fishing-coalition-review>.

Wescott G (2006) The long and winding road: the development of a comprehensive, adequate and representative system of highly protected marine protected areas in Victoria, Australia. *Ocean and Coastal Management* **49**, 905–922. doi:10.1016/j.ocecoaman.2006.08.001.

Index

www.ingramcontent.com/pod-product-compliance
Ingram Content Group UK Ltd.
Pitfield, Milton Keynes, MK11 3LW, UK
UKHW061957041026
15591UKWH00016B/1049

9 781486 301942